AF755447

Advances in

ECOLOGICAL RESEARCH

VOLUME 43

Advances in Ecological Research

Series Editor: GUY WOODWARD
School of Biological and Chemical Sciences
Queen Mary University of London
London, UK

Advances in
ECOLOGICAL RESEARCH

VOLUME 43

INTEGRATIVE ECOLOGY:
FROM MOLECULES TO ECOSYSTEMS

Edited by

GUY WOODWARD

School of Biological and Chemical Sciences,
Queen Mary University of London,
London E1 4NS, UK

ELSEVIER

AMSTERDAM • BOSTON • HEIDELBERG • LONDON
NEW YORK • OXFORD • PARIS • SAN DIEGO
SAN FRANCISCO • SINGAPORE • SYDNEY • TOKYO
Academic Press is an imprint of Elsevier

Academic Press is an imprint of Elsevier

32 Jamestown Road, London NW1 7BY, UK
Linacre House, Jordan Hill, Oxford OX2 8DP, UK
Radarweg 29, PO Box 211, 1000 AE Amsterdam, The Netherlands
30 Corporate Drive, Suite 400, Burlington, MA 01803, USA
525 B Street, Suite 1900, San Diego, CA 92101-4495, USA

First edition 2010

Copyright © 2010 Elsevier Ltd. All rights reserved.

No part of this publication may be reproduced, stored in a retrieval system or transmitted in any form or by any means electronic, mechanical, photocopying, recording or otherwise without the prior written permission of the Publisher.

Permissions may be sought directly from Elsevier's Science & Technology Rights Department in Oxford, UK: phone (+44) (0) 1865 843830; fax (+44) (0) 1865 853333; email: permissions@elsevier.com. Alternatively you can submit your request online by visiting the Elsevier web site at http://elsevier.com/locate/permissions, and selecting *Obtaining permission to use Elsevier material.*

Notice
No responsibility is assumed by the publisher for any injury and/or damage to persons or property as a matter of products liability, negligence or otherwise, or from any use or operation of any methods, products, instructions or ideas contained in the material herein. Because of rapid advances in the medical sciences, in particular, independent verification of diagnoses and drug dosages should be made.

ISBN: 978-0-12-385005-8
ISSN: 0065-2504

For information on all Academic Press publications
visit our website at elsevierdirect.com

Printed and bound in UK

10 11 12 10 9 8 7 6 5 4 3 2 1

Working together to grow
libraries in developing countries

www.elsevier.com | www.bookaid.org | www.sabre.org

ELSEVIER BOOK AID International Sabre Foundation

Contents

Systems Biology for Ecology: From Molecules to Ecosystems

KEVIN J. PURDY, PAUL J. HURD, JORDI MOYA-LARAÑO, MARK TRIMMER, BRIAN B. OAKLEY AND GUY WOODWARD

Assessing the Contribution of Micro-Organisms and Macrofauna to Biodiversity–Ecosystem Functioning Relationships in Freshwater Microcosms

JULIA REISS, R. A. BAILEY, FERNANDA CÁSSIO, GUY WOODWARD AND CLÁUDIA PASCOAL

Environmental Warming and Biodiversity–Ecosystem Functioning in Freshwater Microcosms: Partitioning the Effects of Species Identity, Richness and Metabolism

DANIEL M. PERKINS, BRENDAN G. MCKIE, BJÖRN MALMQVIST, STEVEN G. GILMOUR, JULIA REISS AND GUY WOODWARD

Individual-Based Food Webs: Species Identity, Body Size and Sampling Effects

GUY WOODWARD, JULIA BLANCHARD, RASMUS B. LAURIDSEN, FRANCOIS K. EDWARDS, J. IWAN JONES, DAVID FIGUEROA, PHILIP H. WARREN AND OWEN L. PETCHEY

The Temperature Dependence of the Carbon Cycle in Aquatic Ecosystems

GABRIEL YVON-DUROCHER, ANDREW P. ALLEN, JOSÉ M. MONTOYA, MARK TRIMMER AND GUY WOODWARD

Contributors to Volume 43

ANDREW P. ALLEN, *Department of Biological Sciences, Macquarie University, Sydney, NSW 2109, Australia.*

R.A. BAILEY, *Queen Mary University of London, School of Mathematical Sciences, London E1 4NS, United Kingdom.*

FERNANDA CÁSSIO, *CBMA - Centre of Molecular and Environmental Biology, Department of Biology, University of Minho, Campus de Gualtar, 4710-057 Braga, Portugal.*

FRANCOIS K. EDWARDS, *School of Biological and Chemical Sciences, Queen Mary University of London, London E1 4NS, United Kingdom, and Centre for Ecology and Hydrology, Wallingford, Oxon. United Kingdom.*

DAVID FIGUEROA, *School of Biological and Chemical Sciences, Queen Mary University of London, London E1 4NS, United Kingdom, and Facultad de Recursos Naturales, Universidad Católica de Temuco, Montt 56 - Casilla 15-D, Chile.*

JACK FORSTER, *Queen Mary University of London, School of Biological and Chemical Sciences, London E1 4NS, United Kingdom.*

STEVEN G. GILMOUR, *School of Mathematical Sciences, Queen Mary University of London, London E1 4NS, United Kingdom.*

HELMUT HILLEBRAND, *Institute for Chemistry and Biology of the Marine Environment, Carl-von-Ossietzky University Oldenburg, Schleusenstrasse 1, 26382 Wilhelmshaven, Wilhelmshaven.*

ANDREW G. HIRST, *Queen Mary University of London, School of Biological and Chemical Sciences, London E1 4NS, United Kingdom.*

PAUL J. HURD, *School of Biological & Chemical Sciences, Queen Mary, University of London, Mile End Road, London E1 4NS, United Kingdom.*

J. IWAN JONES, *School of Biological and Chemical Sciences, Queen Mary University of London, London E1 4NS, United Kingdom.*

RASMUS B. LAURIDSEN, *School of Biological and Chemical Sciences, Queen Mary University of London, London E1 4NS, United Kingdom.*

BJÖRN MALMQVIST, *Department of Ecology and Environmental Science, Umeå University, Sweden.*

BRENDAN G. McKIE, *Department of Aquatic Sciences and Assessment, Swedish University of Agricultural Sciences, Uppsala, Sweden.*

JOSÉ M. MONTOYA, *School of Biological & Chemical Sciences, Queen Mary University of London, London E1 4NS. United Kingdom, and Institute of Marine Sciences (ICM-CSIC) Pg. Marítim de la Barceloneta, 37-49E-08003 Barcelona, Spain.*

STEFANIE D. MOORTHI, *Institute for Chemistry and Biology of the Marine Environment, Carl-von-Ossietzky University Oldenburg, Schleusenstrasse 1, 26382 Wilhelmshaven, Wilhelmshaven.*

JORDI MOYA-LARAÑO, *Functional and Evolutionary Ecology, Estación Experimental de Zonas Áridas – CSIC, Carretera de Sacramento s/n, La Cañada de San Urbano, 04120-Almería, Spain.*

CLÁUDIA PASCOAL, *CBMA - Centre of Molecular and Environmental Biology, Department of Biology, University of Minho, Campus de Gualtar, 4710-057 Braga, Portugal.*

DANIEL M. PERKINS, *School of Biological and Chemical Sciences, Queen Mary University of London, London E1 4NS, United Kingdom.*

OWEN L. PETCHEY, *Department of Animal and Plant Sciences, University of Sheffield, Western Bank, Sheffield, S10 2TN, United Kingdom.*

ROBERT PTACNIK, *Institute for Chemistry and Biology of the Marine Environment, Carl-von-Ossietzky University Oldenburg, Schleusenstrasse 1, 26382 Wilhelmshaven, Wilhelmshaven.*

KEVIN J. PURDY, *Department of Biological Sciences, University of Warwick, Coventry, CV4 7AL. United Kingdom.*

JULIA REISS, *School of Biological and Chemical Sciences, Queen Mary University of London, London E1 4NS, United Kingdom.*

REBECCA STEWART, *Queen Mary University of London, School of Biological and Chemical Sciences, London E1 4NS, United Kingdom.*

MARK TRIMMER, *School of Biological & Chemical Sciences, Queen Mary, University of London, Mile End Road, London E1 4NS, United Kingdom.*

PHILIP H. WARREN, *Department of Animal and Plant Sciences, University of Sheffield, Western Bank, Sheffield, S10 2TN, United Kingdom.*

GUY WOODWARD, *School of Biological & Chemical Sciences, Queen Mary, University of London, Mile End Road, London E1 4NS, United Kingdom.*

GABRIEL YVON-DUROCHER, *School of Biological & Chemical Sciences, Queen Mary University of London, London E1 4NS. United Kingdom.*

Preface

This thematic volume of *Advances in Ecological Research* integrates different levels of organisation and biological disciplines, in an attempt to provide a more holistic view of ecological systems. As such, none of these chapters are typical representatives of any single traditional field in ecology, as between them they attempt to forge new links from molecular biology to ecosystems ecology, from microbial to macro-organismal ecology, from individuals to food webs, or from metabolic theory to biodiversity-ecosystem functioning (B-EF) theory. Given that many of the long-established barriers within ecology, and biology as a whole, are rapidly dissolving, this volume reflects as our science as it enters an exciting and evermore interdisciplinary phase. This is being achieved via a combination of new theoretical and empirical perspectives and methodological advances, many of which underpin the chapters presented here.

In recent years, we have witnessed an ever-strengthening link develop between the previously disparate disciplines of community and ecosystems ecology, catalysed in particular by the mushrooming of B-EF and food web research, and the previous companion volume of *Advances in Ecological Research* was devoted to the latter (Volume 42: *Ecological Networks*). These two fields have also started to converge, as vertical and horizontal interactions within communities, and their consequences for ecosystem processes are starting to be considered simultaneously. Further, the fact that interactions occur between individuals operating within metabolic and foraging constraints, is now being recognised and it is becoming increasingly clear that that this level of organisation needs to be considered in B-EF and food web studies, as highlighted in this volume (e.g., Perkins *et al.*, 2010; Reiss *et al.*, 2010a; Woodward *et al.*, 2010). The chapters by Reiss *et al.* (2010a) and Perkins *et al.* (2010), for instance, use the metabolic capacity of individuals to predict ecosystem process rates: the former is one of the first studies to manipulate both microbial and macrofaunal diversity simultaneously in a B-EF experiment and the latter describes a range of potential impacts of climate change on B-EF relationships. In both instances, species richness had no detectable effect on process rates, which were driven primarily by the body size and metabolic demands of the consumers and secondarily by species identity effects *per se*. The chapter by Yvon-Durocher *et al.* (2010) also demonstrates how relatively simple predictions based on individual metabolism can be used to characterise the responses of whole-ecosystem carbon cycling to environmental warming in a long-term field experiment.

The chapter by Woodward *et al.* (2010) includes the first description of a set of food webs constructed from both species- and a size-based perspective simultaneously using individual-level data, and as in the two B-EF papers, this study also revealed size-based redundancy within the system: diet width and the position of feeding links were better explained by body size than by species identity *per se.*, and the allometric diet breadth model predicted up to 84% of the links in a food web correctly, based on body mass data alone.

Within the past two decades, in parallel with the heightened activity in B-EF and food web research, molecular and microbial ecology have also evolved rapidly into mature scientific disciplines in their own right, fuelled by a series of technological breakthroughs, as described in the chapter by Purdy *et al.* (2010). The recent advent of Next Generation Sequencing (NGS) has opened up vast new vistas on a previously hidden world and, perhaps more than any other methodological advance in the last decade, this seems destined to change our view of ecology in radical and unexpected ways for many years to come. Large reservoirs of cryptic biodiversity and even previously unknown ecosystem processes are being unveiled at an astonishing rate, and the first ripples of this are now being felt in general ecology as molecular and microbial ecology become increasingly integrated into the corpus of the wider discipline. This wealth of new information offers novel ways to investigate whether microbial ecology follows general ecological principles, or whether there are inherent differences between the microscopic and macroscopic worlds, both of which are central themes in the chapters of Reiss *et al* (2010b) and Ptacnik *et al* (2010). The latter revisits a longstanding question in ecology – Hutchinson's classic "paradox of the plankton" – from a new perspective and highlights how, even in seemingly homogenous planktonic environments, a large number of species can be supported, and are indeed needed to sustain multiple ecosystem processes.

In summary, the seven chapters in this volume explore themes related to linking structure to functioning and each highlights the role of body size and/or metabolism as playing key roles in achieving this aim. Of course, this does not imply that these are the only variables of importance, rather that they provide a potentially useful means of collapsing a large amount of biologically relevant variation into a small number of dimensions, and this also enables other axes of interest to be explored more easily. In addition, the use of an individual-based perspective can help to bridge different disciplines and levels of organisation, and to provide potentially novel perspectives and insights into both old and new ecological questions.

Guy Woodward

REFERENCES

Perkins, D.M., McKie, B.G., Malmqvist, B., Gilmour, S.G., Reiss, J., and Woodward, G. (2010). Environmental warming and biodiversity-ecosystem functioning in freshwater microcosms: Partitioning the effects of species identity, richness and metabolism. *Adv. Ecol. Res.* **43**, 177–208.

Ptacnik, R., Moorthi, S.D., and Hillebrand, H. (2010). Hutchinson reversed, or why there need to be so many species. *Adv. Ecol. Res.* **43**, 1–43.

Purdy, K.J., Hurd, P.J., Moya-Laraño, J., Trimmer, M., and Woodward, G. (2010). Systems biology for ecology: From molecules to ecosystems. *Adv. Ecol. Res.* **43**, 87–149.

Reiss, J., Bailey, R.A., Cássio, F., Woodward, G., and Pascoal, C. (2010a). Assessing the contribution of micro-organisms and macrofauna to biodiversity-ecosystem functioning relationships in freshwater microcosms. *Adv. Ecol. Res.* **43**, 151–176.

Reiss, J., Cássio, F., Pascoal, C., Forster, J., Stewart, R., and Hirst, A.G. (2010b). When microscopic organisms inform general ecological theory. *Adv. Ecol. Res.* **43**, 45–85.

Woodward, G., Blanchard, J., Lauridsen, R.B., Edwards, F.K., Jones, J.I., Figueroa, D., Warren, P.H., and Petchey, O.L. (2010). Individual-based food webs: Species identity, body size and sampling effects. *Adv. Ecol. Res.* **43**, 209–265.

Yvon-Durocher, G., Allen, A.P., Montoya, J.M., Trimmer, M., and Woodward, G. (2010). The temperature dependence of the carbon cycle in aquatic systems. *Adv. Ecol. Res.* **43**, 267–313.

Hutchinson Reversed, or Why There Need to Be So Many Species

ROBERT PTACNIK, STEFANIE D. MOORTHI AND HELMUT HILLEBRAND

SUMMARY

There is compelling evidence for dispersal limitation among microscopic organisms, including phyto- and zooplankton, especially from studies addressing spatial patterns in taxon richness. This evidence is not in conflict with the widely accepted importance of strong local interactions in the plankton. However, the simultaneous importance of dispersal limitation and strong local interactions can only be understood when taking high temporal turnover rates into account.

Current observational and experimental evidence suggests that biodiversity–ecosystem functioning (B–EF) relationships do not differ systematically from those known from higher organisms. Plankton communities are not saturated by default.

ADVANCES IN ECOLOGICAL RESEARCH VOL. 43 0065-2504/10 $35.00
© 2010 Elsevier Ltd. All rights reserved
DOI: 10.1016/S0065-2504(10)43001-9

Although the pelagial has little spatial structure, it is rich in environmental dimensionality when considering the dimensionality in time and chemical and physical properties, resulting in complex biotic interactions.

We propose a conceptual model explaining B–EF effects in plankton, which contrasts environmental dimensionality with trait dimensionality of the community. This model, which is applicable to ecological communities in general, predicts that positive B–EF relationships depend on sufficient environmental dimensionality. We show how this model can be applied to understand B–EF relationships along gradients of productivity and stoichiometry.

Our major conclusions are that local community dynamics of plankton communities may be better understood when putting them into a wider spatial context, that is, considering regional species pools. Moreover, the framework of environmental and trait dimensionality can be used to make concise predictions for the occurrence and strength of B–EF relationships.

I. INTRODUCTION

The increasing awareness of the accelerating loss of global biodiversity (Worm *et al.*, 2006) has supported a major shift in ecological research in the past decade or so. Initially, researchers were mainly interested in how diversity is regulated in natural communities, and how apparently similar species may coexist, but the focus has now moved towards understanding diversity effects on ecosystem processes and services (Hillebrand and Matthiessen, 2009; Hooper *et al.*, 2005; Reiss *et al.*, 2009).

Starting from Tilman's seminal grassland experiments (Tilman *et al.* 1996), research on biodiversity–ecosystem functioning (B–EF) relationships has progressed rapidly, especially in terrestrial ecology (Hooper *et al.*, 2005). In aquatic habitats, most of the experimental work to date has focused on B–EF relationships in either microbial microcosms (e.g. Petchey *et al.*, 1999) or, more commonly, among the benthic macrofauna (e.g. Perkins *et al.*, 2010), with very few studies including both micro- and macro-organisms (but see Reiss *et al.*, 2010b). Benthic communities are in many ways much more similar to terrestrial communities than are their pelagial counterparts, which have so far received least attention in B–EF research. In fact, only 7 of 84 studies in the synthesis data set assembled by Cardinale *et al.* (2006b) deal with pelagic organisms, and these are all laboratory, rather than field, experiments. Nevertheless, this experimental work with artificial plankton communities played a pivotal role in the process of progressing from the early focus on grassland communities and primary producers into how diversity affects trophic interactions and food web dynamics (McGrady-

Steed *et al.*, 1997; Naeem and Li, 1997). However, while these experiments used planktonic organisms as model communities, they were not specifically designed to address pelagic systems, but to test first principles applicable to ecological communities in general.

The fact that diverse plankton communities exist within a seemingly homogenous environment with only a small number of limiting resources (light and one or few nutrients) has led to the notion of 'The paradox of the plankton', as first proposed in Hutchinson's classic 1961 paper. This apparent paradox implies a high degree of redundancy within these communities, in terms of, for instance, comparable resource requirements, similar uptake mechanisms of resources and similar vulnerability regarding predation, and is based on the intuitive assumption that local diversity of highly mobile organisms largely reflects local dynamics. The biological distinctness of planktonic communities in lakes, as well as the fact that they represent sensitive indicators to environmental stress, such as acidification and eutrophication (e.g. Watson *et al.*, 1997), has supported a 'locally centred' view on plankton communities, implicitly assuming that spatial processes are of secondary importance (see Section II).

This local focus has been further supported by the apparent ubiquitous distribution of many planktonic morphospecies (Fenchel and Finlay 2004). For decades, microscopic organisms have been considered as not being limited by dispersal, implying that local community composition simply reflects local processes. Baas-Becking's tenet 'everything is everywhere' (de Wit and Bouvier, 2006) has apparently been reinforced and fostered by the results of many studies. It has been argued that microbial organisms such as phytoplankton are highly abundant and disperse rapidly and thus are not prone to local extinction; moreover, the local diversity is considered so high that a reduction in ecosystem functioning with the loss of species is not expected, since many species can potentially perform similar roles (Finlay, 2002). These views have been challenged by the more critical evaluation of signs of biogeography in microbes and protists (Green and Bohannan, 2006; Martiny *et al.*, 2006; Smith *et al.*, 2005; Vyverman *et al.*, 2007) and new molecular techniques in particular have challenged the perceived existence of global diaspora (Hurd *et al.*, 2010). Increasingly, recent evidence suggests that despite this seeming 'ubiquity' of micro-organisms biogeographic diversity patterns are indeed manifested among the bacterioplankton (Fuhrman *et al.*, 2008), phytoplankton (Ptacnik *et al.*, 2010; Smith *et al.*, 2005) and zooplankton (Rutherford *et al.*, 1999) and that micro-organism diversity often follows similar patterns found for macro-organisms, for example, in relation to productivity (Irigoien *et al.*, 2004; Smith, 2007) or area (Horner-Devine *et al.*, 2004). Recent meta-analyses suggest that such patterns indeed exist across microbial taxa, even though they may be weaker or responses less steep than for macro-organisms (Drakare *et al.*, 2006; Hillebrand, 2004; Soininen *et al.*, 2007). An increasing number of studies find strong support for regional diversity control in both phyto- and zooplankton (see Section III). At the same time, there is accumulating evidence that comparable scaling relationships between biodiversity and functioning exist both in the microscopic and in the macroscopic world, contradicting the assumption that fundamental differences necessarily exist (see Section IV).

Both the ongoing paradigm shift regarding dispersal limitation in the microscopic world and the increasing awareness about ecosystem functioning relationships in plankton communities motivated us to summarize existing knowledge and to specify the need for further research. Although this chapter started initially as a review, especially its main part developed into a conceptional paper, integrating recent evidence with new ideas into a framework for future B–EF research in plankton ecology. After summarizing characteristics of the pelagic environment and plankton as a group (Section I), we address the ongoing paradigm shift within the diversity of microscopic organisms, which includes most of the plankton (Section II). We then address how competing views of local versus regional diversity control can be reconciled, and summarize the existing evidence for B–EF relationships in the plankton, addressing different functional groups (Section III). The main part of our review addresses the underlying mechanisms, where we emphasize the concepts of environmental dimensionality and trait diversity as central principles for understanding B–EF relationships (Section IV). Finally, we provide an outlook as to how these mechanisms can be applied within the framework of a highly successful ecological concept, ecological stoichiometry (ES; Section V). We conclude our review with an outlook on research needs and further research directions that offer promise for the future (Section VI).

II. PECULIARITIES OF THE PLANKTON

Plankton encompasses all organisms that are largely passively transported in the open water. We focus in our review on the communities inhabiting surface waters of lakes and oceans, which are most relevant in terms of productivity and nutrient cycles. Regarding B–EF relationships in the plankton, most studies so far have addressed freshwater communities, much in contrast to the relative global importance of lakes and oceans—while the oceans cover approximately 71% of the planet's surface and contribute approximately 50% to the global amount of primary production (Falkowski *et al.*, 1998), the quantitative importance of lakes to global nutrient cycling is comparatively small, given that lakes only make up 0.8% of the earth's surface area (Downing *et al.*, 2006).

Community dynamics in pelagic systems differ substantially from community dynamics in benthic or terrestrial systems. In particular, the plankton community of the upper mixed surface layer of the water column (epilimnion in lakes) inhabits a comparatively homogenous environment where ongoing mixing counteracts the emergence of patches. In a well-mixed environment, all organisms potentially interact with each other, much in contrast to terrestrial systems, where many species are either 'sessile' (plants) or have very limited range sizes (most small invertebrates except for flying insects). A further peculiarity is the short generation time of phyto- and zooplankton. Especially on the level of primary producers, aquatic systems are characterized by short-lived microalgae in contrast to annual to perennial species in terrestrial systems (Shurin *et al.*, 2006). This not only changes the dynamics of the system but also allows processes that span several generations to be studied within a relatively short time, which is a key advantage of using these model systems (Reiss *et al.*, 2010a).

These short generation times are related to the small size of most phyto- and zooplankton organisms. While the majority of mesozooplankton organisms (>200 μm) are constituted by metazoa, a major part of bacterivorous and herbivorous nano- and microzooplankton (<200 μm), as well as the entire guild of primary producers (=phytoplankton) consist of unicellular organisms (protists and cyanobacteria). Up to now, the analysis of protistan diversity in the context of B–EF relationships was mainly based on morphospecies distinctions using microscopic and culture-dependent methods. While these methods have contributed valuable groundwork for analysing protistan diversity in pelagic ecosystems, they are restricted to mainly larger organisms (>20 μm) with a distinctive morphology and by the fact that it is still impossible to culture the majority of protists (Moreira and López-García, 2002). A considerable number of protists are therefore likely to have escaped microscopic identification (Dawson and Pace, 2002; Moreira and López-García, 2002). In the past years, molecular biological approaches assessing the diversity of natural microbial assemblages have revealed a tremendous protistan diversity in various marine and freshwater habitats (e.g. Countway *et al.*, 2007; Massana *et al.*, 2004; Moon-van der Staay *et al.*, 2001; Not *et al.*, 2008; Šlapeta *et al.*, 2005), including large numbers of undescribed taxa and even new lineages (Hurd *et al.*, 2010 Massana *et al.*, 2002; Not *et al.*, 2007; Romari and Vaulot, 2004). Furthermore, these studies have demonstrated that in all systems investigated only a few taxa dominate protistan assemblages, while there is a huge number of rare taxa present at extremely small percentages (Caron and Countway, 2009). The extent of this 'unseen' diversity contained in the 'rare biosphere' is still largely unknown, as is its potential ecological importance (Hurd *et al.*, 2010). Caron and

Countway (2009) hypothesized that the members of the 'rare biosphere' not only confer a high level of functional redundancy to a given ecosystem at any given time, but also represent a strong potential for compensatory dynamics under changing environmental conditions. This is based on the fact that microbial species might have different optima regarding specific environmental conditions, but play similar ecological roles and thus can maintain ecosystem functioning (Caron and Countway, 2009; Dolan *et al.*, 2009). Therefore, the tremendous amount of rare taxa may act as a potential biological buffer, ensuring relatively stable community functioning over broad ranges of environmental forcing factors that influence protistan community composition. However, these changes in community structure may affect higher trophic levels substantially by altering resource competition and predator–prey relationships even if rates of elemental cycling and energy flow remain relatively constant (Caron and Countway, 2009).

Fundamental differences between pelagic and terrestrial systems lead to different niche partitioning dynamics among producers and consumers. Instead of being able to selectively locate particular habitats or patches with favourable environmental conditions, plankton organisms are subject to external forces such as wind, water currents and vertical mixing and are more or less passively transported horizontally and vertically. Therefore, they have to cope with a high variability in light, nutrients and other physical and chemical environmental conditions. While in terrestrial and benthic systems niche partitioning among species occurs to a great extent on a spatial scale along an environmental gradient (e.g. consumers are able to locate their feeding patches selectively), plankton organisms partition their niches more on a functional and a temporal scale (e.g. seasonal occurrence of plankton organisms; Wetzel, 2001).

It is not only the physical environment of plankton systems that is different from terrestrial and benthic systems but also the biotic structure (Shurin *et al.*, 2006): for instance, herbivores in pelagic systems differ from terrestrial consumers by ingesting whole 'prey' organisms instead of ingesting plant parts or parts of algal mats (terrestrial and macrophytes-dominated benthic systems, respectively). Also, because structural supporting tissues, such as lignin, are not so prevalent among phytoplankton, an overall larger proportion of plant production is consumed by herbivores in the plankton than in terrestrial ecosystems, and this facilitates top-down effects (Cebrian and Lartigue, 2004).

In contrast to terrestrial systems, in pelagic systems the functional boundary between primary producers and consumers is not as clearly defined, as many plankton organisms exhibit mixotrophy. This term generally refers to organisms which combine different nutritional modes, but is used in a restricted sense for organisms specifically combining photosynthesis and phagotrophy in plankton ecology (e.g. Sanders, 1991; Stickney *et al.*, 2000) and in this review. Mixotrophy has been observed in different groups of planktonic protists such as phytoflagellates, ciliates and sarcodines, and is known from virtually all types of surface waters (Riemann *et al.*, 1995; Sanders, 1991; Stoecker, 1998). Phototrophic and phagotrophic contribution to mixotrophic nutrition varies widely among mixotrophs (Holen and Boraas, 1995; Jones, 1994), ranging from primarily phototrophic protists supplementing their demands in nutrients by ingestion of prey to primarily heterotrophic ones which use heterotrophy to fulfil the majority of their energy requirements (Stoecker, 1998). These diverse nutritional strategies enable planktonic 'super-generalist' organisms to survive suboptimal environmental conditions, such as, for instance, light limitation, low concentrations of dissolved nutrients or low prey abundances.

The global significance of mixotrophy in pelagic systems has been demonstrated in numerous recent studies of bacterivorous phytoflagellates (e.g. Moorthi *et al.*, 2009; Zubkov and Tarran, 2008), mixotrophic dinoflagellates (e.g. Jeong *et al.*, 2005) and ciliates (e.g. Dolan and Perez, 2000). Due to the fact that mixotrophs act on two trophic levels, they increase the complexity of trophic interactions in planktonic food webs, while they also enhance the trophic efficiency and thus the amount of biomass supported at higher trophic levels (Ptacnik *et al.*, 2004; Sanders, 1991). Consequently, mixotrophy and other additional nutritional strategies (omnivory, cannibalism) further enhance the trait diversity in planktonic organisms; as consumer diversity increases, the diversity of nutritional strategies also increases, which can have much more complex consequences on the prey assemblage in terms of biomass, diversity and community structure than purely heterotrophic organisms differing only in feeding preferences and rates. Therefore, alternative nutritional strategies in plankton communities might have a stabilizing effect on ecosystem function by acting as a buffer for the system, by providing alternative pathways that might buffer the effect of species loss, at least in its initial stages.

III. DISPERSAL LIMITATION IN THE PLANKTON

A central prerequisite for B–EF relationships to be manifested is the existence of diversity gradients, which themselves must result from some external process in space or time. Only if diversity is controlled at least partly by independent, external processes, such as the supply rate of new species to a local ecosystem, may we regard it as a true factor driving system dynamics. Dispersal limitation became first evident from the study of island biogeography (MacArthur and Wilson, 1967), which revealed that the richness of an island's flora and fauna is a function of an island's size. However, even well-connected systems situated on large continents are not necessarily 'saturated' by dispersal: for example, species richness in natural assemblages of higher plants often increases when seed dispersal is enhanced experimentally (Turnbull *et al.*, 2000; Vyverman *et al.*, 2007).

It has repeatedly been proposed that communities of organisms < 1 mm are not dispersal limited and that they effectively operate within a global diaspora (Figure 1A; Fenchel and Finlay, 2004; Finlay, 2002). This assumption is based on the apparent enormous population sizes and high dispersal potential, and that many microbial morphospecies appear to have worldwide distributions

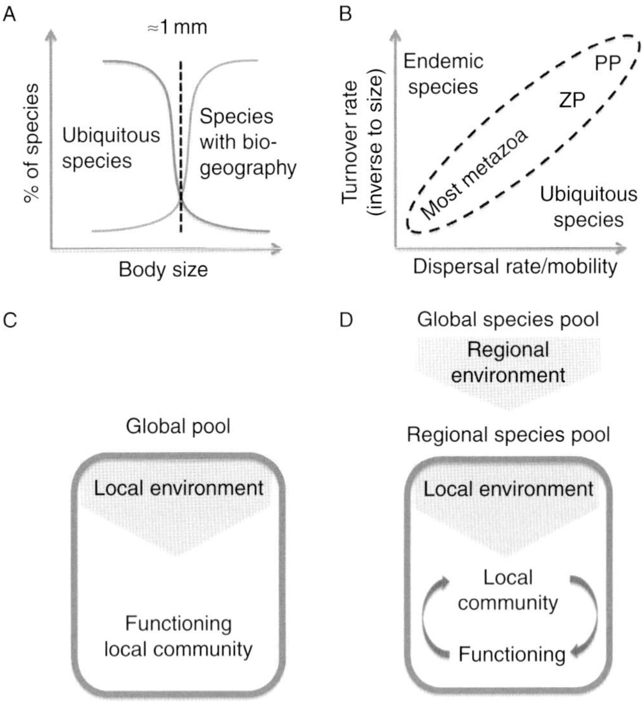

Figure 1 Illustration of the ongoing paradigm shift in plankton ecology. (A) Organisms < 1mm were assumed to be globally dispersed (modified from Finlay, 2002). (B) Recent evidence suggests that both dispersal rate and community turnover rate affect the expression of regional meta-populations (Ptacnik *et al.*, 2010). Since both processes scale with body size, most organisms are assumed to be found along the diagonal, that is, to have meta-populations. Combinations of low dispersal with fast turnover (endemic species) or vice versa (globally dispersed taxa) should be the exception.

(Finlay, 2002). The '*Everything is everywhere*' (de Wit and Bouvier, 2006; Finlay, 2002) hypothesis can be split into two predictions. First, fast dispersal overrides regional speciation patterns, implying there is no biogeography in microbes (i.e. there is only one global species pool). All local communities receive their species from this unique global pool, so community composition reflects solely the result of local sorting. Second, fast dispersal in small organisms implies that microbial communities are always saturated with regard to ecosystem functioning, that is, isolation by distance (e.g. island size) does not affect community saturation in microscopic organisms.

The first prediction is difficult to falsify for two reasons. First, the absence of a given species from a given site is difficult (or impossible) to prove in the microscopic world. Second, it is not possible to prove that its absence is really due to dispersal limitation or whether the given local environment selected against a particular species. However, the evidence for dispersal limitation in plankton organisms comes from highly diverging approaches. Numerous studies address dispersal limitation based on compositional data. Variance partitioning tries to separate the share of variation that can be attributed to either local environment or spatial distance (Peres-Neto *et al.*, 2006). If spatial diversity patterns cannot be attributed to comparable spatial patterns in environmental factors, this is regarded as an indication of regional species pools (aka metacommunities), implying the existence of dispersal limitation (Leibold *et al.*, 2004; Steiner and Leibold, 2004). There are two distinct approaches for addressing whether dispersal affects diversity or not. The analysis of beta diversity evaluates the effect of spatial distance on community composition, and addresses primarily the first prediction. There are several recent studies that have attempted to evaluate the existence of regional species pools in microbes by analysis of beta diversity (Martiny *et al.*, 2006; Nabout *et al.*, 2009; Van der Gucht *et al.*, 2007). While most of these studies point out that part of the spatial variation cannot be attributed to local environmental factors, they often find predominant local control of beta diversity.

A different approach is to study spatial patterns of richness, which since it correlates with functioning in B–EF relationships, directly addresses the second prediction. Dodson (1992) was among the first to study spatial patterns in zooplankton richness in lakes of the temperate zone. Besides finding significant relationships with local parameters, Dodson showed that zooplankton species richness increases with the number of surrounding lakes, pointing at connectivity as a factor that influences taxon richness. Since then, several studies report spatial autocorrelation in terms of taxon richness in phytoplankton (Ptacnik *et al.*, 2010) and benthic diatoms (Telford *et al.*, 2006; Vyverman *et al.*, 2007). Likewise, a number of studies analysing large-scale data on zooplankton diversity found strong regional patterns, pointing at regional control of richness (Hessen *et al.*, 2006; Shurin *et al.*, 2000).

A third line of evidence for dispersal limitation comes from an increasing awareness about the effects of shipping traffic on the dispersal of marine organisms, including microalgae (Kaluza *et al.*, 2010). The true extent of species colonizing new habitats via anthropogenic vectors is almost certainly underestimated since the focus in research of non-indigenous species has been on potentially harmful taxa, such as toxic phytoplankton species. A recent study by Olli and co-workers (in press), analysing a 30-year time-series of phytoplankton community composition in the Baltic Sea, found a tremendous and consistent shift in community composition within this period, especially in the central and eastern basins. Since the Baltic is among the most travelled coastal waters (Kaluza *et al.*, 2010), a relationship with ship traffic seems feasible. Further research will have to show to what extent such ongoing shifts in coastal plankton communities can be attributed to ship traffic etc. (Olli *et al.*, in press; Paavola *et al.*, 2005).

Given the increasingly compelling evidence for dispersal limitation from the different lines of reasoning and evidence, how does it occur in the presence of high abundance and mobility of plankton organisms? We propose here that the existence of dispersal limitation in microscopic organisms, including plankton, can only be understood by taking turnover rates into account. The generation time of organisms and thus the temporal turnover rate of corresponding communities are generally inversely related to body size (Brown *et al.*, 2004; Korhonen *et al.*, 2010). Hence, community turnover rate scales with dispersal rate in most organisms. From that, a conceptual model can be derived with dispersal limitation being a function of both dispersal and turnover rate (Figure 1B). In organisms with a low turnover rate, a low dispersal rate is sufficient to allow 'saturation' of niche occupancy, whereas at higher turnover rates, higher dispersal is necessary to provide species for rapidly emerging niches. Thus, for a community with rapid temporal species turnover, dispersal limitation can occur despite rapid dispersal. In fact, communities along the diagonal (Figure 1B) are expected to have similar-sized metacommunities. Exceptions from this general rule are given either by fast growing organisms with limited dispersal (endemic species; upper left) or by slowly growing species with high dispersal, which are therefore ubiquitous. Taking turnover into account reverses Hutchinson's classic paradox of the plankton—instead of asking why there are so many species (Hutchinson, 1961), we have to start analysing how often and under which conditions pelagic communities lack species.

Prevailing regional control of species richness has previously been reported for various higher organisms in both terrestrial and aquatic environments (examples in Cornell and Karlson, 1997). A recurrent pattern is a positive scaling between local and regional species richness. Irrespective of the actual mechanism affecting regional species pool size, such patterns generally point at a functional link between local and regional species pool size, which then has been shown to correlate with total area (Island or continent; Shurin *et al.*, 2000; Vyverman *et al.*, 2007), regionally averaged productivity (Ptacnik *et al.*, 2010) and pH (Telford *et al.*, 2006). From the ubiquity of positive correlations between local and regional diversity, Cornell and Karlson (1997) concluded that local dynamics have secondary effects on local community assembly and that regional dynamics (dispersal limitation) constrain local species richness. However, there is little doubt that local dynamics are very important in shaping plankton communities: for example, Shurin (2000) showed experimentally that they dominate pond zooplankton communities. Shurin *et al.* (2000) also found a linear relationship between regional and local richness in lake zooplankton. They concluded that zooplankton is ecologically saturated at the local level with regard to the respective regional species pool, while regional pools may be historically unsaturated. Similarly, Ptacnik *et al.* (2009) found that local community composition (but not richness) correlates strongly with local total phosphorus in Norwegian lakes, indicating that regional control of richness does not contradict strong local sorting (Ptacnik *et al.*, 2010). Rather, local richness reflects the dynamic local colonization–extinction equilibrium. With increasing dispersal (i.e. increasing size of regional species pools), a balance is reached at a higher richness level, but ultimately, it is the local dynamics that select taxa from the regional pool. Corroborating this finding, He *et al.* (2005) showed analytically that the regional effect on local richness does not preclude local species sorting. In metacommunities, linear relationships between regional and local richness can occur despite strong local interactions (Hillebrand, 2005). The accumulating evidence for the existence of metacommunities in phyto- and zooplankton lends much support to empirical studies linking phytoplankton resource use and community stability to the diversity of local communities (see Section IV).

The conventional view of plankton communities being largely controlled by local environment and dynamics (Figure 1C) is increasingly eroding. The diversity seen in natural communities reflects neither pure local coexistence nor entire regional control of plankton richness. Rather, it reflects a dynamic colonization–extinction equilibrium, integrating local and spatial processes (MacArthur and Wilson, 1967; Vandvik and Goldberg, 2006). With increasing dispersal, this equilibrium is reached at a higher level. Becoming aware of dispersal limitation in the plankton, we must acknowledge that plankton diversity is strongly impacted by regional processes. This implies that plankton diversity is not a function but a driver of system dynamics at the local scale (Figure 1D).

IV. PRESENT EVIDENCE FOR B–EF RELATIONSHIPS IN THE PLANKTON

A. Primary Production and Resource Use

Following Tilman's experiments with grassland communities, the first studies to show B–EF relationships with plankton communities came from experiments with artificial food webs (McGrady-Steed *et al.*, 1997; Naeem and Li, 1997). These studies manipulated plankton diversity at levels clearly below those seen in natural communities, and revealed that essential functions such as primary and secondary production, as well as stability of those processes, scale with the diversity of the communities. However, such positive B–EF relationships are not ubiquitous: other phytoplankton studies have found neutral (Gamfeldt *et al.*, 2005) or negative B–EF relationships (Schmidtke *et al.*, 2010) or more complex patterns across time (Weis *et al.*, 2007). Whereas these studies dealt with diversity levels far below natural communities in order to test general relationships, very few studies have systematically analysed the relationship between phytoplankton richness and ecosystem functioning at richness levels observed in the field. Ptacnik *et al.* (2008) and Striebel *et al.* (2009a), however, have shown that phytoplankton resource-use efficiency (RUE), measured as either algal chlorophyll or algal carbon per unit limiting nutrient (total phosphorus), scales with taxon richness in natural communities. Moreover, Striebel *et al.* (2009a) have shown that the relationship between richness and RUE scales consistently from artificially assembled, species poor communities to species rich field communities. These studies were the first to confirm that there seem to be general B–EF relationships among all primary producers, including microbial organisms.

Diversity does not only affect average resource use. In agreement with findings from higher plants, Ptacnik *et al.* (2008) showed that temporal variability in resource use (RU) and community composition were inversely related with richness, but increased with productivity. Likewise, Steiner (2005) showed that temporal variability of pond zooplankton biomass is inversely related with richness, but increases with system productivity. By contrast, Downing *et al.* (2008) found no evidence for a stabilizing effect of diversity in plankton systems when large synchronizing forces (i.e. strong seasonality) prohibited strong compensatory dynamics.

B. Resource Use in Heterotrophic Bacteria

We are unaware of direct evidence for B–EF scaling relationships from field studies with aquatic bacteria. However, positive B–EF relationships are evident from communities of soil bacteria (e.g. Griffiths *et al.*, 2000).

Moreover, Langenheder and co-workers (2006) studied the performance of heterotrophic bacteria in the pelagic zone of humic lakes and found that resource use correlates with changes in community composition along a vertical depth gradient, indicating that diversity affects functioning of bacterial communities.

C. Secondary Production and Trophic Interactions

In recent years, studies investigating the relationship between biodiversity and ecosystem functioning have begun to incorporate trophic-level effects in multitrophic systems (see reviews by Duffy *et al.*, 2007; Srivastava *et al.*, 2009). All natural systems contain multiple trophic levels; furthermore, species loss in higher trophic levels appears to occur more frequently than at lower trophic levels (Petchey *et al.*, 1999, 2004), which can have serious effects on ecosystems (Hughes and Connell, 1999; Jackson *et al.*, 2001) that can even be larger than the comparable loss of primary producers (Duffy *et al.*, 2003).

The majority of B–EF studies have been conducted using terrestrial or benthic aquatic systems, with only a few of them investigating consumer effects in planktonic food webs. Steiner *et al.* (2005) investigated zooplankton effects on algal prey in a freshwater system and showed a positive effect of zooplankton diversity on zooplankton biomass. Even though no zooplankton diversity effect on total algal biomass was detected, increased zooplankton diversity significantly altered the size structure of algae, increasing the relative abundance of large, grazer-resistant algae. However, Gamfeldt *et al.* (2005) both manipulated consumer (ciliates) and prey (algae) richness and identity independently and showed clear biodiversity effects of both consumers and prey, within and across trophic levels. Consumer richness reduced prey and increased consumer biomass, with the most diverse prey assemblage supporting the highest biomass of consumers at their highest richness. Thus, enhanced energy transfer was associated with simultaneous increases in the richness of consumers and prey. Dzialowski and Smith (2008) tested whether the effects of consumer identity and diversity are nutrient dependent in a freshwater food web (cladoceran consumers and algal prey) and found greater total zooplankton and lower algal biomass with increasing zooplankton diversity only under nutrient enrichment. This suggested that diversity effects were related to biological mechanisms, such as resource-use complementarity, that were enhanced at high nutrient concentrations. As consumer diversity *per se* did not have consistent effects on prey (no effect, strong effect, nutrient-dependent effect), other mechanisms, including specific consumer interactions with other consumers and their prey, must be relevant when regarding biodiversity effects, indicating that consumer identity and traits are important in this context. Worsfold *et al.* (2009) demonstrated in a microcosm experiment with three trophic levels that the presence of a specialist predator, though rare, significantly altered the effects of a generalist predator on the total biomass of prey species, indicating that the impact of species loss from high trophic levels could be very context dependent. These results corroborate previous results from a number of terrestrial and benthic aquatic studies (e.g. Bruno and O'Connor, 2005; Finke and Denno, 2005; Snyder *et al.*, 2006; Straub and Snyder, 2006), showing that changes in diversity on higher trophic levels can substantially affect ecosystem function in natural systems. However, the direction of the change (+/−) depended on species identity and thus on consumer traits in relation to other consumers (interference competition, intra-guild predation) and their prey (ingestion rates, feeding preferences). Theory suggests that herbivores can have very different effects on biomass, diversity and composition of their prey, depending on their specialization (generalist vs. specialist, food web connectivity; Thébault and Loreau, 2003), which consequently affects the complementary use of resources and prey abundances through direct and indirect interactions. The first empirical tests of this model are now underway through experimental manipulation of ciliate diversity and their degree of specialization (Filip *et al.*, in preparation). Finally, a distinctly different aspect of functional protist diversity affecting secondary production comes from feeding experiments with marine copepods. These studies show that monospecific diets (especially diatoms) often result in reduced reproductive success in copepods, while mixed diets (especially diatoms with flagellated algae and/or the heterotrophic protozoa) result in high reproductive success ('trophic upgrading'; Kleppel, 1993; Klein Breteler *et al.*, 1999).

D. Underyielding and Superspecies

The studies highlighted above show that B–EF relationships in plankton communities show rather divergent results and are also very diverse in their empirical approaches. Moreover, a number of papers suggest 'underyielding' and the presence of 'superspecies' dominating the ecosystem process across a broad range of conditions, that is, either species mixtures performed worse than single species or mixture effects were purely reflecting the dynamics of a single species. In laboratory experiments with eight phytoplankton species from different algal groups, consistent underyielding was detected, which was caused by a negative dominance effect (Schmidtke *et al.*, 2010). One fast-growing species (*Monoraphidium*) pre-empted the resources in mixtures but had lower final yields than slow-growing species. However, these slower species, which were responsible for the high biovolume in monocultures, were unable to grow fast enough before *Monoraphidium* had consumed most resources. Thus, underyielding was mainly based on a trade-off between growth rate and biomass production. In a subsequent experiment, Schmidtke (2009) found overyielding and transgressive overyielding in a different phytoplankton community, which showed a positive correlation between growth rate and biomass production. From the two experiments, Schmidtke and co-workers concluded that overyielding in phytoplankton communities very much depended on the monopolization of resources by fast-growing species and the correlation between growth rate and biomass production. Other plankton B–EF studies concur with this example, insofar as the results often depended on the performance of single species, which dominated irrespective of different environmental conditions or community composition (Norberg, 2000; Weis *et al.*, 2008). We will devote the next section to propose mechanisms unifying the divergent results of pelagic B–EF experiments and the seemingly paradox occurrence of 'superspecies' and underyielding.

V. MECHANISMS UNDERLYING PELAGIC B–EF RELATIONSHIPS

We propose here that the key to understand the differentiation between the outcome of simple laboratory experiments and B–EF relationships in field studies is the variability in environmental conditions and the variance in traits established in the plankton community. We will develop the argument in the following section, representing a conceptual model to understand B–EF relationships in plankton, but with general relevance to other systems. We will then highlight the applicability of this model through two examples, 'productivity–trait dimensionality' and 'spectral coexistence and stoichiometry'. Finally, we will broaden the view on 'stoichiometric constraints of B–EF relationships', ending with predictions for the role of biodiversity in the face of anthropogenically changed biogeochemical cycles.

A. Environmental and Trait Dimensionality

The concept of niche dimensionality has been used to explain the change of plant biodiversity following the addition of multiple nutrients (Harpole and Tilman, 2007). The number of plant species in a grassland decreased when progressively more nutrient types were added and as such the number of potentially limiting resources (=niche dimension) declined. Similarly, a survey of lakes provided evidence that phytoplankton diversity decreases in line with the number of potentially limiting resources (Grover and Chrzanowski,

2004; Interlandi and Kilham, 2001). The interpretation of this effect of the number of limiting resources on species coexistence is based on resource-ratio theory stating that a higher number of limiting nutrients allows more species to coexist (Tilman, 1982).

Niche dimensionality can, however, be envisioned much more broadly than just the number of limiting resources (Figure 2A). For instance, the diversity of pelagic communities is strongly affected by both temporal (Beisner, 2001; Flöder *et al.*, 2002; Litchman, 1998) and spatial variation (Barnett and Beisner, 2007) of resource supply. Thus, variability in time and space and the number of limiting resources are all part of resource dimensionality, which represents one axis of 'environmental dimensionality' which is broader

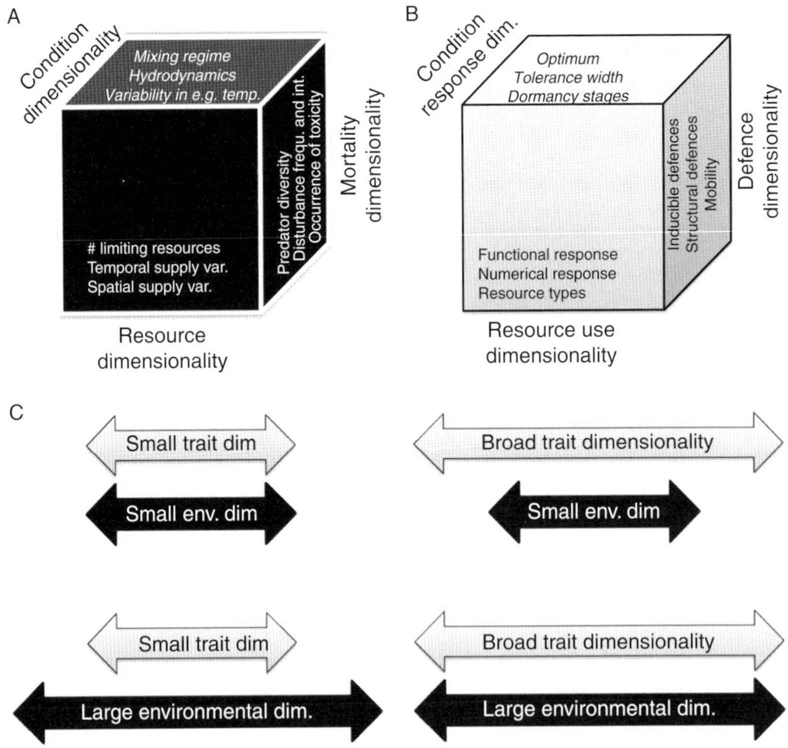

Figure 2 (A) Environmental dimensionality describes the potential for species coexistence through differential responses to resources, abiotic conditions and sources of mortality. (B) Trait dimensionality describes the functional diversity of the assemblage in response to resources, conditions and mortality sources. (C) Strong positive B–EF relationships are predicted only if a large environmental dimensionality allows for species coexistence and a broad trait dimensionality allows for complementarity.

than niche dimensionality (*sensu* Harpole and Tilman, 2007). A second axis of environmental dimensionality is represented by the variation in abiotic conditions (excluding resources), which can include physical and chemical fluctuations. Phytoplankton diversity has been shown to increase through fluctuations in temperature (Descamps-Julien and Gonzalez, 2005; Jiang and Morin, 2007), although this effect is not ubiquitous (Burgmer, 2009; see below for reasons for this deviation). In a large survey of lake zooplankton data, species richness also increased with temperature fluctuations, but decreased with increasing variance in chemical variables (Shurin *et al.*, 2010). Thus, abiotic fluctuations *per se* do not necessarily enhance the dimensionality of environmental conditions (see below).

A third axis comprises the dimensionality of mortality sources, such as increasing diversity of consumer types, the frequency and intensity of disturbance effects or the occurrence of toxic substances (Figure 2). Again, a broader dimensionality on this axis increases the potential for higher diversity if, for example, predation enhances prey diversity by facilitating trade-offs between prey responses (Chesson and Kuang, 2008). Also, abiotic mortality sources may enhance environmental dimensionality, as diversity is strongly affected by the frequency and magnitude of disturbance events (Shea *et al.*, 2004). These mortality effects are strongly related to resource dimensionality, as altering resource supply and ratios alter the relationship between disturbance and diversity (Cardinale *et al.*, 2006a; Kondoh, 2001). The use of allelopathic substances toxic to competitors might alter the diversity of responding communities, for example, by altering sign and intensity of competition between phytoplankton species (Declerck *et al.*, 2007; Hulot and Huisman, 2004).

Thus, the box representing environmental dimensionality (Figure 2A) is a box of 'niche opportunities', as it describes in different axes (which are not necessarily exhaustive) the ways organisms can partition their environment. Decreasing the width of these axes also restricts the opportunities for species to coexist, whereas 'a larger box' translates to larger chance for coexistence. However, it is critical to emphasize the terms opportunity and chance here, as the potential for coexistence also depends on the variability of traits in the community responding to the environmental dimensionality. In the experiment by Burgmer and Hillebrand described above (Burgmer, 2009), increasing the variance of temperature did not alter competitive interactions, but favoured an already dominant species in a species-poor-community. Thus, high environmental dimensionality has to be met by high 'trait dimensionality' (Figure 2B) in order to allow for a large diversity response and large diversity effects on ecosystem processes.

The consideration of trait-based approaches to ecology has recently been fostered by fundamental synthesis efforts for phytoplankton (Litchman and Klausmeier, 2008; Litchman *et al.*, 2007) and zooplankton (Barnett *et al.*, 2007). In analogy to environmental dimensionality, trait dimensionality can be addressed for different axes. Organisms may differ in the way they partition resources by consuming different species, by differing in their uptake rates (functional response) and their resource requirements for growth (numerical response). If species differ more in these traits and show higher trade-offs along these axes, they are more likely to coexist (Chesson, 2000). The same is true for the mortality axis, as has been shown for predation, which increases coexistence if it allows stronger trade-offs in responses among prey species (Chesson and Kuang, 2008). The same argument—at least conceptually—holds as an explanation for the intermediate disturbance hypothesis, where very frequent/strong or very infrequent/weak disturbances select for the same type of responses across all species and thus constrain diversity, whereas intermediate levels maximize the mortality niche as species can exhibit differential responses such as evasion, tolerance, recovery or resistance. Thus, intermediate disturbances allow for a maximum mortality niche and enhance diversity if the species in the community show a broad width of responses (i.e. trade-offs between response types).

This latter argument highlights the important fact that the box axes in Figure 2 are not the axes of increasing 'intensity' or increasing 'magnitude', but of maximizing differences (dimensions) in the environment (Figure 2A) or in the community (Figure 2B). Thus, also the response to variation in other-abiotic conditions depends on the width of the response axis, that is, the trade-off in species' optimal conditions, the widths of their physiological tolerance and their difference in tolerating adverse conditions, for example, by dormancy. Thus, in the face of large disturbances, the mortality dimensionality might decline. In comparison of zooplankton composition between disturbed and undisturbed lakes, the disturbed communities consisted of more closely related species (Helmus *et al.*, 2010). This disturbance-induced reduction in phylogenetic diversity occurred independent of the species richness, evenness or abundance and indicated that the disturbance was strong enough to reduce the number of successful trait combinations.

We have argued before that a trait-based approach is needed to understand B–EF relationships in general (Hillebrand and Matthiessen, 2009). We extend this argument here in proposing that strong B–EF relationships require the occurrence of both high environmental and high trait dimensionality (Figure 2C). If the environment is invariant, increasing diversity—which in these cases is also often artificially maintained by the experimentalist or modeller—has no dimension to allow for higher functional process rates. If the trait dimensionality is low, the species are too similar to make effective use of whatever small or large variation exists in the environment.

These considerations might explain seemingly counterintuitive results on 'superspecies' in pelagic B–EF experiments. Single species are apt to outperform mixtures under stable conditions, whereas mixture can show higher process rates under variable conditions (Norberg *et al.*, 2001). Thus, constant conditions are more likely to promote species identity effects rather than B–EF relationships. Broadening the environmental dimensionality alone does not necessarily provide stronger B–EF results if the trait dimensionality is low. In a microcosm study with pelagic algae, no stronger diversity effects were observed when algae were grown in heterogeneous (different N:P supply in patches) compared to homogeneous patches (Weis *et al.*, 2008). An impact of heterogeneity on biodiversity effects was prevented by the fact that the same algal species dominated the biomass in both heterogeneous and homogeneous conditions.

Evidence for the tenet that higher trait diversity makes a difference to B–EF effects comes from a study on rock pool communities: when assembling zooplankton from a broader spatial scale (as a proxy for sampling more divergent zooplankton), both zooplankton productivity and consumption of algae increased (Naeslund and Norberg, 2006). Also, the effect of consumer diversity on phytoplankton biomass strongly depends on the presence of differences in the prey community leading to a broader resource dimensionality, which, for example, can comprise differences in edibility of prey species (Norberg, 2000; Steiner *et al.*, 2005). Simpler experiments using only one prey type are more prone to miss these interactions.

We are not aware of more direct tests of increasing trait or environmental dimensionality in plankton, but general support for these ideas comes from studies in other ecosystems. Increasing the spatial heterogeneity of a habitat (i.e. increasing environmental dimensionality) strengthened B–EF relationships in terrestrial (Tylianakis *et al.*, 2008) and marine benthic (Griffin *et al.*, 2009) ecosystems. Environmental dimensionality is also bound to increase with longer duration of experiments, which has been shown to enhance B–EF relationships (Cardinale *et al.*, 2007; van Ruijven and Berendse, 2009). The link to trait dimensionality has been demonstrated clearly in a recent study (Otto *et al.*, 2008), where enhanced predation with higher predator richness depended on temporal niche separation, which in this case was given by non-overlapping phenology of predators which created a form of temporal niche complementarity.

Even the recent interest in ecosystem multifunctionality (Gamfeldt *et al.*, 2008; Hector and Bagchi, 2007; Reiss *et al.*, 2009; Zavaleta *et al.*, 2010), where stronger B–EF relationships are proposed if multiple functions are considered in an ecosystem, reflects considerations of environmental dimensionality. Multiple functions deal with different resources and therefore form a larger resource dimension axis than is the case for single functions, and thus enable larger trait dimensionality to play out.

By contrast, strong synchronizing forces, such as high latitude seasonality, can reduce both environmental and trait dimensionality, as they select for certain traits and inhibit the functional consequences of trade-offs.

Consequently, synchronization can decrease the potential for compensatory dynamics between zooplankton taxa and thus preclude diversity effects on the temporal stability of the entire community (Downing *et al.*, 2008). The occurrence of compensatory dynamics may be masked by large seasonal signals which can synchronize population dynamics. The stabilizing force of compensatory dynamics between species in a functionally rich community may thus be restricted to sub-annual time scales (Vasseur and Gaedke, 2007), or put another way, synchronized and compensatory dynamics may occur on different time scales (Vasseur *et al.*, 2005). Such synchronization might also affect the relationship between richness and compositional turnover. Shurin *et al.* (2007) showed that species turnover exhibited qualitatively different associations with the total number of zooplankton species in different biogeographic regions: at temperate latitudes, higher zooplankton diversity resulted in lower turnover, whereas polar lakes showed both low turnover and low richness. An accompanying model suggested that feedbacks from diversity on colonization or extinction rates can produce this empirical pattern of reduced turnover with increasing species pool size (Shurin, 2007). Thus, extreme seasonality at high latitudes might prevent richness–stability relationships.

In summary, our conceptual model highlights some important aspects of B–EF relationships, which are proposed to hold beyond pelagic communities. First, B–EF relationships cannot be seen independent of the mechanisms constraining coexistence (Hillebrand and Matthiessen, 2009), the potential for which increases if niche overlap decreases and fitness differences are minimized (Chesson, 2000; Chesson and Kuang, 2008). Increasing trait and increasing environmental dimensionality decrease niche overlap and increase the potential for trade-offs leading to fitness equality. Thus, the same mechanisms facilitating coexistence lead to a higher potential for positive B–EF relationships.

B. Productivity–Environmental and Trait Dimensionality

After temperature, productivity is arguably the next major structuring gradient of aquatic environments. Especially in lakes, productivity may vary over several orders of magnitude (Wetzel, 2001). Though this productivity gradient has increased with anthropogenic eutrophication, tremendous ranges can also be observed in non-impacted lakes: for example, in Fennoscandia, lakes not impacted by eutrophication range from $<2\ \mu g$ TP L^{-1} in Norway to $>50\ \mu g\ L^{-1}$ in Finland (Henriksen *et al.*, 1998).

The variability of phytoplankton community composition diversifies with increasing productivity (Figure 3A). Both temporal (within site) and spatial (among sites) variabilities of phytoplankton community composition

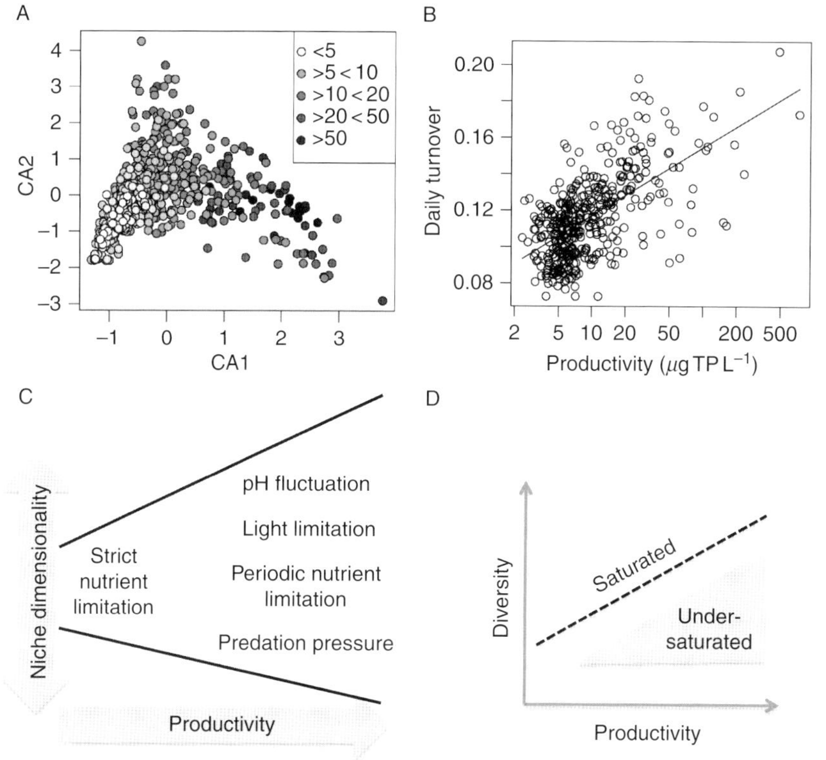

Figure 3 Productivity and environmental dimensionality. (A) Variability among phytoplankton samples from different lakes (=spatial turnover) increases with productivity [legend; μg TP L^{-1}; data from Norwegian lakes (Ptacnik *et al.*, 2010); correspondence analysis on cubic-root transformed genus data]. (B) Temporal community turnover in lake phytoplankton as a function of productivity; every dot gives the turnover averaged over one season (Norwegian lakes; see Ptacnik *et al.*, 2008 for details). (C) Various biotic and abiotic factors become more important as productivity increases. (D) As a consequence, diversity for maintenance of ecosystem functions increases with productivity.

increase with productivity (Ptacnik *et al.*, 2010; Figure 3A,B). Since phytoplankton community composition gives a sensitive bioindicator of nutrient conditions (e.g. Ptacnik *et al.*, 2009), the observed diversification seen in Figure 3A does not merely reflect random fluctuations, but indicates that the number of potential community configurations increases with productivity. Ptacnik *et al.* (2010) have not included hypertrophic lakes in their study, but it seems arguable that this positive effect of productivity on community turnover extends into hypertrophic lakes.

The pattern points at a relationship between productivity and environmental dimensionality and can be confirmed by considering the potential constraints acting on phytoplankton communities at different productivity levels (Figure 3C). At very low productivity, mineral nutrients are limiting to growth and species compete within a constrained total niche. With increasing productivity, the number of potential interactions and constraints increases. Adaptation to changing physical conditions (especially reduced light conditions) and vertical migration represent 'niches' opening in productive systems. Grazing pressure and top–down control increases, opening larger mortality dimensionality (see Figure 2A) as different taxa may exhibit grazer avoidance by either maximizing growth rates, or by expressing grazing resistance by colony formation or expression of defence structures (Reynolds *et al.*, 2002). Allelopathy and toxicity are additional traits that are more often encountered in productive systems (Huisman and Hulot, 2005; Scheffer *et al.*, 1997).

In the context of dispersal limitation, increasing niche space implies an increasing risk for communities to be unsaturated. We argue that the destabilizing effect of elevated productivity (=enrichment), known as 'the paradox of enrichment' (Rosenzweig, 1972), can be better understood considering the interaction between productivity and environmental dimensionality (Figure 3D). As a proof of concept, we tested this hypothesis by analysing phytoplankton RUE (Ptacnik *et al.*, 2008) as a function of productivity (ln (total phosphorus)) and diversity (genus richness) in lakes with low (southwestern Norway) and high (Southern Finland) beta diversity (see Ptacnik *et al.*, 2008 for details on the data and analysis of RUE). In both subsets, RUE increases with diversity (Figure 4). Moreover, variability of RUE tends to increase with productivity, confirming the generality of the destabilizing enrichment effect (Rosenzweig, 1972). However, variability is generally higher in Western Norway (Figure 4A), where species richness is lower. Moreover, in Western Norway, RUE is high only at low productivity, but decreases with increasing productivity. On the contrary, average RUE is high throughout the productivity gradient in Finland (Figure 4B). The data confirm our expectation that symptoms of under-saturation, such as low and unstable resource use, should apply especially to productive systems with species poor communities.

By affecting beta diversity, productivity also affects the size of regional species pools. In a detailed study on regional diversity patterns in lake phytoplankton, Ptacnik *et al.* (2010) showed that phytoplankton genus richness exhibits a scale-dependent productivity–diversity relationship, similar to what has been demonstrated previously for amphibians and macro-invertebrates (Chase and Leibold, 2002; Chase and Ryberg, 2004). The underlying mechanism is that temporal and spatial community turnover both scale with productivity (Figure 3A and B), resulting in a positive correlation between

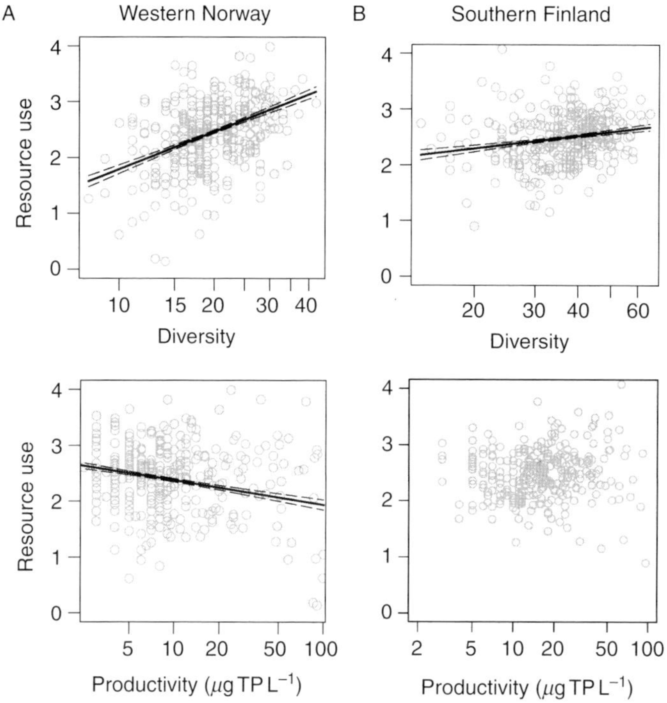

Figure 4 Phytoplankton resource-use efficiency [RUE; ln(μg chlorophyll *a* L^{-1}/μmol TP L^{-1})] as a function of productivity in species-poor (Western Norway, *A, left*) and species rich lakes (Southern Finland, *B, right*). The analysis corresponds to the analysis shown in Ptacnik *et al.* (2008), Table 1 therein, except that two more confined subsets (Western Norway and Southern Finland) with a similar range in TP concentrations (≤ 100 μg L^{-1}) were chosen here in order not to confound local with regional effects of TP (Ptacnik *et al.*, 2010) (see Appendix for regression statistics).

productivity and beta diversity. The scale-dependent productivity–diversity relationship highlights that local and regional dynamics influence diversity patterns at local and regional scale interactively. Local productivity ultimately affects local dynamics and species sorting, but the size of the regional pool, which is fueling each of its local pools, is an integrated function of connectivity and beta diversity (Chase and Ryberg, 2004).

The interactive effect of local and regional processes implies that differences between local and regional environmental parameters may provide important insight into local ecosystem functioning. This is particularly true for productivity as a direct driver of niche space. Sites with high productivity surrounded by low-productive lakes may be particularly 'isolated'. An example for such a situation is outlined in Figure 4—productive lakes in Western Norway (Figure 4A) are situated in a regional pool mostly comprising oligotrophic systems, while the productive lakes in Finland (Figure 4B) are surrounded by many other productive lakes. Though these relationships are derived from observational data sets of lake phytoplankton, they are likely to have wider applicability. For example, Steiner (2005) showed that temporal variability of zooplankton biomass increases with productivity, but is inversely related with its taxon richness.

Resource use is considered a major driver behind positive B–EF relationships (Cardinale *et al.*, 2006b). Low and unstable resource use opens windows of opportunity for single species to invade systems and to monopolize resources. Harmful algal blooms are the most conspicuous events of resource monopolization in aquatic environments which coincide with low algal diversity. It remains to be seen, however, whether low diversity results from blooms, or whether diversity patterns play a role in bloom initiation.

C. Spectral Coexistence and Stoichiometry

Compared with terrestrial plants, phytoplankton has a much higher phylogenetic diversity, and diversity in pigments and light use strategies (Gragham *et al.*, 2010). Pigment diversity facilitates coexistence in phytoplankton because it reduces overlap in the wavelength spectrum used (Stomp *et al.*, 2004, 2007). In addition to the pigments of green land plants with rather similar absorption spectra (mainly chlorophyll *a* and *b* and carotenoids), different algal groups comprise other chlorophyll and carotenoid types including an array of xanthophylls, but also phycoerythrin and phycocyanin (Figure 5). Thus, increasing diversity of algae is reflected by an increasing diversity of pigments (Striebel *et al.*, 2009a).

In contrast to algae, terrestrial plants show high trait dimensionality for mineral resource uptake (Figure 5). Plants differ in the spatial, temporal and chemical characteristics of the soil nitrogen (N) pools they use (Kahmen *et al.*, 2006; McKane *et al.*, 2002; von Felten *et al.*, 2009). Likewise, there are ample possibilities to partition different soil phosphorus (P) pools (Turner, 2008), especially through the diversity of mycorrhizal fungi involved in P-sequestration (Van der Heijden *et al.*, 1998). In pelagic communities, there is much less ability to show niche complementarity in mineral resource use. On one hand, the entrainment of algae in the water column reduces the chance to differentiate horizontal spatial and temporal resource gradients compared with the localised competition for mineral nutrients in soils. On the other hand, algae mainly sequester inorganic N and P by direct uptake from the water column, with other sources being present but less predominant (but see Section II, for the role and occurrence of mixotrophy in the phytoplankton).

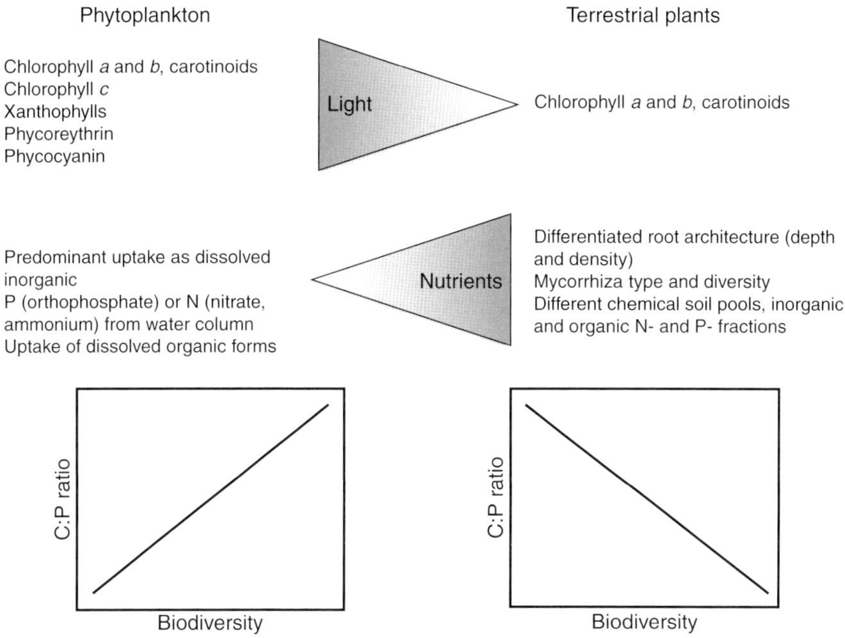

Figure 5 Comparison of trait dimensionality between phytoplankton and terrestrial plants. Phytoplankton species host a broader variety in light harvesting pigments, whereas plants command over a broader variety of nutrient uptake mechanisms. Therefore, we expect differential effects of biodiversity on C:P ratios of primary producers at sea than at land (see text for details).

This divergence between pelagic and terrestrial primary producers can be described as much larger trait dimensionality in phytoplankton with regard to light use, relative to larger trait dimensionality for nutrient uptake in terrestrial plants. These differences have potentially major consequences for the stoichiometry of B–EF relationships, as the magnitude of the biodiversity effects corresponds to the variance in traits in the community (Hillebrand and Matthiessen, 2009). In pelagic communities, we expect higher trait variance in light acquisition than in mineral resource acquisition traits, so higher algal diversity should maximize light use and thus carbon fixation (Figure 5). In terrestrial communities, light acquisition traits show far less spectral divergence between species, thus higher plant diversity should maximize nutrient uptake rather than carbon fixation. This leads us to predict that increasing diversity in phytoplankton communities should increase C:P ratios of the autotrophs, whereas increasing diversity in terrestrial plants should decrease C:P ratios.

At least the first part of our tenet has been formally tested. In an elegant combination of field and laboratory investigations, Striebel *et al.* (2009a) elegantly showed that high algal diversity correlated with higher pigment diversity, which then led to higher carbon fixation (=net primary production) at higher diversity due to light use complementarity. As a consequence, the C-fixation efficiency increased more rapidly with diversity than the P-uptake efficiency, which then leads to increasing C:P ratios with increasing phytoplankton diversity (Striebel *et al.* 2008, 2009b).

The potential scaling of the stoichiometry of ecosystem processes to phytoplankton diversity has major consequences for the trophic transfer of nutrients in pelagic systems. As aquatic herbivores feed on prey comparably rich in carbon and low in N and P, a further increase in C:nutrient ratios with increasing diversity might reduce food quality further, leading to an increasing stoichiometric mismatch and thus lower rates of prey removal (Hillebrand *et al.*, 2009a). These stoichiometric consequences of shifts in plankton diversity for trophic interactions have rarely been assessed so far, but recent evidence suggests that trophic propagations are likely, but context dependent. Urabe and Waki (2009) showed that higher phytoplankton diversity mitigated the effects of altered CO_2 availability on algal food quality for primary consumers.

D. Stoichiometry of Ecosystem Functioning

ES has been one of the most successful ecological frameworks in recent decades. It considers the demand of organisms for and the availability of multiple elements to make predictions about autotroph and heterotroph growth, trophic interactions and ecosystem dynamics (Sterner and Elser, 2002). Recent synthesis efforts have highlighted that autotroph growth is rarely limited by a single element only; instead, the predominance of co-limitation allows much larger producer responses to multiple additions than to the addition of single nutrients (Elser *et al.*, 2007). Such co-limitation can occur through physiological coupling of uptake processes within individuals, different growth status between individuals within populations, and different resource requirements between species within communities (Danger *et al.*, 2008). It is a central tenet of ES that producers are more flexible than their consumers in elemental composition. While this is true as a general principle, there is considerable variation across types of ecosystem and organisms (Persson *et al.*, 2010). Therefore, nutrient regeneration from animals consuming primary producers has strong stoichiometric constraints (Hillebrand *et al.*, 2008). Biotic components of ES thus may have strong regulating influence on the large-scale coupling of biogeochemical cycles for multiple elements (Lenton and Klausmeier, 2007; Riebesell *et al.*, 2007; Woodward *et al.*, 2010).

Given such tight linkages between biology and biogeochemistry, it should be obvious that ES should also play an important role in determining B–EF relationships, but this has rarely been assessed (Woodward, 2009). In a grazer–periphyton study, however, Hillebrand *et al.* (2009b) found consumer diversity to affect the stoichiometry of nutrient recycling. The same was suggested in a study simulating the loss of fish species from Lake Tanganyika and a neotropical river system (McIntyre *et al.*, 2007).

Based on the conceptual model described above (Figure 2) and the knowledge on light:nutrient resource efficiency in phytoplankton of different diversity (Figure 5), we will discuss two connected hypotheses below addressing ES as a constraint of B–EF relationships as well as a response to B–EF relationships. First, we propose that the ES of resource availability constrains the strength of B–EF relationships as it affects the environmental dimensionality. Second, we detail how trait dimensionality couples processes within ecosystems stoichiometrically, leading to novel predictions about the magnitude and variance of multifunctionality.

Considering species competing for two limiting resources, classical resource competition theory predicts that—depending on resource supply ratios—at maximum two species will coexist at equilibrium (Tilman, 1982) (Figure 6A). The identity of these species depends on the position of their zero-net-growth requirements and their consumption vectors. The stable coexistence points are all characterized by each species being limited by another resource due to trade-offs in critical resource requirements. Increasing the concentration of resource 1 (R1) while keeping resource 2 (R2) constant thus shifts the community from being exclusively limited by R1 (at very low supply) to all species being limited by R2 (Figure 6B). Coexistence of species is thus restricted to intermediate resource supply levels of a resource or—if both resources vary in concentrations—to intermediate resource ratios. Thus, the ratio of R1:R2 constrains the probability of co-limitation and the number of limiting resources (Figure 6C). Therefore, the supply ratio also constrains environmental dimensionality, given that both very low and very high ratios leave limited space for trade-offs to play out. A better competitor for R2 will not contribute much to overall ecosystem productivity if R1 is limiting productivity in all species and vice versa.

The same argument holds for more than two limiting resources. Only if these resources are supplied in a balanced way (balanced with regard to the average critical resource requirements), trait complementarity leads to higher coexistence and to stronger B–EF relationships (Figure 6D). Thus, the environmental dimensionality (the potential number of limiting resources) is maximized at a multivariate resource supply in the centre of the triangle (Figure 6D). If resource supply is characterized by imbalances such that only one resource is limiting, both coexistence and B–EF relationship are strongly constrained. We propose that a greater distance from the centre, that is, a

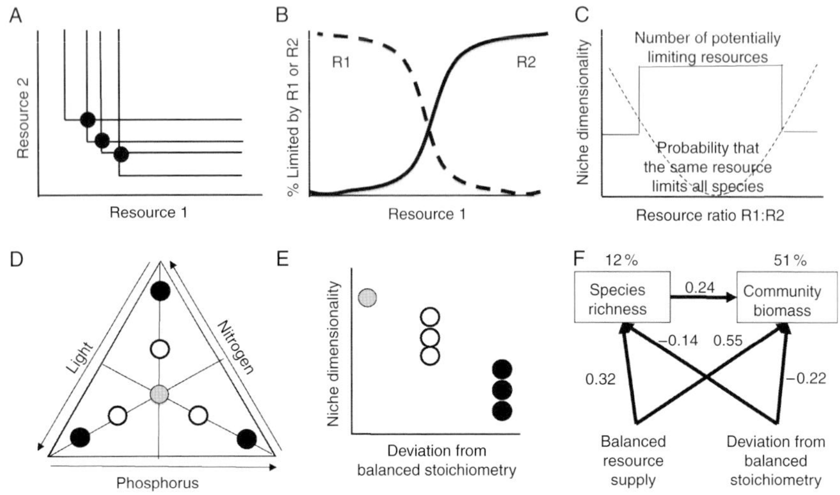

Figure 6 (A) Three species (lines give zero-net-growth-isoclines, ZNGI) competing for two resources according to Tilman's (1982) resource competition model. (B) Along a supply gradient of resource 1 the probability of being limited by this resource decreases. (C) At intermediate resource ratios coexistence is enhanced by enhancing environmental (niche) dimensionality as there is low probability that all species are limited by the same resource. (D) The same argument holds for >2 resources. At very unbalances supply ratios (black dots), most to all species are limited by the same resource, whereas at balanced supply ratios, there is a high probability of multiple resources being limiting. (E) This transforms to the prediction that with stronger deviation from balanced stoichiometry, environmental dimensionality declines and therewith the scope for positive B–EF relationships. (F) This prediction is tested by a phytoplankton data set, showing that deviation from balanced stoichiometry decreases richness, and thereby indirectly reduces community biomass (adapted from Cardinale *et al.*, 2009).

stronger deviation from balanced stoichiometry) reduces environmental dimensionality (Figure 6E) and hence the potential for positive B–EF relationships to be manifested.

Two lines of evidence indirectly support this prediction. In an elegant modelling approach, Gross and Cardinale (2007) showed that diversity of primary producers drove primary production in metacommunities, but only at intermediate supply ratios of two limiting resources. The analysis of a large-scale phytoplankton data set (Ptacnik *et al.*, 2008) then revealed that both richness and community biomass depended on the balanced supply (positively) and on the deviation in multiple resource stoichiometry from this balance (negatively), whereas increasing phytoplankton richness increased phytoplankton biomass (Figure 6F, derived from Cardinale *et al.*, 2009).

These studies suggest that a fruitful way to foster our understanding of pelagic B–EF relationships is to consider the resource supply stoichiometry explicitly in theoretical and empirical approaches. Such an approach is useful also in the light of human-dominated global nutrient cycles. The overall amount of biologically available nutrients has changed dramatically for all major elements needed by algae (Falkowski *et al.*, 1998; Vitousek *et al.*, 1997), which also alters the stoichiometry of biogeochemical cycles (Ptacnik *et al.*, 2005) and the predominantly limiting element (Elser *et al.*, 2009). Thus, environmental dimensionality might increase or decrease under globally changing element cycles, which allows the prediction of non-linear (synergistic or antagonistic) interactions between multiple stressors of global change such as altered biogeochemistry and biodiversity.

ES is not only a constraint of environmental dimensionality but also constrains trait dimensionality. Important processes in pelagic ecosystems involving different elements such as C, N and P are tightly linked by the stoichiometry of the involved organisms. Assuming that autotrophs are more flexible in their body composition than their consumers (Persson *et al.*, 2010), the stoichiometric constraints increase with increasing trophic level in pelagic food webs. Thus, the uptake of different resources by phytoplankton can be more uncoupled within individuals than the trophic transfer of different elements through herbivorous and carnivorous consumers. Stoichiometric considerations can thus inform our understanding of ecosystem multifunctionality, that is, the sum of biogeochemical processes in an ecosystem (Gamfeldt *et al.*, 2008).

While the relationship between environmental conditions and stoichiometry variations mediated by physiological processes has been studied extensively (Sterner and Elser, 2002) and has been extrapolated to the level of biogeochemical cycling (Oschlies *et al.*, 2008), our understanding about the relationship between changes in biodiversity and stoichiometry is still in its infancy. A few studies have included stoichiometric considerations in B–EF research, indicating that the body stoichiometry of algae (Striebel *et al.*, 2009b), the food quality for herbivorous consumers (Urabe and Waki, 2009) and the elemental recycling ratios from consumers (Hillebrand *et al.*, 2009b; McIntyre *et al.*, 2007) scale to the biodiversity of algae and consumers, respectively. A recent modeling study additionally suggests that consumer-mediated nutrient recycling feeds back on algal coexistence (Kato *et al.*, 2007). However, these studies represent only some aspects of the potential dynamics in these systems. Therefore, we will present a few general predictions as to how B–EF relationships should change depending on the stoichiometric functional ecology of the species involved.

Two aspects of organismal stoichiometry will affect B–EF relationship. First, the correlation between species contributions to the processing of different elements or resources will define the potential for ecosystem multifunctionality (Figure 7A–C). If the potential to convert resource 1 into biomass (RUE) is strongly coupled to the RUE for resource 2 (Figure 7A), processes are physiologically strictly coupled within individuals and any

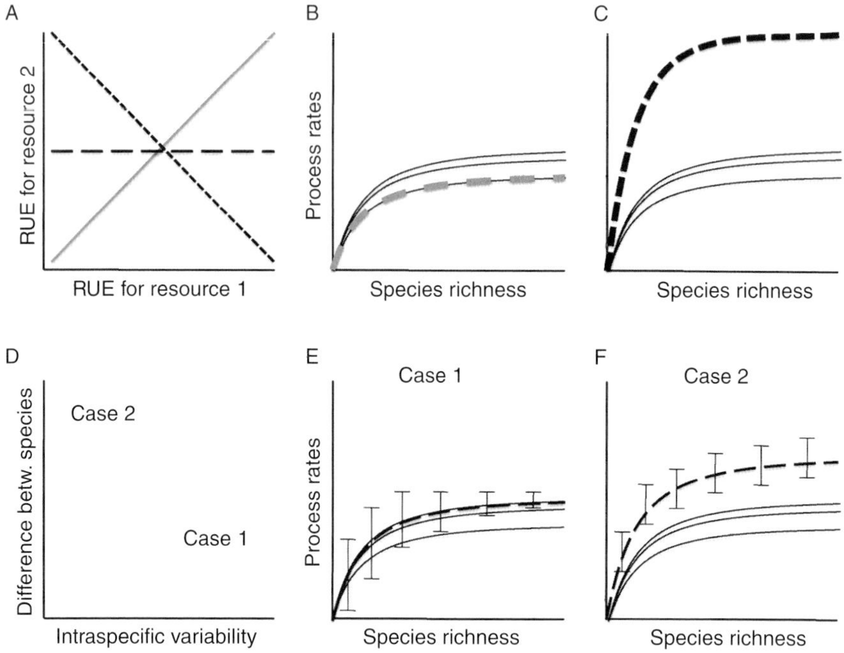

Figure 7 A stoichiometric consideration of ecosystem multifunctionality (*sensu* Gamfeldt *et al.*, 2008). (A–C) The correlation between species contributions to the processing of different elements or resources defines the potential for ecosystem multifunctionality. (A) Correlations between efficiency to process resources 1 and 2. (B) B–EF in the case of positive correlations. (C) B–EF in the case of negative correlations. (D) In the case of uncorrelated RUEs for different elements, the outcome depends on the flexibility within populations relative to the difference between species. (E) Species showing a broad variability in functional and numerical responses to different resources but differing little in their average RUE show negligible multifunctionality. (F) Species differing in their capacities to transform different resources into biomass with little flexibility foster a high degree of multifunctionality (see text for details).

species' impact on the processing of different elements will be positively correlated. In that case, the effect of biodiversity on multifunctionality will not differ from effects on single functions, except that it will be the single process with slowest relative rate that determines overall functioning (Figure 7B). If, however, the RUE of different elements requires strong trade-offs, such that a single species contributes to RUE either for R1 or for R2, the processing of multiple elements will be much more sensitive to biodiversity than the processing of single elements (Figure 7C).

In the case of uncorrelated RUEs for different elements, the outcome depends on the flexibility within populations relative to the difference between species (Figure 7D). If populations show a broad variability in functional and numerical responses to different resources but differ little in their average RUE, then we expect weak changes by considering multiple processes compared to a single process (Figure 7E). Higher species richness will, however, lead to a lower variability for the overall multifunctionality, given that multiple processes are more strongly buffered against temporal fluctuations. If species differ in their capacities to transform different resources into biomass but show little flexibility, there will be a high degree of multifunctionality, that is, more diverse assemblages will strongly outperform monocultures (Figure 7F).

These predictions are yet untested, but provide a framework for including stoichiometric constraints into the consideration of how altered biodiversity affects biogeochemical processes in pelagic environments. The inclusion of these constraints is without doubt of utmost importance to disentangle multiple stressor effects in a globally changing ocean (Riebesell *et al.*, 2009). Effects of ocean acidification on global carbon fluxes are often generalized across the breadth of coccolithophorid calcifying algae, but they show strong differences in their sensitivity to reduced pH (Langer *et al.*, 2006) and their effect on carbon flux depends on carbon:nutrient stoichiometry (Riebesell *et al.*, 2007), which opens up for a strong interaction between biogeochemically and biodiversity-mediated changes in such a globally important ecosystem process.

VI. OUTLOOK AND CONCLUSIONS

Two major conclusions can be drawn from this review of conceptual and empirical approaches to B–EF in plankton. First, biodiversity in microscopic organisms including phyto- and zooplankton cannot be understood from the local perspective alone. Rather, taxon richness depends on dispersal in much the same way as what is known from higher organisms. While it appears obvious that microscopic organisms disperse faster and that some morphospecies may have global dispersal, high dispersal is also required to sustain species rich communities in presence of fast turnover rates. Second, in contrast to intuition, we argue here that niches are potentially diverse in pelagic ecosystems, though distinctly different from spatial niches in terrestrial and benthic ecosystems. This large environmental dimensionality reflects the potential of pelagic organisms to differ in their resource use, in their response to sources of mortality, and to the rapid change in environmental conditions. It thus appears that fast turnover and strong interactions indeed depend on species rich communities. Interestingly, a recent meta-analysis on terrestrial and aquatic ecosystems highlighted that diversity scales with strength of various biotic interactions (predation, parasitism, mutualistic interactions) as shown by the fact that both diversity and interaction strength increase towards low latitudes (Schemske *et al.*, 2009).

Recognizing that dimensionality of both environment and traits can be quantified will provide significant progress for a mechanistic understanding of B–EF relationships in the plankton and elsewhere (Hillebrand and Matthiessen, 2009; Reiss *et al.*, 2009). We have shown that general principles such as length of environmental gradients and stoichiometric constraints can be combined into a promising framework for modelling dynamic interactions.

Furthermore, spatial processes should be taken into account when evaluating B–EF relationships. From what has emerged from only a few systematic studies analysing richness patterns across wider geographic areas, it appears that dispersal limitation must be considered. The regional scale may actually critically enhance our understanding of how diversity of planktonic communities links to ecosystem functions under given environmental drivers and anthropogenic stressors. Of particular topical interest is how regional pools affect the adaptability of ecosystems under a changing climate (Woodward *et al.*, 2010).

Most of the evidence discussed in this review is derived from limnological research. This is not because we consider lakes to be more important systems than marine ones, but because most of the data available on dispersal limitation and B–EF relationships come from lentic systems, which are particularly suitable for the study of diversity patterns, since they from distinct patches separated by land ('inverted islands'; Turner, 1999). A particular challenge will be to find how species pool dynamics function in more open systems. Recent work (Barton *et al.*, 2010) has shown, for instance, that horizontal advection has critical effects on phytoplankton diversity in the oceans. Diversity gradients may also be assumed for transitional systems such as brackish estuaries and fjords, which are particularly impacted by eutrophication and pollution, and seem to be very prone to invasion by non-indigenous species at the same time (Paavola *et al.*, 2005).

Due to increasing environmental problems and loss of species in surface waters on one hand and the increasing dependency on food from aquatic habitats on the other hand, there is an urgent need for a better understanding of the pelagial and its components. We have shown here that its microscopic representatives, that is, the plankton, may not be excluded from the study of B–EF scaling relationships in lakes and oceans. It is time to shift our focus from wondering about diversity in the plankton towards asking when and under which conditions plankton communities lack species, and what the implications of this may be.

ACKNOWLEDGEMENTS

Comments by G. Woodward and two anonymous reviewers are gratefully acknowledged. H. H. was financed through the German Science Foundation (DFG Hi 848/7-1).

APPENDIX. PTACNIK, MOORTHI AND HILLEBRAND: HUTCHINSON REVERSED OR WHY THERE NEED TO BE SO MANY SPECIES

Regression summary for multiple linear regressions predicting phytoplankton RUE (=ln(chlorophyll.a/TP), mol/mol) from genus richness (G) and ln (TP, μg L^{-1}) in Norwegian lakes. The analysis corresponds to Table 1 in Ptacnik *et al.* (2008), except that two more confined subsets (Western Norway and Southern Finland) with a similar range in TP concentrations (≤ 100 μg L^{-1}) were chosen here in order not to confound local with regional effects of TP (Ptacnik *et al.*, 2010).
For both regressions overall *p*, R^2, number of observations (*n*) and the estimated coefficients are given.

	p	R^2	*n*	Intercept (SD)	*G* (SD)	ln(TP) (SD)
Western Norway	<0.01	0.19	341	0.09 (0.33)	0.93 (0.11)***	−0.19 (0.04)***
Southern Finland	<0.01	0.03	297	1.3 (0.37)***	0.29 (0.09)**	0.05 (0.04)

*** = <0.001; ** = <0.01.

REFERENCES

Barnett, A., and Beisner, B.E. (2007). Zooplankton biodiversity and lake trophic state: Explanations invoking resource abundance and distribution. *Ecology* **88**, 1675–1686.
Barnett, A.J., Finlay, K., and Beisner, B.E. (2007). Functional diversity of crustacean zooplankton communities: Towards a trait-based classification. *Freshwater Biol.* **52**, 796–813.
Barton, A.D., Dutkiewicz, S., Flierl, G., Bragg, J., and Follows, M. (2010). Patterns of diversity in marine phytoplankton. *Science*.
Beisner, B.E. (2001). Plankton community structure in fluctuating environments and the role of productivity. *Oikos* **95**, 496–510.
Brown, J.H., Gillooly, J.F., Allen, A.P., Savage, V.M., and West, G.B. (2004). Toward a metabolic theory of ecology. *Ecology* **85**, 1771–1789.

Bruno, J.F., and O'Connor, M.I. (2005). Cascading effects of predator diversity and omnivory in a marine food web. *Ecol. Lett.* **8**, 1048–1056.
Burgmer, T. (2009). *Effects of Climate Variability and Physical Forcing on the Diversity of Aquatic Organisms*. Botanical Institute. University of Cologne Köln, PhD thesis. http://kups.ub.uni-koeln.de/volltexte/2009/2892/.
Cardinale, B.J., Hillebrand, H., and Charles, D.F. (2006a). Geographic patterns of diversity in streams are predicted by a multivariate model of disturbance and productivity. *J. Ecol.* **94**, 609–618.
Cardinale, B.J., Srivastava, D.S., Emmett, Duffy J., Wright, J.P., Downing, A.L., Sankaran, M., and Jouseau, C. (2006b). Effects of biodiversity on the functioning of trophic groups and ecosystems. *Nature* **443**, 989–992.
Cardinale, B.J., Wright, J.P., Cadotte, M.W., Carroll, I.T., Hector, A., Srivastava, D. S., Loreau, M., and Weis, J.J. (2007). Impacts of plant diversity on biomass production increase through time because of species complementarity. *Proc. Natl. Acad. Sci. USA* **104**, 18123–18128.
Cardinale, B.J., Hillebrand, H., Harpole, W.S., Gross, K., and Ptacnik, R. (2009). Separating the influence of resource 'availability' from resource 'imbalance' on productivity-diversity relationships. *Ecol. Lett.* **12**, 475–487.
Caron, D.A., and Countway, P.D. (2009). Hypotheses on the role of the protistan rare biosphere in a changing world. *Aquat. Microb. Ecol.* **57**, 227–238.
Cebrian, J., and Lartigue, J. (2004). Patterns of herbivory and decomposition in aquatic and terrestrial ecosystems. *Ecol. Monogr.* **74**, 237–259.
Chase, J.M., and Leibold, M.A. (2002). Spatial scale dictates the productivity-biodiversity relationship. *Nature* **416**, 427–430.
Chase, J.M., and Ryberg, W.A. (2004). Connectivity, scale-dependence, and the productivity-diversity relationship. *Ecol. Lett.* **7**, 676–683.
Chesson, P. (2000). Mechanisms of maintenance of species diversity. *Annu. Rev. Ecol. Syst.* **31**, 343–366.
Chesson, P., and Kuang, J.J. (2008). The interaction between predation and competition. *Nature* **456**, 235–238.
Cornell, H.V., and Karlson, R. (1997). Local and regional processes as controls of species richness. In: *Spatial Ecology: The Role of Space in Population Dynamics and Interspecific Interactions* (Ed. by D. Tilman, P. Kareiva, D. Tilman and P. Kareiva), pp. 250–268. Princeton University Press, New Jersey.
Countway, P.D., Gast, R.J., Dennett, M.R., Savai, P., Rose, J.M., and Caron, D.A. (2007). Distinct protistan assemblages characterize the euphotic zone and deep sea (2500 m) of the western North Atlantic (Sargasso Sea and Gulf Stream). *Environ. Microbiol.* **9**, 1219–1232.
Danger, M., Daufresne, T., Lucas, F., Pissard, S., and Lacroix, G. (2008). Does Liebig's law of the minimum scale up from species to communities? *Oikos* **117**, 1741–1751.
Dawson, S.C., and Pace, N.R. (2002). Novel kingdom-level eukaryotic diversity in anoxic environments. *Proc. Natl. Acad. Sci. USA* **99**, 8324–8329.
de Wit, R., and Bouvier, T. (2006). 'Everything is everywhere, but, the environment selects'; what did Baas Becking and Beijerinck really say? *Environ. Microbiol.* **8**, 755–758.
Declerck, S., Vanderstukken, M., Pals, A., Muylaert, K., and de Meester, L. (2007). Plankton biodiversity along a gradient of productivity and its mediation by macrophytes. *Ecology* **88**, 2199–2210.
Descamps-Julien, B., and Gonzalez, A. (2005). Stable coexistence in a fluctuating environment: An experimental demonstration. *Ecology* **86**, 2815–2824.

Dodson, S. (1992). Predicting crustacean zooplankton species richness. *Limnol. Oceanogr.* **37**, 848–856.
Dolan, J.R., and Perez, M.T. (2000). Costs, benefits and characteristics of mixotrophy in marine oligotrichs. *Freshwater Biol.* **45**, 227–238.
Dolan, J.R., Ritchie, M.E., Tunin-Ley, A., and Pizay, M.D. (2009). Dynamics of core and occasional species in the marine plankton: Tintinnid ciliates in the north-west Mediterranean Sea. *J. Biogeogr.* **36**, 887–895.
Downing, J.A., Prairie, Y.T., Cole, J.J., Duarte, C.M., Tranvik, L.J., Striegl, R.G., McDowell, W.H., Kortelainen, P., Caraco, N.F., and Melack, J.M. (2006). The global abundance and size distribution of lakes, ponds, and impoundments. *Limnol. Oceanogr.* **51**, 2388–2397.
Downing, A.L., Brown, B.L., Perrin, E.M., Keitt, T.H., and Leibold, M.A. (2008). Environmental fluctuations induce scale-dependent compensation and increase stability in plankton ecosystems. *Ecology* **89**, 3204–3214.
Drakare, S., Lennon, J.J., and Hillebrand, H. (2006). The imprint of the geographical, evolutionary and ecological context on species-area relationships. *Ecol. Lett.* **9**, 215–227.
Duffy, J.E., Richardson, J.P., and Canuel, E.A. (2003). Grazer diversity effects on ecosystem functioning in seagrass beds. *Ecol. Lett.* **6**, 637–645.
Duffy, J.E., Cardinale, B.J., France, K.E., McIntyre, P.B., Thébault, E., and Loreau, M. (2007). The functional role of biodiversity in ecosystems: Incorporating trophic complexity. *Ecol. Lett.* **10**, 522–538.
Dzialowski, A.R., and Smith, V.H. (2008). Nutrient dependent effects of consumer identity and diversity on freshwater ecosystem function. *Freshwater Biol.* **53**, 148–158.
Elser, J.J., Bracken, M.E.S., Cleland, E.E., Gruner, D.S., Harpole, W.S., Hillebrand, H., Ngai, J.T., Seabloom, E.W., Shurin, J.B., and Smith, J.E. (2007). Global analysis of nitrogen and phosphorus limitation of primary producers in freshwater, marine and terrestrial ecosystems. *Ecol. Lett.* **10**, 1135–1142.
Elser, J.J., Andersen, T., Baron, J.S., Bergstrom, A.K., Jansson, M., Kyle, M., Nydick, K.R., Steger, L., and Hessen, D.O. (2009). Shifts in lake N:P stoichiometry and nutrient limitation driven by atmospheric nitrogen deposition. *Science* **326**, 835–837.
Falkowski, P.G., Barber, R.T., and Smetacek, V. (1998). Biogeochemical controls and feedbacks on ocean primary production. *Science* **281**, 200–206.
Fenchel, T., and Finlay, B.J. (2004). The ubiquity of small species: Patterns of local and global diversity. *Bioscience* **54**, 777–784.
Filip, J., Bauer B., Beniermann A., Gaedke U., Hillebrand, H., and Moorthi S.D. Consumer diversity and specialization affect prey biomass, diversity and composition. To be submitted to Ecology Letters.
Finke, D.L., and Denno, R.F. (2005). Predator diversity and the functioning of ecosystems: The role of intraguild predation in dampening trophic cascades. *Ecol. Lett.* **8**, 1299–1306.
Finlay, B.J. (2002). Global dispersal of free-living microbial eukaryote species. *Science* **296**, 1061–1063.
Flöder, S., Urabe, J., and Kawabata, Z. (2002). The influence of fluctuating light intensities on species composition and diversity of natural phytoplankton communities. *Oecologia* **133**, 395–401.
Fuhrman, J.A., Steele, J.A., Hewson, I., Schwalbach, M.S., Brown, M.V., Green, J.L., and Brown, J.H. (2008). A latitudinal diversity gradient in planktonic marine bacteria. *Proc. Natl. Acad. Sci. USA* **105**, 7774–7778.

Gamfeldt, L., Hillebrand, H., and Jonsson, P.R. (2005). Species richness changes across two trophic levels simultaneously affect prey and consumer biomass. *Ecol. Lett.* **8**, 696–703.

Gamfeldt, L., Hillebrand, H., and Jonsson, P.R. (2008). Multiple functions increase the importance of biodiversity for overall ecosystem functioning. *Ecology* **89**, 1223–1231.

Gragham, L.E., Graham, J.M., and Wilcox, L.W. (2010). *Algae*, 2nd ed. Benjamin Cummings (Pearson), San Francisco, CA.

Green, J., and Bohannan, B.J.M. (2006). Spatial scaling of microbial biodiversity. *Trends Ecol. Evol.* **21**, 501–507.

Griffin, J.N., Jenkins, S.R., Gamfeldt, L., Jones, D., Hawkins, S.J., and Thompson, R.C. (2009). Spatial heterogeneity increases the importance of species richness for an ecosystem process. *Oikos* **118**, 1335–1342.

Griffiths, B.S., Ritz, K., Bardgett, R.D., Cook, R., Christensen, S., Ekelund, F., Sørensen, S.J., Bååth, E., Bloem, J., de Ruiter, P.C., Dolfing, J., and Nicolardot, B. (2000). Ecosystem response of pasture soil communities to fumigation-induced microbial diversity reductions: An examination of the biodiversity-ecosystem function relationship. *Oikos* **90**, 279–294.

Gross, K., and Cardinale, B.J. (2007). Does species richness drive community production or vice versa? Reconciling historical and contemporary paradigms in competitive communities. *Am. Nat.* **170**, 207–220.

Grover, J.P., and Chrzanowski, T.H. (2004). Limiting resources, disturbance, and diversity in phytoplankton communities. *Ecol. Monogr.* **74**, 533–551.

Harpole, W.S., and Tilman, D. (2007). Grassland species loss resulting from reduced niche dimension. *Nature* **446**, 791–793.

He, F., Gaston, K.J., Connor, E.F., and Srivastava, D.S. (2005). The local-regional relationship: Immigration, extinction, and scale. *Ecology* **86**, 360–365.

Hector, A., and Bagchi, R. (2007). Biodiversity and ecosystem multifunctionality. *Nature* **448**, 188–190.

Helmus, M.R., Keller, W., Paterson, M.J., Yan, N.D., Cannon, C.H., and Rusak, J.A. (2010). Communities contain closely related species during ecosystem disturbance. *Ecol. Lett.* **13**, 162–174.

Henriksen, A., Skjelvåle, B.L., Mannio, J., Wilander, A., Harriman, R., Jensen, J.P., Fjeld, E., and Moiseenko, T. (1998). Northern European Lake Survey, 1995—Finland, Norway, Sweden, Denmark, Russian Kola, Russian Karelia, Scotland and Wales. *AMBIO* **27**, 80–91.

Hessen, D.O., Faafeng, B.A., Smith, V.H., Bakkestuen, V., and Walseng, B. (2006). Extrinsic and intrinsic controls of zooplankton diversity in lakes. *Ecology* **87**, 433–443.

Hillebrand, H. (2004). On the generality of the latitudinal gradient. *Am. Nat.* **163**, 192–211.

Hillebrand, H. (2005). Regressions of local on regional diversity do not reflect the importance of local interactions or saturation of local diversity. *Oikos* **110**, 195–198.

Hillebrand, H., and Matthiessen, B. (2009). Biodiversity in a complex world: Consolidation and progress in functional biodiversity research. *Ecol. Lett.* **12**, 1405–1419.

Hillebrand, H., Frost, P., and Liess, A. (2008). Ecological stoichiometry of indirect grazer effects on periphyton nutrient content. *Oecologia* **155**, 619–630.

Hillebrand, H., Borer, E.T., Bracken, M.E.S., Cardinale, B.J., Cebrian, J., Cleland, E.E., Elser, J.J., Gruner, D.S., Harpole, W.S., Ngai, J.T., Sandin, S., Seabloom, E.W., *et al.* (2009a). Herbivore metabolism and stoichiometry each constrain herbivory at different organizational scales across ecosystems. *Ecol. Lett.* **12**, 516–527.

Hillebrand, H., Gamfeldt, L., Jonsson, P.R., and Matthiessen, B. (2009b). Consumer diversity indirectly changes prey nutrient content. *Mar. Ecol. Prog. Ser.* **380**, 33–41.

Holen, D.A., and Boraas, M.E. (1995). Mixotrophy in chrysophytes. In: *Chrysophyte Algae* (Ed. by C.D. Sandgren, J.P. Smol and J.J. Kristiansen), pp. 119–140. Cambridge University Press, Cambridge.

Hooper, D.U., Chapin, F.S., Ewel, J.J., Hector, A., Inchausti, P., Lavorel, S., Lawton, J.H., Lodge, D.M., Loreau, M., Naeem, S., Schmid, B., Setala, H., *et al.* (2005). Effects of biodiversity on ecosystem functioning: A consensus of current knowledge. *Ecol. Monogr.* **75**, 3–35.

Horner-Devine, M.C., Lage, M., Hughes, J.B., and Bohannan, B.J.M. (2004). A taxa-area relationship for bacteria. *Nature* **432**, 750–753.

Hughes, T.P., and Connell, J.H. (1999). Multiple stressors on coral reefs: A long term perspective. *Limnol. Oceanogr.* **44**, 932–940.

Huisman, J., and Hulot, F.D. (2005). Population dynamics of harmful cyanobacteria. In: *Harmful cyanobacteria* (Ed. by J. Huisman, H.C.P. Matthijs and P.M. Visser), pp. 143–176. Springer, Dordrecht, The Netherlands.

Hulot, F.D., and Huisman, J. (2004). Allelopathic interactions between phytoplankton species: The roles of heterotrophic bacteria and mixing intensity. *Limnol. Oceanogr.* **49**, 1424–1434.

Hurd, P., Woodward, G., Trimmer, M., and Purdy, K. (2010). Systems biology for ecology: From molecules to ecosystems. *Adv. Ecol. Res.* **43**, 87–149.

Hutchinson, G.E. (1961). The paradox of the plankton. *Am. Nat.* **95**, 137–145.

Interlandi, S.J., and Kilham, S.S. (2001). Limiting resources and the regulation of diversity in phytoplankton communities. *Ecology* **82**, 1270–1282.

Irigoien, X., Huisman, J., and Harris, R.P. (2004). Global biodiversity patterns of marine phytoplankton and zooplankton. *Nature* **429**, 863–867.

Jackson, J.B.C., Kirby, M.X., Berger, W.H., Bjorndal, K.A., Botsford, L.W., Bourque, B.J., Bradbury, R.H., Cooke, R., Erlandson, J., Estes, J.A., Hughes, T.P., Kidwell, S., *et al.* (2001). Historical overfishing and the recent collapse of coastal ecosystems. *Science* **293**, 629–638.

Jeong, H.J., Yoo, Y.D., Park, J.Y., Song, J.Y., Kim, S.T., Lee, S.H., Kim, K.Y., and Yih, W.H. (2005). Feeding by phototrophic red-tide dinoflagellates: Five species newly revealed and six species previously known to be mixotrophic. *Aquat. Microb. Ecol.* **40**, 133–150.

Jiang, L., and Morin, P.J. (2007). Temperature fluctuation facilitates coexistence of competing species in experimental microbial communities. *J. Anim. Ecol.* **76**, 660–668.

Jones, R.I. (1994). Mixotrophy in planktonic protists as a spectrum of nutritional strategies. *Mar. Microb. Food Webs* **8**, 87–96.

Kahmen, A., Renker, C., Unsicker, S.B., and Buchmann, N. (2006). Niche complementarity for nitrogen: An explanation for the biodiversity and ecosystem functioning relationship? *Ecology* **87**, 1244–1255.

Kaluza, P., Kölzsch, A., Gastner, M.T., and Blasius, B. (2010). The complex network of global cargo ship movements. *J. R. Soc. Interface* **7**, 1093–1103.

Kato, S., Urabe, J., and Kawata, M. (2007). Effects of temporal and spatial heterogeneities created by consumer-driven nutrient recycling on algal diversity. *J. Theor. Biol.* **245**, 364–377.

Klein Breteler, W.C.M., Schogt, N., Baas, M., Schouten, S., and Kraay, G.W. (1999). Trophic upgrading of food quality by protozoans enhancing copepod growth: Role of essential lipids. *Mar. Biol.* **135**, 191–198.

Kleppel, G.S. (1993). On the diets of calanoid copepods. *Mar. Ecol. Prog. Ser.* **99**, 183–195.

Kondoh, M. (2001). Unifying the relationships of species richness to productivity and disturbance. *Proc. R. Soc. B Biol. Sci.* **268**, 269–271.

Korhonen, J.J., Soininen, J., and Hillebrand, H. (2010). A quantitative analysis of temporal turnover in aquatic species assemblages across ecosystems. *Ecology* **91**, 508–517.

Langenheder, S., Sobek, S., and Tranvik, L. (2006). Changes in bacterial community composition along a solar radiation gradient in humic waters. *Aquat. Sci. Res. Across Bound.* **68**, 415–424.

Langer, G., Geisen, M., Baumann, K.H., Kläs, J., Riebesell, U., Thoms, S., and Young, J.R. (2006). Species-specific responses of calcifying algae to changing seawater carbonate chemistry. *Geochem. Geophys. Geosyst.* **7,** Q09006, doi:10.1029/2005GC001227.

Leibold, M.A., Holyoak, M., Mouquet, N., Amarasekare, P., Chase, J.M., Hoopes, M.F., Holt, R.D., Shurin, J.B., Law, R., Tilman, D., Loreau, M., and Gonzalez, A. (2004). The metacommunity concept: A framework for multi-scale community ecology. *Ecol. Lett.* **7**, 601–613.

Lenton, T.M., and Klausmeier, C.A. (2007). Biotic stoichiometric controls on the deep ocean N:P ratio. *Biogeosciences* **4**, 353–367.

Litchman, E. (1998). Population and community responses of phytoplankton to fluctuating light. *Oecologia* **117**, 247–257.

Litchman, E., and Klausmeier, C.A. (2008). Trait-based community ecology of phytoplankton. *Ann. Rev. Ecol. Evol. Syst.* **39**, 615–639.

Litchman, E., Klausmeier, C.A., Schofield, O.M., and Falkowski, P.G. (2007). The role of functional traits and trade-offs in structuring phytoplankton communities: Scaling from cellular to ecosystem level. *Ecol. Lett.* **10**, 1170–1181.

MacArthur, R.H., and Wilson, E.O. (1967). *The Theory of Island Biogeography.* Princeton University Press, Princeton, NJ.

Martiny, J.B.H., Bohannan, B.J.M., Brown, J.H., Colwell, R.K., Fuhrman, J.A., Green, J.L., Horner-Devine, M.C., Kane, M., Krumins, J.A., Kuske, C.R., Morin, P.J., Naeem, S., *et al.* (2006). Microbial biogeography: Putting microorganisms on the map. *Nat. Rev. Microbiol.* **4**, 102–112.

Massana, R., Balagué, V., Guillou, L., and Pedrós-Alió, C. (2004). Picoeukaryotic diversity in an oligotrophic coastal site studied by molecular and culturing approaches. *FEMS Microbiol. Eco.* **50**, 231–243.

Massana, R., Guillou, L., Díez, B., and Pedrós-Alío, C. (2002). Unveiling the organisms behind novel eukaryotic ribosomal DNA sequences from the ocean. *Appl. Environ. Microbiol.* **68**, 4554–4558.

McGrady-Steed, J., Harris, P.M., and Morin, P.J. (1997). Biodiversity regulates ecosystem predictability. *Nature* **390**, 162–164.

McIntyre, P.B., Jones, L.E., Flecker, A.S., and Vanni, M.J. (2007). Fish extinctions alter nutrient recycling in tropical freshwaters. *Proc. Natl. Acad. Sci. USA* **104**, 4461–4466.

McKane, R.B., Johnson, L.C., Shaver, G.R., Nadelhoffer, K.J., Rastetter, E.B., Fry, B., Giblin, A.E., Kielland, K., Kwiatkowski, B.L., Laundre, J.A., and Murray, G. (2002). Resource-based niches provide a basis for plant species diversity and dominance in arctic tundra. *Nature* **415**, 68–71.

Moon-van der Staay, S.Y., De Wachter, R., and Vaulot, D. (2001). Oceanic 18S rDNA sequences from picoplankton reveal unsuspected eukaryotic diversity. *Nature* **409**, 607–610.

Moorthi, S.D., Caron, D.A., Gast, R.J., and Sanders, R.W. (2009). Mixotrophy: A widespread and important ecological strategy for planktonic and sea-ice nanoflagellates in the Ross Sea. *Antarctica Aquat. Microb. Ecol.* **54**, 269–277.

Moreira, D., and López-García, P. (2002). The molecular ecology of microbial eukaryotes unveils a hidden world. *Trends Microbiol.* **10**, 31–38.

Nabout, J.C., Siqueira, T., Bini, L.M., and Nogueira, I.d.S. (2009). No evidence for environmental and spatial processes in structuring phytoplankton communities. *Acta Oecol.* **35**, 720–726.

Naeem, S., and Li, S. (1997). Biodiversity enhances ecosystem reliability. *Nature* **390**, 507–509.

Naeslund, B., and Norberg, J. (2006). Ecosystem consequences of the regional species pool. *Oikos* **115**, 504–512.

Norberg, J. (2000). Resource-niche complementarity and autotrophic compensation determines ecosystem-level responses to increased cladoceran species richness. *Oecologia* **122**, 264–272.

Norberg, J., Swaney, D.P., Dushoff, J., Lin, J., Casagrandi, R., and Levin, S.A. (2001). Phenotypic diversity and ecosystem functioning in changing environments: A theoretical framework. *Proc. Natl. Acad. Sci. USA* **98**, 11376–11381.

Not, F., Latasa, M., Scharek, R., Viprey, M., Karleskind, P., Balagué, V., *et al.* (2008). Protistan assemblages across the Indian Ocean, with a specific emphasis on the picoeukaryotes. *Deep-Sea Res. I* **55**, 1456–1473.

Not, F., Valentin, K., Romari, K., Lovejoy, C., Massana, R., Tobe, K., *et al.* (2007). Picobiliphytes: A marine picoplanktonic algal group with unknown affinities to other eukaryotes. *Science* **315**, 253–255.

Olli, K., Klais, R., Tamminen, T., Ptacnik, R., Andersen, T. (in press) Long term changes in the Baltic sea phytoplankton community. *Boreal Environ. Res.*

Oschlies, A., Schulz, K.G., Riebesell, U., and Schmittner, A. (2008). Simulated 21st century's increase in oceanic suboxia by CO_2-enhanced biotic carbon export. *Global Biogeochem. Cycles* GB4008, doi:10.1029/2007GB003147.

Otto, S.B., Berlow, E.L., Rank, N.E., Smiley, J., and Brose, U. (2008). Predator diversity and identity drive interaction strength and trophic cascades in a food web. *Ecology* **89**, 134–144.

Paavola, M., Olenin, S., and Leppäkoski, E. (2005). Are invasive species most successful in habitats of low native species richness across European brackish water seas? *Estuar. Coast. Shelf Sci.* **64**, 738–750.

Peres-Neto, P.R., Legendre, P., Dray, S., and Borcard, D. (2006). Variation partitioning of species data matrices: Estimation and comparison of fractions. *Ecology* **87**, 2614–2625.

Perkins, D.M., McKie, B.G., Malmqvist, B., Gilmour, S.G., Reiss, J., and Woodward, G. (2010). Environmental warming and biodiversity-ecosystem functioning in freshwater microcosms: Partitioning the effects of species identity, richness and metabolism. *Adv. Ecol. Res.* **43**, 177–209.

Persson, J., Fink, P., Goto, A., Hood, J.M., Jonas, J., and Kato, S. (2010). To be or not to be what you eat: Regulation of stoichiometric homeostasis among autotrophs and heterotrophs. *Oikos* **119**, 741–751.

Petchey, O.L., McPhearson, P.T., Casey, T.M., and Morin, P.J. (1999). Environmental warming alters food-web structure and ecosystem function. *Nature* **402**, 69–72.

Petchey, O.L., Downing, A.L., Mittelbach, G.G., Persson, L., Steiner, C.F., Warren, P.H., and Woodward, G. (2004). Species loss and the structure and functioning of multitrophic aquatic systems. *Oikos* **104**, 467–478.

Ptacnik, R., Sommer, U., Hansen, T., and Martens, V. (2004). Effects of microzooplankton and mixotrophy in an experimental planktonic food web. *Limnol. Oceanogr.* **49**, 1435–1445.

Ptacnik, R., Jenerette, D.G., Verschoor, A.M., Huberty, A.F., and Solimini, A.G. (2005). Applications of ecological stoichiometry for sustainable acquisition of ecosystem services. *Oikos* **109**, 52–62.

Ptacnik, R., Solimini, A.G., Andersen, T., Tamminen, T., Brettum, P., Lepisto, L., Willen, E., and Rekolainen, S. (2008). Diversity predicts stability and resource use efficiency in natural phytoplankton communities. *Proc. Natl. Acad. Sci. USA* **105**, 5134–5138.

Ptacnik, R., Solimini, A., and Brettum, P. (2009). Performance of a new phytoplankton composition metric along a eutrophication gradient in Nordic lakes. *Hydrobiologia* **633**, 75–82.

Ptacnik, R., Andersen, T., Brettum, P., Lepistö, L., and Willén, E. (2010). Regional species pools control community saturation in lake phytoplankton. *Proc. R. Soc. B Biol. Sci.* doi: 10.1098/rspb.2010.1158.

Reiss, J., Bridle, J.R., Montoya, J.M., and Woodward, G. (2009). Emerging horizons in biodiversity and ecosystem functioning research. *Trends Ecol. Evol.* **24**, 505–514.

Reiss, J., Cássio, F., Forster, J., Hirst, A., Pascoal, C., and Stewart, R. (2010a). When micrscopic organisms inform general ecological theory. *Adv. Ecol. Res.* **43**, 45–85.

Reiss, J., Bailey, R.A., Woodward, G., Cássio, F., and Pascoal, C. (2010b). Assessing the contribution of micro-organisms and macrofauna to biodiversity-ecosystem functioning relationships in freshwater microcosms. *Adv. Ecol. Res.* **43**, 151–176.

Reynolds, C.S., Huszar, V., Kruk, C., Naselli-Flores, L., and Melo, S. (2002). Towards a functional classification of the freshwater phytoplankton. *J. Plankton Res.* **24**, 417–428.

Riebesell, U., Schulz, K.G., Bellerby, R.G.J., Botros, M., Fritsche, P., Meyerhofer, M., Neill, C., Nondal, G., Oschlies, A., Wohlers, J., and Zollner, E. (2007). Enhanced biological carbon consumption in a high CO_2 ocean. *Nature* **450**, 545–548.

Riebesell, U., Körtzinger, A., and Oschlies, A. (2009). Sensitivities of marine carbon fluxes to ocean change. *Proc. Natl. Acad. Sci. USA* **106**, 20602–20609.

Riemann, B., Havskum, H., Thingstad, F., and Bernard, C. (1995). The role of mixotrophy in pelagic environments. In: *Molecular Ecology of Aquatic Microbes* (Ed. by I. Joint), pp. 87–114. Springer-Verlag, Berlin.

Romari, K., and Vaulot, D. (2004). Composition and temporal variability of picoeukaryote communities at a coastal site of the English Channel from 18S rDNA sequences. *Limnol. Oceanogr.* **49**, 784–798.

Rosenzweig, M.L. (1972). Paradox of enrichment: Destabilization of exploitation ecosystems in ecological time. *Science* **171**, 385–387.

Rutherford, S., D'Hondt, S., and Prell, S. (1999). Environmental controls on the geographic distribution of zooplankton diversity. *Nature* **400**, 749–753.

Sanders, R.W. (1991). Mixotrophic protists in marine and freshwater ecosystems. *J. Protozool.* **38**, 76–81.

Scheffer, M., Rinaldi, S., Gragnani, A., Mur, L.R., and Nes, E.H. (1997). On the dominance of filamentous cyanobacteria in shallow, turbid lakes. *Ecology* **78**, 272–282.

Schemske, D.W., Mittelbach, G.G., Cornell, H.V., Sobel, J.M., and Roy, K. (2009). Is there a latitudinal gradient in the importance of biotic interactions? *Ann. Rev. Ecol. Evol. Syst.* **40**, 245–269.

Schmidtke, A. (2009). Biodiversity Effects on the Performance of Terrestrial Plant and Phytoplankton Communities. Institute for Biochemistry and Biology, University of Potsdam, Potsdamp. 138.

Schmidtke, A., Gaedke, U., and Weithoff, G. (2010). A mechanistic basis for underyielding in phytoplankton communities. *Ecology* **91**, 212–221.

Shea, K., Roxburgh, S.H., and Rauschert, E.S.J. (2004). Moving from pattern to process: Coexistence mechanisms under intermediate disturbance regimes. *Ecol. Lett.* **7**, 491–508.

Shurin, J.B. (2000). Dispersal limitation, invasion resistance, and the structure of pond zooplankton communities. *Ecology* **81**, 3074–3086.

Shurin, J.B. (2007). How is diversity related to species turnover through time? *Oikos* **116**, 957–965.

Shurin, J.B., Havel, J.E., Leibold, M.A., and Pinel-Alloul, B. (2000). Local and regional zooplankton species richness: A scale-independent test for saturation. *Ecology* **81**, 3062–3073.

Shurin, J.B., Gruner, D.S., and Hillebrand, H. (2006). All wet or dried up? Real differences between aquatic and terrestrial food webs. *Proc. R. Soc. B Biol. Sci.* **273**, 1–9.

Shurin, J.B., Arnott, S.E., Hillebrand, H., Longmuir, A., Pinel-Alloul, B., Winder, M., and Yan, N.D. (2007). Diversity-stability relationship varies with latitude in zooplankton. *Ecol. Lett.* **10**, 127–134.

Shurin, J.B., Winder, M., Adrian, R., Keller, W., Matthews, B., Paterson, A.M., Paterson, M.J., Pinel-Alloul, B., Rusak, J.A., and Yan, N.D. (2010). Environmental stability and lake zooplankton diversity—Contrasting effects of chemical and thermal variability. *Ecol. Lett.* **13**, 453–463.

Šlapeta, J., Moreira, D., and López-García, P.The extent of protist diversity: Insights from molecular ecology of freshwater eukaryotes. *Proc. R. Soc. B* **272**, 2073–2081.

Smith, V.H. (2007). Microbial diversity-productivity relationships in aquatic ecosystems. *FEMS Microbiol. Ecol.* **62**, 181–186.

Smith, V.H., Foster, B.L., Grover, J.P., Holt, R.D., Leibold, M.A., and deNoyelles, F. (2005). Phytoplankton species richness scales consistently from laboratory microcosms to the world's oceans. *Proc. Natl. Acad. Sci. USA* **102**, 4393–4396.

Snyder, W.E., Snyder, G.B., Finke, D.L., and Straub, C.S. (2006). Predator biodiversity strengthens herbivore suppression. *Ecol. Lett.* **9**, 789–796.

Soininen, J., McDonald, R., and Hillebrand, H. (2007). The distance decay of similarity in ecological communities. *Ecography* **30**, 3–12.

Srivastava, D.S., Cardinale, B.J., Downing, A.L., Duffy, J.E., Jouseau, C., Sankaran, M., and Wright, J.P. (2009). Diversity has stronger top-down than bottom-up effects on decomposition. *Ecology* **90**, 1073–1083.

Steiner, C.F. (2005). Temporal stability of pond zooplankton assemblages. *Freshwater Biol.* **50**, 105–112.

Steiner, C.F., and Leibold, M.A. (2004). Cyclic assembly trajectories and scale-dependent productivity-diversity relationships. *Ecology* **85**, 107–113.

Steiner, C.F., Darcy-Hall, T.L., Dorn, N.J., Garcia, E.A., Mittelbach, G.G., and Wojdak, J.M. (2005). The influence of consumer diversity and indirect facilitation on trophic level biomass and stability. *Oikos* **110**, 556–566.

Sterner, R.W., and Elser, J.J. (2002). Ecological Stoichiometry. Princeton University Press, Princeton.

Stickney, H.L., Hood, R.R., and Stoecker, D.K. (2000). The impact of mixotrophy on planktonoc marine ecosystems. *Ecol. Model.* **125**, 203–230.

Stoecker, D.K. (1998). Conceptual models of mixotrophy in planktonic protists and some ecological and evolutionary implications. *Eur. J. Protistol.* **34**, 281–290.

Stomp, M., Huisman, J., de Jongh, F., Veraart, A.J., Gerla, D., Rijkeboer, M., Ibelings, B.W., Wollenzien, U.I.A., and Stal, L.J. (2004). Adaptive divergence in pigment composition promotes phytoplankton biodiversity. *Nature* **432**, 104–107.

Stomp, M., Huisman, J., Voros, L., Pick, F.R., Laamanen, M., Haverkamp, T., and Stal, L.J. (2007). Colourful coexistence of red and green picocyanobacteria in lakes and seas. *Ecol. Lett.* **10**, 290–298.

Straub, C.S., and Snyder, W.E. (2006). Species identity dominates the relationship between predator biodiversity and herbivore suppression. *Ecology* **87**, 277–282.

Striebel, M., Sporl, G., and Stibor, H. (2008). Light-induced changes of plankton growth and stoichiometry: Experiments with natural phytoplankton communities. *Limnol. Oceanogr.* **53**, 513–522.

Striebel, M., Behl, S., Diehl, S., and Stibor, H. (2009a). Spectral niche complementarity and carbon dynamics in pelagic ecosystems. *Am. Nat.* **174**, 141–147.

Striebel, M., Behl, S., and Stibor, H. (2009b). The coupling of biodiversity and productivity in phytoplankton communities: Consequences for biomass stoichiometry. *Ecology* **90**, 2025–2031.

Telford, R.J., Vandvik, V., and Birks, H.J.B. (2006). Dispersal limitations matter for microbial morphospecies. *Science* **312**, 1015.

Thébault, E., and Loreau, M. (2003). Food-web constraints on biodiversity–ecosystem functioning relationships. *Proc. Natl. Acad. Sci. USA* **100**, 14949–14954.

Tilman, D. (1982). Resource Competition and Community Structure. Princeton University Press, Princeton, NJ.

Tilman, D., Wedin, D., and Knops, J. (1996). Productivity and sustainability influenced by biodiversity in grassland ecosystems. *Nature* **379**, 718–720.

Turnbull, L.A., Crawley, M.J., and Rees, M. (2000). Are plant populations seed-limited? A review of seed sowing experiments. *Oikos* **88**, 225–238.

Turner, G.F. (1999). Explosive speciation of African cichlid fishes. In: *Evolution of Biological Diversity* (Ed. by A.E. Magurran and R.M. May), pp. 113–129. Oxford University Press, Oxford.

Turner, B.L. (2008). Resource partitioning for soil phosphorus: A hypothesis. *J. Ecol.* **96**, 698–702.

Tylianakis, J.M., Rand, T.A., Kahmen, A., Klein, A.M., Buchmann, N., Perner, J., and Tscharntke, T. (2008). Resource heterogeneity moderates the biodiversity-function relationship in real world ecosystems. *PLoS Biol.* **6**, 947–956.

Urabe, J., and Waki, N. (2009). Mitigation of adverse effects of rising CO_2 on a planktonic herbivore by mixed algal diets. *Global Change Biol.* **15**, 523–531.

Van der Gucht, K., Cottenie, K., Muylaert, K., Vloemans, N., Cousin, S., Declerck, S., Jeppesen, E., Conde-Porcuna, J.-M., Schwenk, K., Zwart, G., Degans, H., Vyverman, W., *et al.* (2007). The power of species sorting: Local factors drive bacterial community composition over a wide range of spatial scales. *Proc. Natl. Acad. Sci. USA* **104**, 20404–20409.

Van der Heijden, M.G.A., Klironomos, J.N., Ursic, M., Moutoglis, P., Streitwolf, Engel R., Boller, T., Wiemken, A., and Sanders, I.R. (1998). Mycorrhizal fungal diversity determines plant biodiversity, ecosystem variability and productivity. *Nature* **396**, 69–72.

van Ruijven, J., and Berendse, F. (2009). Long-term persistence of a positive plant diversity-productivity relationship in the absence of legumes. *Oikos* **118**, 101–106.

Vandvik, V., and Goldberg, D.E. (2006). Sources of diversity in a grassland metacommunity: Quantifying the contribution of dispersal to species richness. *Am. Nat.* **168**, 157–167.

Vasseur, D.A., and Gaedke, U. (2007). Spectral analysis unmasks synchronous and compensatory dynamics in plankton communities. *Ecology* **88**, 2058–2071.

Vasseur, D.A., Gaedke, U., and McCann, K.S. (2005). A seasonal alternation of coherent and compensatory dynamics occurs in phytoplankton. *Oikos* **110**, 507–514.

Vitousek, P.M., Aber, J.D., Howarth, R.W., Likens, G.E., Matson, P.A., Schindler, D.E., Schlesinger, W.H., and Tilman, D. (1997). Human alterations of the global nitrogen cycle: Sources and consequences. *Ecol. Appl.* **7**, 737–750.

von Felten, S., Hector, A., Buchmann, N., Niklaus, P.A., Schmid, B., and Scherer-Lorenzen, M. (2009). Belowground nitrogen partitioning in experimental grassland plant communities of varying species richness. *Ecology* **90**, 1389–1399.

Vyverman, W., Verleyen, E., Sabbe, K., Vanhoutte, K., Sterken, M., Hodgson, D.A., Mann, D.G., Juggins, S., Vijver, B.V.d., Jones, V., Flower, R., Roberts, D., *et al.* (2007). Historical processes constrain patterns in global diatom diversity. *Ecology* **88**, 1924–1931.

Watson, S.B., McCauley, E., and Downing, J.A. (1997). Patterns in phytoplankton taxonomic composition across temperate lakes of differing nutrient status. *Limnol. Oceanogr.* **42**, 487–495.

Weis, J.J., Cardinale, B.J., Forshay, K.J., and Ives, A.R. (2007). Effects of species diversity on community biomass production change over the course of succession. *Ecology* **88**, 929–939.

Weis, J.J., Madrigal, D.S., and Cardinale, B.J. (2008). Effects of algal diversity on the production of biomass in homogeneous and heterogeneous nutrient environments: A microcosm experiment. *PLoS ONE* **3**, e2825.

Wetzel, R.G. (2001). Limnology. 3rd ed. Academic Press, San Diego, CA.

Woodward, G. (2009). Biodiversity, ecosystem functioning and food webs in freshwaters: Assembling the jigsaw puzzle. *Freshwater Biol.* **54**, 2171–2187.

Woodward, G., Benstead, J.P., Beveridge, O.S., Blanchard, J., Brey, T., Brown, L., Cross, W.F., Friberg, N., Ings, T.C., Jacob, U., Jennings, S., Ledger, M.E., *et al.* (2010). Ecological networks in a changing climate. *Adv. Ecol. Res.* **42**, 71–138.

Worm, B., Barbier, E.B., Beaumont, N., Duffy, J.E., Folke, C., Halpern, B.S., Jackson, J.B., Lotze, H.K., Micheli, F., Palumbi, S.R., Sala, E., Selkoe, K.A., *et al.* (2006). Impacts of biodiversity loss on ocean ecosystem services. *Science* **314**, 787–790.

Worsfold, N.T., Warren, P.H., and Petchey, O.L. (2009). Context-dependent effects of predator removal from experimental microcosm communities. *Oikos* **118**, 1319–1326.

Zavaleta, E.S., Pasari, J.R., Hulvey, K.B., and Tilman, G.D. (2010). Sustaining multiple ecosystem functions in grassland communities requires higher biodiversity. *Proc. Natl. Acad. Sci. USA* **107**, 1443–1446.

Zubkov, M.V., and Tarran, G.A. (2008). High bacterivory by the smallest phytoplankton in the North Atlantic Ocean. *Nature* **455**, 224–226.

When Microscopic Organisms Inform General Ecological Theory

JULIA REISS, JACK FORSTER, FERNANDA CÁSSIO, CLÁUDIA PASCOAL, REBECCA STEWART AND ANDREW G. HIRST

SUMMARY

Free-living protists and very small metazoans, such as meiofauna or mesozooplankton, are extremely abundant in aquatic habitats and they play a key role in numerous ecosystem processes, including nutrient cycling and supplying energy to higher trophic levels. Assessing their ecological roles is needed to understand natural systems in their own right, but their study can also be used to inform and test general ecological theories.

Organisms smaller than 2 mm in size are ideal candidates for laboratory or field studies with a theoretical focus, and we illustrate this point with published examples of pioneering and recent work in which the use of protozoans, aquatic fungi or small metazoans has informed theories within population and community ecology and also more synthetic ecological theories that span different levels of biological organisation (e.g. biodiversity–ecosystem functioning relationships and the metabolic theory of ecology). These literature examples highlight how ecological theory has been stimulated by this research and vice versa.

ADVANCES IN ECOLOGICAL RESEARCH VOL. 43 0065-2504/10 $35.00
© 2010 Elsevier Ltd. All rights reserved
DOI: 10.1016/S0065-2504(10)43002-0

Studies of small species might be sufficiently universal that they can also be applied to the ecology of larger organisms and by extension they could offer insights into how ecosystems operate. In the main part of this chapter, we discuss whether these small worlds can represent the larger world by giving three case studies related to three ecological concepts and topics: the temperature-size rule (TSR), allometric scaling of population abundance and biodiversity–ecosystem functioning relationships. In the first two case studies, we discuss if and how protists differ from small metazoans; in all three case studies we explore if patterns observed for small organisms vary from those detected for larger organisms and the consequences this might have for detecting unifying ecological principles.

We used a combination of our own and published data to test predictions related to the TSR and allometric scaling. In the first case, we found that the mechanisms which generate the phenomenon of larger adult size at low temperatures are likely to be fundamentally different for protozoans and multicellular animals. In our second case study, we regressed body size of protists and small metazoans from a large data set against their population abundance. We found that single-celled organisms also differed from small metazoans with regard to their allometries and that both groups differed from larger metazoans. In both these cases, there is evidence that extrapolations from the small to the larger world need to be treated with some caution. In the final case study, we explore to what extent biodiversity–ecosystem functioning relationships among small organisms (aquatic fungi) are similar to those observed for larger organisms.

I. INTRODUCTION

The invention of the microscope opened a new world for early investigators of small-sized organisms. Being rich in fascinating features and exhibiting enormous variety, ciliates ("infusoria") and rotifers ("wheel animalcules") were among the first microscopically small animals to be described by Antonie van Leeuwenhoek in 1674 (published in 1702) and John Harris in 1696 (published in Hudson and Gosse, 1886). The beauty and species richness of small fauna has continued to occupy biologists and taxonomists over the following two centuries (e.g. Foissner and Berger, 1996; Haeckel, 1904; Kahl, 1935; Ruttner-Kolisko, 1977; Streble and Krauter, 1988), and vast numbers of taxonomic descriptions have been made of unicellular organisms and small metazoans in the open water, surfaces and sediments of aquatic systems. Ecologists are still trying to quantify the true importance of theses ubiquitous, abundant and diverse members of natural ecosystems. Microscopic organisms can be found in all aquatic and wet habitats and are typically divided into groups based on their size, taxonomy or functional

Table 1 Classification of microscopic eukaryotes in freshwater and marine habitats

	Microscopic, eukaryotic plankton	Microscopic, eukaryotic benthos
Size groups	Picoplankton are < 2 μm in size, nanoplankton are 2–20 μm in size, microplankton are 20–200 μm in size (all three groups include heterotrophic organisms, but also others) and mesozooplankton describes all heterotrophic organisms 200–2000 μm in length (Sieburth *et al.*, 1978)	Nano-, micro- and meiofauna, where nano- and microfauna have the same size as their pelagic counterparts. Meiofauna refers to micro-metazoans that pass a 500 μm sieve (Reiss and Schmid-Araya, 2008). This definition makes meiofauna comparable to mesozooplankton and distinguishes them from unicellular organisms
Taxonomic groups	Single-celled and multicellular eukaryotes can be assigned to one of eight major groups in the (most recent) phylogenetic tree (Baldauf, 2003). Within the protozoans and micro-metazoans (metazoans smaller than 2 mm), representatives of many major phyla can be found and in some phyla individuals never grow beyond a very small size. For example, all species within the Rotifera (wheel animals), Gastrotricha, Gnathostomulida (jaw worms), Kinorhyncha (mud dragons) or Priapulida (priapulid worms) are all microscopic. Further, most groups are extremely species rich (examples for taxonomy and information on taxonomic keys: Rundle, Robertson and Schmid-Araya, 2002; Hausmann *et al.*, 2003)	

attributes (Table 1). The taxa involved include bacteria and single-celled algae, but as these groups have been considered in detail in other recent ecological reviews (e.g. Ptacnik *et al.*, 2010; Purdy *et al.*, 2010), we will limit our focus to protozoans, aquatic fungi and microscopic metazoans [meiofauna and mesozooplankton (Table 1) which we refer to as "micro-metazoans"]. Species belonging to these groups have been widely used in experiments addressing general ecological theory (e.g. Bonsall and Hassell, 2005; Cadotte *et al.*, 2005; Jessup *et al.*, 2004; Lawton, 1995; Pascoal *et al.*, 2010; Petchey *et al.*, 2002, Yoshida, 2005).

The high species richness of these small organisms highlights their ecological importance as repositories of biodiversity. The notion that there are more small than there are large species was confirmed over twenty years ago by meta-analyses of species richness within size classes (May, 1986). In his seminal paper "The search for patterns in the balance of nature: advances and retreats", May (1986) showed that species richness peaks within the 0.5–1 cm body size range of animals, but pointed out that the lower species richness within the size classes less than 0.5 cm could be an artefact of sampling effort and poor taxonomic resolution: "The most serious problem is our current uncertainty, by a factor 10 or more, of the total number of species on the globe. This uncertainty centres on insects and other small creatures, and could well raise the totals in the categories below 1 cm ... by almost an order of magnitude" (p. 1124; May, 1986). Others have argued that species less than 1 cm that have resting stages and high dispersal show high local diversity but low global diversity (Fenchel and Finlay, 2004; Finlay, 2002). Whether this is true or not (May, 1999; Ptacnik *et al.*, 2010), the global species richness of micro-metazoans is undoubtedly very high: for example, there are more than 11,000 species of free living nematodes and over 13,000 copepod species (Boxshall and Defaye, 2008; Traunspurger, 2000). While the species concept is problematic for some protists and prokaryotes due, for instance, to the prevalence of asexual reproduction and/or relatively open genomes (Purdy *et al.*, 2010), biodiversity within these even smaller groups is also substantial [e.g. estimated ciliate abundance of 3,000 (Finlay, 2002) to 8,000 (Hillebrand *et al.*, 2001); bacterial diversity in the oceans of up to 2 Mio taxa (Curtis *et al.*, 2002)].

Small organisms are not only species rich, but they also comprise a wide diversity of life strategies, feeding and locomotory modes, trophic positions and other functional traits. Further, the range of body size within protists and micro-metazoans spans many orders of magnitude, thus they are comparable to larger organisms (e.g. the insects). Even within the same taxonomic group (e.g. phylum or even genus), individual taxa often cannot be ascribed to a single trophic group, but may comprise representatives from various trophic levels: for example, nematode species can be generalists or specialists which can feed on detritus, bacteria, algae and/or be predatory (Traunspurger, 2000), as is also true of the larger metazoans.

The wide range of traits (e.g. body size and trophic position), their high species diversity and small size mean that microscopic organisms are useful candidates for testing general ecological theory (Jessup *et al.*, 2004). Their fast reproduction means that multiple generations can be observed over a relatively short study period, and their small size implies that arenas in controlled laboratory experiments can be replicated to a high degree. These properties have allowed them to be used for testing a wide range of ecological and evolutionary theories and ideas (Table 2) and microscopic organisms assembled in micro- or mesocosms (e.g. chemostats) are commonly used as "model systems" [i.e. as "simplified representations of more complex systems" (p. 189; Jessup *et al.*, 2004)]. Early laboratory experiments with protozoans showed that microscopic organisms can be used to inform general biological theories (e.g. interspecific competition; Gause, 1934) and since the mid-1970s in particular ecologists have increasingly discovered the potential of laboratory experiments with these organisms to extract ecological

Table 2 Examples for laboratory studies that have used small organisms to test ecological theory

Study	Focus	Organisms and set-up	Results
Population dynamics and food chains			
Gause (1934)	Competitive exclusion and coexistence	Example for one of the described experiments: one bacterial species (*Bacillus subtilis*); two protozoan species (*Paramecium caudatum* and *Stylonychia mytilus*); three treatm. (2 mono-1 diculture); (d) 4 days	Mathematical equations for competition and coexistence
Jost *et al.* (1973)	Competitive exclusion and predator–prey oscillations	Two bacterial species, one protist species; (c) batch and continuous; (d) up to approximately 28 days	Presence of predator allows coexistence of the two prey species
Kaunzinger and Morin (1998)	Relationship between productivity and food chain length	One bacterial, two protozoan species; (a) up to 24 treatm.; (b) up to 100 ml; (c) 12 nutrient levels; (d) up to 40 days	Productivity controls food chain length
Morin (1999)	Predator–prey dynamics with and without intra-guild predation	Three bacterial, two protozoan (ciliate) species; (b) 100 ml; (c) two productivity levels; (d) 60 days	The response of food webs to resource enrichment can depend on the amount of omnivory
Bohannan and Lenski (1999)	Effect of resource enrichment on predator–prey dynamics, evolution	One bacteria, one phage; (b) 30 ml chemostat; (c) different glucose levels; (d) 8.3 days	Decrease in the stability of both the predator and prey populations
Diehl and Feissel (2000)	Effects of enrichment on three-level food chains, omnivory	Unknown bacteria, two protozoan (ciliate) species; (a) five treatm.; (b) 50 ml; (d) 150 days	Mathematical model with six predictions, for example predicting successful invasion of an omnivore

(continued)

Table 2 *(continued)*

Study	Focus	Organisms and set-up	Results
Diehl and Feissel (2001)	Effects of enrichment on three-level food chains, omnivory	Unknown bacteria, two protozoan (ciliate) species; (a) 3×4 treatm.; (b) 50 ml; (d) 150 days	Enrichment of the shared resource (on which the intra-guild prey is the superior competitor) reduces the population size of the intra-guild prey
Becks *et al.* (2005)	Chaotic and stable state population dynamics of coexisting species	Two bacterial, one protozoan (ciliate) species; (a) 185 ml cultures; (c) chemostat; (d) up to 55 days duration	Resource supply determines chaotic or stable population dynamics
Community assembly			
Holyoak and Lawler (1996)	Dispersal, predator–prey metapopulation dynamics	Two protozoan species; (b) 30 ml with unknown bacteria; (c) some microcosms linked for dispersal	Confirmation of three predictions, for example, predator:prey ratios are greatest at high dispersal rates
Fox *et al.* (2000)	Local species saturation	Bacterial, protist and micro-metazoan species; (a) 10×5 treatm.; (b) 100 ml; (c) supplemented nutrients; (d) 7 weeks	Species richness does not set an upper limit to number of local species
Law *et al.* (2000)	Invasion	Six protist species; (a) 55 treatm.; (b) 70 ml with unidentified bacteria; (d) 9.5 days	Species richness does not affect invasion
McGrady-Steed and Morin (2000)	Species richness, size and stability of populations	9 autotrophic protist, 20 heterotrophic protist, 3 micro-metazoan species; (a) five treatm.; (b) 100 ml with four bacterial species; (d) 42 days duration	Wide range of patterns

Warren *et al.* (2003)	Coexistence and invasion	Six protist species; (a) 63 treatm.; (b) 70 ml with unknown bacteria; (d) 2 months	Persistent communities with coexisting species are species poor. Some predators act as catalytic species: they invade the resident community and then go extinct
Fukami (2004)	Assembly history, ecosystem size and species diversity	12 protist, 2 micro-metazoan (rotifer) species; (a) 16 treatm.; (b) up to 850 ml with four bacterial species and unidentified flagellates; (d) 125 days	Assemblage history affects diversity more strongly in smaller ecosystems
Fox (2007)	Dispersal and source-sink dynamics on competitive outcome	Three protozoan (ciliate) species; (a) $6 \times 2 \times 2$ treatm.; (b) 40 ml with three bacterial species; (c) some microcosms linked for dispersal; (d) 35 days	Dispersal alters competitive outcome
Fox (2008)	Level of productivity and assembly (invasion and extinction)	25 protist, 3 micro-metazoan (rotifer) species, 1 group of unidentified flagellates; (a) 2×4 treatm.; (b) 60 ml; (c) four levels of carbon; (d) 105 days	Level of productivity does not determine community assembly. Communities are comprised out of species with appropriate traits
Food web			
Neill (1975)	Role of competitive exclusion, coexistence, predation and resource use on stable state	18 algal, 12 micro-metzoan (crustacean) species, 1 fish species; (b) 1500 ml; (d) 35 weeks	Predation, life history and resource use affect equilibrium of populations and communities
Smith and Cooper (1982)	Competition	Three micro-metazoan (cladoceran) species; (a) up to 21 treatm.; (b) 1 l; (c) addition of five invertebrates	Competition depends on resource levels and the age (size) of competitors

(*continued*)

Table 2 (*continued*)

Study	Focus	Organisms and set-up	Results
Lawler and Morin (1993)	Food chain length, omnivory	Five protist species; (b) 100 ml with unknown bacteria; (d) 45 days duration	Long food chains and those with abundant omnivory are less likely to be stable than shorter, simpler food webs
Balciunas and Lawler (1995)	Relative strength of top-down and bottom-up effects in food chains	(a) 12 treatm.	When intermediate species has a size refuge, more resource sustains its population, but presence of other intermediate prey make it more vulnerable to predation
Schrag and Mittler (1996)	Coexistence	Bacterial species and bacteriophages; (a) three treatm.; (b) 11–50 ml; (c) 250 h to 13 days (depending on culture treatment)	The role of spatial refuges in stabilising bacteria–phage interactions. Wall populations of bacteria decreased the stability of the phage species
Weatherby *et al.* (1998)	Coexistence	Bacterial assemblage, six protist species; (a) 63 treatm.; (b) 70 ml; (d) 101 days	Extinction rates are higher in more species-rich systems; collapse can be predicted from initial community
Worsfold *et al.* (2009)	Increased extinction frequency, decreased population abundance and biodiversity loss with removal of large generalist predator	Unidentified bacteria and flagellate species, 11 protozoan, 1 micro-metazoan species; (a) four treatm.; (b) 100 ml	Species loss from high trophic levels can be context-dependent, as rare species can have disproportionately large effects in food webs
Biodiversity and ecosystem functioning			
McGrady-Steed *et al.* (1997)	Diversity–stability, insurance hypothesis	9 autotrophic protist, 20 heterotrophic protist, 3 micro-metazoan species; (a) 8 treatm.; (b) 100 ml; (c) addition of ciliate *Euplotes* sp. as invader; (d) 42 days duration	More diverse assemblages show less variation in CO_2 flux and are more resistant to invasion

Naeem and Li (1997)	Species richness and community stability	3 bacterial, 3 autotrophic protist, 13 heterotrophic protist species; (b) 50 ml; (c) three nutrient and three light levels; (d) 57 days duration	Species richness within functional group promotes community stability (measured as abundance)
Petchey *et al*. (1999)	Warming, food web structure and ecosystems functions; insurance hypothesis	At least 7 autotroph species, at least 19 protozoan species, 2 micro-metazoans; (a) eight treatm.; (b) 100 ml with three bacterial species; (c) warming of microcosms over time; (d) 7 weeks	Warming affects food web structure and changes ecosystem processes as a result. More diverse communities are more robust than species poor ones
Hulot *et al*. (2000)	Effects of functional diversity at different trophic levels and nutrient enrichment on final biomass	25 algal, 20 micro-metazoan species; (c) nutrient enrichment; (d) 2 month	Mathematical model "intermediate complexity" explains biomass within trophic levels and functional groups, linear food web models do not
Kassen *et al*. (2000)	Productivity and biodiversity are positively related	*Pseudomonas fluorescens*; (a) 16 nutrient levels; (c) homogeneous versus heterogeneous environments; (d) 2 days	Diversity peaks at intermediate productivity
Naeem *et al*. (2000)	Producer–decomposer co-dependency influences biodiversity effects	12 heterotroph bacteria, 8 green algae species; (a) five algae diversity levels, six bacterial diversity levels; (b) 50 ml; (c) 20 °C, 14:10 h light:dark cycle; (d) 4 weeks	Ecosystem responses to changing biodiversity depend on producer–decomposer co-dependency

(*continued*)

Table 2 (*continued*)

Study	Focus	Organisms and set-up	Results
Fox and McGrady-Steed (2002)	Stability and complexity	Bacteria, protist, micro-metazoan species; (a) 24 treatm.; (b) 100 ml; (c) 23 °C, 14:10 h light: dark cycle; (d) 6 weeks	Trophic interactions cause non-random extinctions
Morin and McGrady-Steed (2004)	Relationship between species richness and stability	9 autotrophic and bacterial species, 20 protozoan, 3 micro-metazoan species; (a) eight treatm.; (d) 6 weeks	More diverse assemblages show less variation in CO_2 flux
Bell *et al*. (2005a)	Species richness and composition; ecosystem functioning	72 heterotrophic bacterial species; (a) 585 treatm. (combinations of 12 diversity levels, 3 time levels); (b) 10 ml; (d) up to 28 days	Synergistic interactions between species and composition determine community respiration
Gamfeldt and Hillebrand (2005)	Species richness and trophic position, consumer and prey biomass	Three algal, four protozoan (ciliate) species; (a) 56 treatm.; (c) 18 °C; (d) 3 weeks	Biodiversity effects on consumer and prey biomass
Duarte *et al*. (2006)	Species diversity and identity; ecosystem processes	Four aquatic fungal species; (a) 15 treatm.; (b) 100 ml; (c) 18 °C; (d) 27 days	Diversity and identity affect litter decomposition, biomass production and reproductive output
Warren and Weatherby (2006)	Energy input and species diversity patterns	Bacterial assemblage, 12 protist species (8 bacterivorous and 4 omnivorous or predators); (a) gradient of energy (six levels); (b) 70 ml; (d) up to 421 days	Effects of energy input on species diversity patterns varied along time
Replansky and Bell (2009)	Species diversity and productivity; environmental complexity	Three yeast species; (a) 182 treatm. (three diversity levels, up to eight carbon sources); (b) 5 ml; (d) 200 generations, 20 days	Positive diversity effects on productivity due to complementarity; effects of environmental complexity are unclear

Salles *et al*. (2009)	Community niche and ecosystem functioning	16 denitrifying bacterial species; (a) combinations of five diversity levels; (b) 50 ml; (c) medium with six carbon sources; (d) 8 days (13 generations)	Larger community niches account for higher denitrification and CO_2 production
Wittebolle *et al*. (2009)	Richness and evenness, ecosystem processes; stress	18 denitrifying bacterial species; (a) 84 treatm. possible; (b) multiwell plates; (c) temperature or salt stress; (d) 20 h	Highly uneven communities show lower denitrification when faced with stress
Pascoal *et al*. (2010)	Diversity and functional stability	Four aquatic fungal species; (a) 27 treatm. (combinations of four diversity levels, three metal levels); (b) 80 ml; (c) 17 °C; (d) 21 days	Diversity increases functional stability, mainly under metal stress
Environmental warming and assemblage patterns			
Delaney (2003)	Growth rate response of bacterial and protist communities to warming and turbulence	Bacterial, one protist species; (a) 4 × 2 × 4 treatm. (four temperatures and two levels of turbulence); (b) 130 ml; (d) 70–120 h	Growth of populations is temperature dependent. Turbulence influences grazing activity
Jiang and Morin (2004)	Warming and species interactions	Two protist (ciliate) species; (a) 3 × 5 treatm.; (b) 100 ml; (c) 10% media replacement every week; (d) 44 days	Temperature dependent interactions
Newsham and Garstecki (2007)	Food web responses to warming, single species response to warming	Bacterial species, six protist (flagellate) species; (a) 4 × 2 treatm.; (b) 300 ml SESM; (d) 252 h	Interactive effects of flagellate species loss and warming on bacterial abundance, evidence for apparent complementarity

(*continued*)

Table 2 *(continued)*

Study	Focus	Organisms and set-up	Results
Beveridge *et al.* (2010)	Interaction between trophic structure and temperature. Effects of warming on food chain length	Two protozoan species; (a) $2 \times 4 \times 6$ treatm.; (b)100 ml (c) six temperature levels, four food chain length levels; (d) 6 weeks	Population density and ecosystem functioning depends on community structure and temperature and the interaction between the two. Positive trend between decomposition and temperature
Disturbance hypotheses			
Buckling *et al.* (2000)	Biodiversity, disturbance frequency, heterogeneous environment	Genotypes within bacterium *Pseudomonas fluorescens*; (a) 20 treatments; (b) 6 ml; (d) 16 days	Unimodal relationship between diversity and disturbance in heterogeneous, but not in homogeneous environments
Garstecki and Wickham (2003)	Biodiversity, disturbance intensity and frequency	Protozoan (amoebae) species; (b) 2.6 l; (c) grid oscillation frequencies; (d) 24 days	Abundance and species richness are not affected by disturbance intensity

The publications are ordered by topic and year of publication. This table gives the authors, the focus of the study, the methods and the results. Four details are given for the methods when they were available: (a) the number of treatments (treatm.); (b) millilitre culture medium; (c) other conditions and (d) duration of the experiment.

principles related to theories (Table 2). This has included population, assemblage and community theories [e.g. population or food chain dynamics (e.g. Gause, 1934; Morin, 1999) and theories based in community ecology (e.g. food web theory; Lawler and Morin, 1993)]. Further, higher level and synergistic theories (e.g. the metabolic theory of ecology (MTE) (e.g. DeLong *et al.*, 2010) or biodiversity and ecosystem functioning relationships (e.g. Pascoal *et al.*, 2010) have also been tested with microscopic organisms. In addition to laboratory experiments, field observations on microscopic organisms have also been used, albeit less frequently, to inform general theories (e.g. Burns and Schallenberg, 1996; Jürgens *et al.*, 1996), including those related to allometric scaling of body mass and abundance and food web structure (e.g. Schmid *et al.*, 2002; Schmid-Araya *et al.*, 2002).

A number of literature reviews reflect the large number and wide topical range of these studies, but whilst many have addressed the justification of using these organisms in microcosms for testing theories (e.g. Bonsall and Hassell, 2005; Cadotte *et al.*, 2005; Jessup *et al.*, 2004; Lawton, 1995; Petchey *et al.*, 2002), there has been less focus on how small organisms might differ fundamentally from larger ones and what this might mean for extracting general ecological rules from observations of microscopic species.

In this chapter, we have two broad aims: firstly, to provide examples for key ecological theories and concepts that have profited from laboratory and field studies on small organisms and secondly, to give more detailed background for three of these concepts while discussing the suitability of using small organisms for testing the theory. We do the latter for three case studies in the main part of this chapter, where we consider the temperature-size rule (TSR), body size–abundance scaling, and biodiversity–ecosystem functioning relationships, respectively. Here, we use previously published and/or new data to assess whether there are differences between groups of small organisms (we focus on the difference between single-celled and multicellular organisms) and whether small organisms are different from larger organisms with respect to the particular theory under consideration. Given the wide range of possible ecological fields we could discuss in this context, the choice of topics for our case studies is necessarily somewhat subjective, but each represents an important current field in ecology today.

II. EXAMPLES FROM THE LITERATURE: WHEN SMALL ORGANISMS HAVE INFORMED GENERAL ECOLOGICAL THEORY

Many previous literature reviews have especially focussed on experiments conducted in micro- or mesocosms set-ups (see reviews by Bonsall and Hassell, 2005; Cadotte *et al.*, 2005; Jessup *et al.*, 2004; Lawton, 1995;

Petchey *et al.*, 2002; Yoshida, 2005). Petchey *et al.* defined micro- and mesocosms as "spatially delimited artificially constructed model ecosystems" (p. 127; Petchey *et al.*, 2002) and they can vary from small vials holding less than one or a few millilitres (e.g. Bell *et al.*, 2005) to large tanks and artificial ponds or enclosures (e.g. Moss *et al.*, 2003; Yvon-Durocher *et al.*, 2010a) holding many hundreds of litres.

While there is a general debate in ecological research about the pros and cons of using micro- or mesocosms versus field studies (Benton *et al.*, 2007; Carpenter, 1996), it is clear that highly controlled experiments can be more than just a stimulus to further research (Benton *et al.*, 2007). For instance, they can produce the parameters that can be fitted in mathematical models (i.e. they inform modelling) which can in turn be tested on natural assemblages (Bonsall and Hassell, 2005). For example, Becks *et al.* (2005) used a combination of laboratory experiments (with bacteria and a ciliate) and mathematical modelling (Lyapunov exponents) to demonstrate how chaos, stable limit cycles and coexistence at equilibrium may occur in this two-prey, one-predator system (see also Bonsall and Hassell, 2005 for a discussion of modelling chaos). Also, experiments using microcosms can draw the focus on particular mechanisms that might operate in natural assemblages but which would have gone undetected as a consequence of the high complexity of natural communities (Benton *et al.*, 2007; Jessup *et al.*, 2004). The main arguments for the use of microcosms are: replication, simplicity and ease of experimental manipulation (including a control of spatial and temporal factors), although the lack of complexity can also restrict the ability to extrapolate to natural systems (Woodward, 2009). These aspects, and the relevance of using micro- and mesocosms for ecological research, have been discussed in more detail elsewhere (Bonsall and Hassell, 2005; Cadotte *et al.*, 2005; Jessup *et al.*, 2004; Lawton, 1995; Petchey *et al.*, 2002; Yoshida, 2005).

Here, we focus on giving examples of laboratory and field research that have either used small organisms to test existing ecological theories or even put new concepts forward by confirming ideas with small organisms. We do this by focussing on the broad topics mentioned above: (1) population, assemblage and community theories and (2) higher level and synthetic theories. The separation of theories into these coarse groups is somewhat artificial, as there are many overlapping aspects between them. For example, we class the metabolic theory of ecology as a "synthetic theory", but hypotheses on individuals and populations are essential for its formulation. Throughout this paper, we will focus primarily on the ecology of protists (protozoans), aquatic fungi and micro-metazoans and have excluded the discussion of evolution *per se* in order to retain a coherent focus on the main topics of interest, whilst recognising that other microscopic organisms have also informed general ecological theory (e.g. plankton: see references in Ptacnik *et al.*, 2010; bacteria: Langenheder and Ragnarsson, 2007; Purdy *et al.*, 2010).

Further, the study of protists and micro-metazoans, as well as other microscopic organisms (especially yeasts and bacteria), has also informed evolutionary theories (e.g. Replansky et al., 2008; Bell and Gonzalez, 2009).

A. Population, Assemblage and Community Theories

Since Gause's pioneering work using ciliates and yeast to demonstrate general concepts of population dynamics (intra- and interspecific competition; Gause, 1934), small organisms have been used in the field of population and community ecology. Two reviews by Bonsall and Hassell (2005) and Holyoak and Lawler (2005) provide an excellent overview of the main concepts in population and metapopulation ecology, in which they give numerous examples of studies that have used microscopic organisms to address these concepts (see topic "population dynamics and food chains" in Table 2). These include competition (Gause, 1934; Jost *et al.*, 1973), population control and effects of environmental variation (Petchey, 2000), omnivory, predation and stability–complexity relationships (Holyoak, 2000; Morin, 1999). These studies have in common that they typically addressed population dynamics within short food chains. Microscopic organisms remain important study objects in population ecology with recent studies addressing topics such as the TSR (see case study below) or chaotic food chain dynamics (Becks *et al.*, 2005). Models and theories on population dynamics are also testable using experiments or observations of larger animals, but their assembly in more complex experiments is logistically challenging and small organisms are often more suitable for complex manipulations, especially when studying intergenerational dynamics.

At the higher levels of organisation, community ecology theories, such as those related to assembly rules, metapopulation ecology or food web dynamics (Table 2), have also profited from the use of microscopic organisms. Cadotte *et al.* (2005) reviewed some of these studies (e.g. the studies by Diehl and Feissel, 2000; Fox and McGrady-Steed, 2002; Holyoak, 2000; Lawler and Morin, 1993; Warren *et al.*, 2003; see Table 2) within the broad topic of community assembly and highlight why microcosm communities are so useful for understanding community ecology: "We see laboratory microcosms as the link between models and the real world" (p. 348; Cadotte *et al.*, 2005). Community assembly experiments have typically tested rules that govern the assembly of "stable" communities (Table 2). For example, a recent study by Fox (2008) tested how productivity affects stable states of communities and he found that species composition varied with productivity, but that the frequency of successful invasions and of post-invasion extinctions did not. These assembly experiments and those addressing food web theory (and food chain length) typically all use a wider range of species. In a review of multispecies community studies, Lawton (1995) summarised many of the findings of those that used small organisms for testing food web theory: for instance, he discusses that theory predicts that long food chains and those with abundant omnivory are less likely to be stable than short food chains, but that only the former has been confirmed by laboratory studies with microscopic organisms. To date, we still have very few empirical data from other systems available to explore intergenerational food web dynamics in this way (Woodward *et al.*, 2010). Traditionally, food web experiments with microscopic species focussed on competition, food chain length or coexistence (Table 2) and the most recent studies have widened this focus (Table 2). For example, the importance of species loss within trophic levels was addressed by Worsfold *et al.* (2009) who used a food web consisting of bacteria, protists and one micro-metazoan. The authors tested whether the loss of a large omnivorous species would affect individual prey species and the assemblage as a whole and found that loss of the omnivore only changed total community biomass when a specialist predator was present in the remaining assemblage.

Compared with the rich literature addressing population and community dynamics in microcosm experiments, field observations are less often used to infer patterns at this level of organisation via the study of small organisms. However, studies on lakes for example have been instrumental in developing and understanding some pivotal theories in ecology (Jürgens *et al.*, 1996). For instance, important concepts such as bottom-up and top-down control and trophic cascades have been shown in studies on lake communities where the taxonomic composition of the microbial food web is determined by feeding interactions within the food web and the top-down control of the mesozooplankton (Burns and Schallenberg, 1996; Jürgens *et al.*, 1996). A seminal example is that dense populations of the cladoceran *Daphnia* can control lower trophic levels and thus suppress populations of smaller zooplankton (Jürgens, 1994). Generally, the study of small zooplankton has underscored how trophic cascades (e.g. from fish to phytoplankton) and alternative equilibria can arise and persist in the plankton (Carpenter *et al.*, 1985; Wickham, 1995).

B. Higher Level and Synthetic Theories

More synthetic theories that integrate multiple levels of organisation (e.g. from individuals to ecosystems) have also been addressed in experiments and empirical field studies using microscopic organisms. We focus specifically on two examples in our case studies below [allometric scaling and biodiversity and ecosystem functioning (B–EF) experiments]. However, they are not the only higher level or synthetic theories that have been informed by studying small organisms. Another example is the "biological insurance hypothesis"

(BIH) which is linked to B–EF theory. It proposes that biodiversity provides a buffer against environmental fluctuations, because species respond differentially to these changes, leading to more predictable aggregate community or ecosystem properties (Yachi and Loreau, 1999). Consequently, species might appear functionally redundant for a given ecosystem process over a certain time frame but they are no longer redundant when this process is measured for longer (and while the environmental conditions change). Studies on both plants and microscopic organisms have provided evidence for the BIH (Leary and Petchey, 2009; Loreau *et al.*, 2002b; Naeem and Li, 1997). Similarly, the intermediate disturbance hypothesis, which predicts that intermediate levels of disturbance (e.g. temperature fluctuations or physical disturbance of the environment) favour species-rich communities, has also been tested using microscopic organisms (Garstecki and Wickham, 2003; Table 2).

Small organisms have also informed synthetic theories related to exploring the effects of environmental warming (Table 2). For instance, Beveridge *et al.* (2010) investigated the interacting effects of temperature and food chain length on species abundance and one ecosystem process (decomposition) in a microcosm experiment with bacteria and two ciliate species. They concluded that in order to predict the response of an organism to a change in temperature both the organism's trophic position and the effects of temperature on species interactions must be taken into account. Warming has also been shown to influence the size structure of small organism assemblages in experimental mesocosms, causing a shift in the community towards smaller body size within the phytoplankton assemblage (Yvon-Durocher *et al.*, 2010c), and this general pattern (which is in line with predictions made by the MTE (e.g. Brown *et al.*, 2004) has also been reported from a wide range of aquatic systems (Woodward *et al.*, 2010 and references therein). In general, experiments with small organisms might be very suitable when more complex and higher level ecological questions are addressed, because measurements can be made within a short period of absolute time.

III. CASE STUDIES: DO SMALL WORLDS REPRESENT LARGER WORLDS?

So far, we have highlighted when small organisms have usefully informed general ecological theory, rather than questioning the general approach. However, if there are fundamental differences between small and large organisms (see Jessup *et al.*, 2004) this could have implications for the generalisations we can make from the study of the former.

Clearly, small and large organisms share many features, but some small organisms might be more suitable for testing certain theories than others.

An obvious difference to larger metazoans is that many protists and micrometazoans can form resting stages (in a short time frame) and can lie dormant until the environmental conditions allow the expression of population abundances. These characteristics mean that laboratory experiments that run over longer time periods have to be monitored in this regard.

In the following three sections, we will consider, using specific examples, patterns and processes exhibited by small organisms in the context of three ecological topics. We examine whether the theories or hypotheses hold for both small and large organisms. In other words, we ask to what extent small worlds can resemble large worlds. We highlight that in some cases there are important differences between unicellular and multicellular organisms and also between microscopic organisms and larger ones, which in our view should be considered when using microscopic organisms in model systems.

A. Temperature-Size Rule

1. The Theory

When ectothermic species develop at cooler temperatures they tend to grow more slowly, but reach a larger size at maturity, than individuals grown at warmer temperatures. This common phenomenon, known as the "temperature-size rule" (Atkinson, 1994), has been detected in over 83% of studies examining organism size at a given developmental stage and at different temperatures (Atkinson, 1994). It is widespread, and has been demonstrated in a bacterium, protists, plants and animal groups including molluscs, arthropods, amphibians and reptiles (Atkinson, 1994). These changes in size can be marked: for example, a meta-analysis of protist data has demonstrated that for each 1 °C increase in temperature, cell volume decreased by around 2.5% of their volume at 15 °C (Atkinson *et al.*, 2003). In metazoa, changes in size with temperature can also be large: for example, in a rotifer species body volume changed by almost 50% over an 8 °C temperature range (Stelzer, 2002).

The TSR is important when in the context of global warming, and the expected decrease in organism size with increasing temperature has been called the "third ecological rule", along with latitudinal/altitudinal shifts and changes in phenology, which are predicted under a warming climate (Daufresne *et al.*, 2009). Shifts in the size distribution within populations could affect predator–prey dynamics in natural systems and even crop yields in agro-ecosystems (Atkinson, 1994). Consequently, if we are to make accurate predictions of future effects associated with a warmer world we need to understand what drives the TSR.

Despite extensive evidence for the existence of the TSR, a general explanation remains elusive (see Angilletta and Dunham, 2003; de Jong, 2010). van der Have and de Jong (1996) proposed an explanation based on the unequal thermal sensitivities of growth (changes in mass) and development rate (changes in life stage). They show that increased temperatures must cause a greater increase in development than in growth rates and suggest that this is because the former is limited by enzymatic processes ($Q_{10} \sim 2$), whereas the latter is limited by diffusion (Q_{10} close to 1). They suggest that their proposed mechanism is universal and thus applies to all organisms, including unicellular and multicellular organisms. Although the universality of their suggested mechanism has been questioned (Angilletta and Dunham, 2003), separating the effects of temperature into growth and development is a useful framework for developing ideas of the TSR (Kingsolver and Huey, 2008), and microscopic organisms offer particular promise in this regard.

Examining growth, development and size form the essential components in determining the cause of the TSR. Short generation times, and relatively easy culturing under controlled-temperature laboratory conditions, make small unicellular organisms particularly appealing subjects for studying the TSR experimentally. Development rates can easily be estimated from doubling times of cell division, and changes in cell biovolumes can be used to make estimates of growth rates. In contrast, in multicellular organisms development rates are complicated by multiple life stages, and longer generation times. Further insights may thus rely heavily on data collected from experiments conducted on unicellular organisms to inform general theories about the mechanism(s) behind the TSR.

2. *Is It Different for Small Organisms?*

Using microscopic organisms to make predictions about the cause of the TSR across all ectothermic groups should be conducted with caution. There might be fundamental differences between how unicellular and multicellular organisms are able to respond to temperature change, and in turn how the mechanism of the TSR might act. Here, we will highlight how some of these issues may have wider implications for how the use of unicells to make broader predictions of processes and life history rates.

Firstly, let us begin by examining the inter-relations between growth and development in the context of the TSR. This is best explored using a simple growth model which assumes individual organism mass increases linearly with time:

$$\frac{g}{D} = \frac{M_A}{M_P}, \tag{1}$$

where g is the average growth rate (increase in individual organism mass, d^{-1}), D is the average development rate (inverse of total development time from progeny to adult stage, d^{-1}), M_A is the mass at achieving adulthood and M_P is the mass of progeny. In unicells, progeny can be considered as a newly divided individual, whereas in metazoans progeny is an individual egg or newly hatched individual offspring. We shall consider a theoretical species of metazoan or unicellular organism in two different thermal environments, one cool and the other warm. As the unicell divides by binary fission, adult mass would be approximately twice that of progeny mass, thus $M_A/M_P=2$ in both environments (Figure 1). However, data on the mass of adults and their progeny in a range of metazoans (including butterflies, copepods and flies) show that progeny mass changes much less with temperature than adult mass, that is, the TSR is more pronounced in the latter life stage. We therefore hypothesise that for a single metazoan species the ratio of adult to progeny mass tends to be greater in cool environments than in warm (i.e. M_A/M_P (cool) $>$ M_A/M_P (warm)).

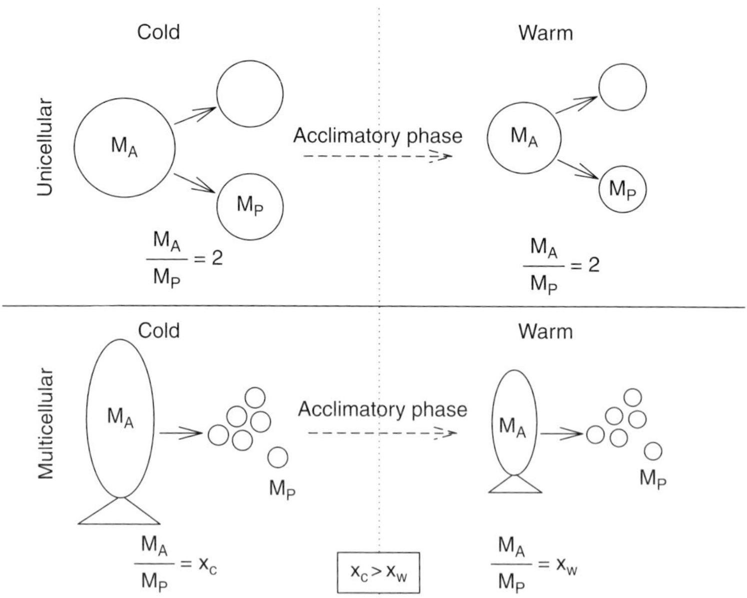

Figure 1 An illustration of how individual unicellular and multicellular species can differ in their size response to temperature. M_A = mass of adult, M_P = mass of progeny, x_c = ratio of M_A/M_P in a multicellular organism at cool temperature, x_w = ratio of M_A/M_P in a multicellular organism at warm temperature. In unicellular organisms, the ultimate ratio of M_A/M_P is fixed as 2 due to binary fission, whereas this ratio is variable in multicellular organisms.

It follows from Eq. (1) that for a single metazoan species the ratio of growth to development must also be greater in cool than in warm environments (i.e. $g/D(\text{cool}) > g/D(\text{warm})$), thus development rate is more sensitive to temperature change than growth rate. Now let us compare this with the situation for unicellular organisms which reproduce by binary fission. Given that the ratio of mass of adult to progeny is the same in cool and warm environments (i.e. a ratio of two) then substituting these values into Eq. (1) shows that for the unicell g/D is fixed in different thermal environments.

The contrasting reproductive strategies of unicellular and multicellular organisms results in fundamental differences in their responses of growth and development rates to temperature. This is counter to the universal explanation proposed by van der Have and de Jong (1996). Although changes in mass do occur in unicells when exposed to a new temperature, this can only occur during an acclimatory phase, and once the new adult size is established at steady state (size-acclimation), M_A/M_P becomes a fixed ratio once more. As progeny size need not be a fixed proportion of adult size in metazoans, they are not constrained in this way; they are therefore able to maintain differences in adult and progeny mass in different thermal environments and thus different rates of growth and development (Forster *et al.*, in review).

A theoretical example for size-acclimated unicellular and multicellular organisms is shown in Figure 2. As the ratio M_A/M_P is fixed for unicellular organisms, the slopes applied to the data for progeny and adult mass are parallel (Figure 2A). In contrast, these slopes differ for metazoa, with progeny mass having a much shallower slope than adult mass (Figure 2B). Applying this to Eq. (1) results in a fixed g/D in the unicellular data but a decrease in g/D in the metazoan data (Figure 2C).

Growth and development rates are key aspects of organism life history, therefore the differences that the TSR highlights between unicells and metazoans are very important. This case study of the TSR shows that multicellular organisms might be able to decouple these rates from one another, whilst this might not be possible in unicells that divide by binary fission. This suggests that extrapolating the population dynamics of multicellular species from single-celled organism models (e.g. the classic predator–prey dynamics of Gause, 1934) might not be reasonable, as the differing reproductive strategy in metazoans allows increased flexibility to other environmental factors. Critically, the difference in reproductive modes of unicellular and multicellular organisms is key here.

The impact of a variable adult to progeny size ratio in multicellular organisms in comparison to unicells has been explored recently with regard to the TSR (Forster *et al.*, in review), but there are also other important implications related to these differences. Growth and development rates are

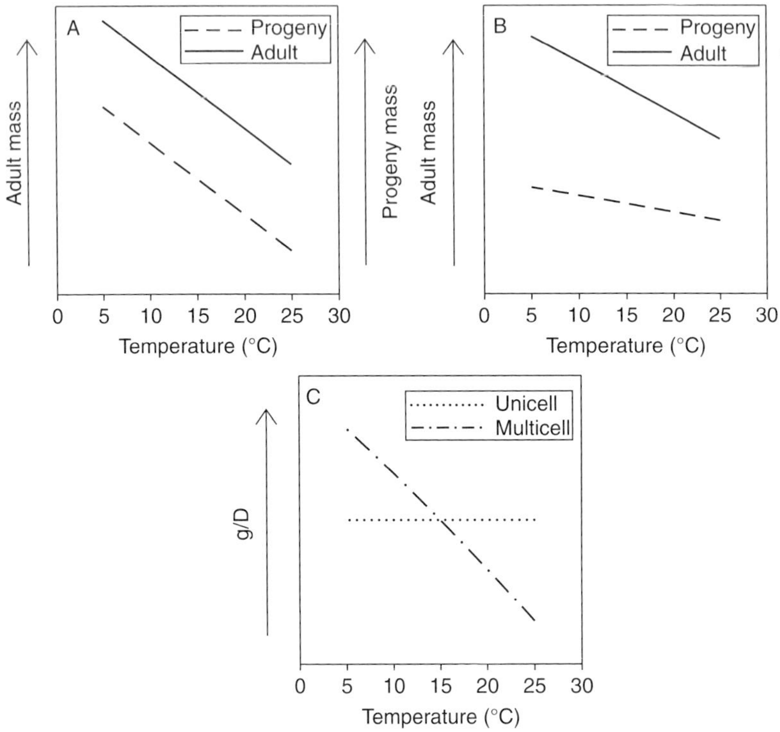

Figure 2 Example of the relationship between adult and progeny mass in (A) unicellular organisms and (B) multicellular organisms, at different environmental temperatures when at steady state. (C) Resultant relationship of growth rate (g) to development rate (D) in different organisms. Adult and progeny mass axes in (A) and (B) are equivalent in relative magnitude, thus an equal gradient indicates the same relative change in mass, whilst an unequal gradient indicates a differing change in mass with temperature.

often treated as being coupled: for example, according to the MTE, these rates are both expected to scale with mass$^{-1/4}$ (Brown *et al.*, 2004). When comparing across different sized unicellular species, this may well be the case, as g/D must be a fixed ratio of 2 in all these species dividing by binary fission (Figure 3). However, this might not hold for all multicellular organisms, as the ratio of egg to adult mass is not always fixed across species of different sizes. As an example, individual egg mass for different sized adults was estimated for sac-spawning and broadcaster species of marine epipelagic copepods, using equations calculated from a meta-analysis of egg to adult sizes in these groups (Bunker and Hirst, 2004). Development rates were calculated at 15 °C for the same range of adult masses using the equation calculated by Gillooly (2000) for zooplankton generation times, in line with

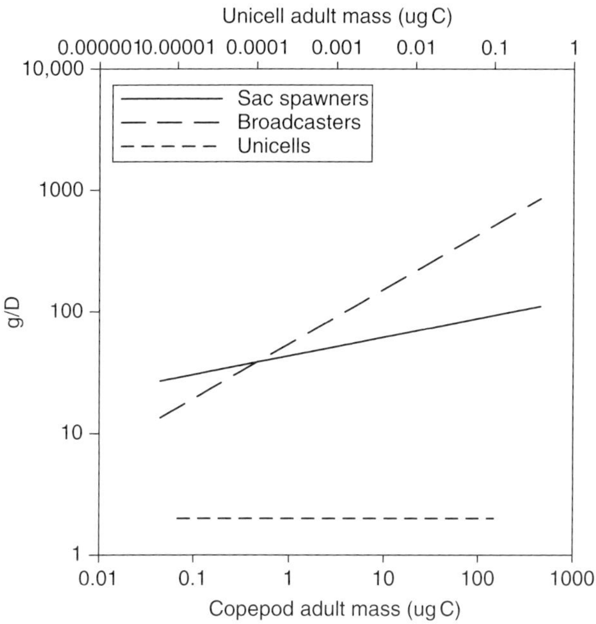

Figure 3 Changes in g/D (growth rate/development rate) associated with adult body mass in unicellular and multicellular organisms. C = carbon. Multicellular data calculated from data on sac-spawning and broadcaster marine epipelagic copepod species (Bunker and Hirst, 2004).

the predictions of the MTE. Adult masses and development rates can then be used to calculate growth rates using Eq. (1). Plotting g/D for these two groups shows that this ratio is not constant with increasing body size but rather differs (Figure 3). Clearly, unicellular species differ from multicellular species in these regards. Therefore, when using microscopic organisms to inform ecological theory, reproductive mode may create a fundamental difference of particular relevance to population-level studies.

B. Allometric Scaling of Abundance: Do Protozoans and Metazoans Show the Same Patterns?

1. The Theory

Body mass is regarded as a key feature of organisms, determining properties and processes on an individual level. Life-history traits and physiological processes, such as life span, growth and reproduction rate, relate to the body mass of an individual, the relationship often following a simple scaling law (power law) of the form

$$Y = aW^b,$$

where Y is the characteristic examined, W is body mass, a is the intercept and b is the scaling exponent (Brown *et al.*, 2004; Peters, 1983).

Above the individual level, ecological characteristics of populations are hypothesised to be determined by the body size of their species. Population properties, such as abundance or production, have frequently been linked with body mass of the species concerned. As with the scaling relationship between body mass and life-history traits, these higher-level relationships have also been suggested to follow power laws (e.g. Damuth, 1981; Enquist, 2002; Jennings *et al.*, 2002). The fundamental questions arising from the described scaling relationships are whether population-, community- and ecosystem-level processes are linked with individual-level processes and with body mass. Several frameworks have been put forward since Damuth's pioneering work (Brown *et al.*, 2004; Damuth, 1981; Gillooly *et al.*, 2001; Glazier, 2010) and among those is the MTE (*sensu* Brown *et al.*, 2004; Gillooly *et al.*, 2001) which assumes that individual body mass determines metabolism which in turn determines the expression of population properties. Community and ecosystem characteristics reflect the combined population properties of all species and can therefore be traced back to the body size distribution of its constituent individuals.

The most frequently described example for how body mass and population characteristics are connected is the so-called *body mass–abundance* (*MN*) scaling relationship, stemming from the early observation that small species are more abundant than larger ones (Damuth, 1981; Mohr, 1940). A simple, inverse linear correlation between body mass and abundance on a log–log axis has been suggested by many authors (Damuth, 1987), for which the MTE predicts a scaling exponent of -0.75 if the rate of resource supply is held constant (Brown *et al.*, 2004). The latter means that this scaling could be observed for species (MTE predictions are for individuals) when all species use the same amount of energy (energetic equivalence rule; Damuth, 1987; Nee *et al.*, 1991). Given that metabolism is the underlying cause of the *MN* relationship, then other ecological characteristics, such as biomass and production, should also conform to quarter scaling laws, as has indeed been widely reported (Brown *et al.*, 2004). The most important external factors are the energy available for all individuals and temperature (as it determines metabolic rates; Brown *et al.*, 2004).

The MTE has come under scrutiny and criticisms, and the study of small organisms has contributed greatly to this debate: it has, for example, been shown recently that the metabolic rate of prokaryotes is different from eukaryotes and, further, that single-celled eukaryotes differ from multicellular ones (DeLong *et al.*, 2010). DeLong *et al.* (2010) suggest that body mass scales with metabolism with a slope of approximately 0.75 on logarithmic scales only in metazoans, but that protists and prokaryotes show much steeper slopes (around 1.06 and 1.96, respectively). Importantly, this different body mass–metabolic rate scaling should be reflected in the body mass–abundance scaling, as they are connected to each other through the energy equivalence rule. If this is true, then small single-celled and small multicellular organisms should obey different rules (e.g. when their body mass is regressed against abundance). The implications would be that the application of MTE (e.g. for modelling community size structure; see Yvon-Durocher *et al.*, in press and references therein) may have to be modified accordingly.

Studies addressing body mass–abundance relationships mainly follow two approaches when visualising the relationship on a species level: body mass of species can be plotted against population abundance or size bins can be created and the abundance of similar sized species is summed per size bin (Blanco *et al.*, 1994). The MTE requires that abundance of all coexisting individuals within a trophic group and body mass category should be summed, as the theory predicts how many individuals of a given size range and trophic group can be supported by a given resource (Brown *et al.*, 2004; Dinmore and Jennings, 2004). Computing data like this should in theory confirm the energetic equivalence rule, but for most studies data are compiled for species rather than trophic groups (e.g. Griffiths, 1998; Schmid *et al.*, 2000). A different approach is to form size bins and to use an individual-based approach, where individuals are binned and abundance is related to these bins. This can be done irrespective of the individual's or species trophic status, taxonomy or any other characteristics, or this information can be included (Petchey and Belgrano, 2010). The term size spectra is commonly used in such studies and they are computed either for mean abundance or for summed abundance of one size bin (e.g. Blanco *et al.*, 1994; Gasol *et al.*, 1991, Woodward *et al.*, 2010).

Most data sets used for demonstrating body mass–abundance scaling have been restricted to specific taxa and many have encompassed mammals, fish or macro-invertebrates (e.g. Damuth, 1981; Enquist, 2002; Jennings *et al.*, 2002). Including small organisms in these types of studies is likely to reveal a more representative body size–abundance scaling pattern, because they span a wide range of body sizes and use energy available in the ecosystem. Small invertebrates such as protozoans and micro-metazoans are an intriguing group in this context because they represent a transitional zone where body sizes of single-celled and multicelled organisms overlap and where many traits, such as reproduction and feeding modes, change fundamentally. Placing these small organisms in the context of allometric scaling theories might help to reveal whether there is indeed a universal phenomenon, which strengthens the generality of ecological predictions, for example, in relation to energy flux (see Yvon-Durocher *et al.*, in press and references therein).

However, allometric studies on microscopic organisms are sparse (but see Finlay, 2002; Schmid *et al.*, 2000, 2002; Stead *et al.*, 2005; and studies on microbial size spectra, e.g. Cavender-Bares *et al.*, 2001; Gasol *et al.*, 1991; Li, 2002). Only a few studies have considered Protozoa or micro-metazoans (Finlay, 2002; Schmid *et al.*, 2000, 2002; Stead *et al.*, 2005).

2. *Is It Different for Small Organisms?*

In the following, we will focus on the body mass–abundance scaling and assume that the general assumption of power-law scaling is correct. The MTE gives precise values for the exponent of the *MN* scaling relationship (−0.75 on log–log scales) and predicts that the main factors affecting population-level characteristics are body mass, trophic level, the amount of available energy and temperature.

In line with the classic MTE predictions (i.e. ignoring the new results of DeLong *et al.*, 2010 for now), we firstly consider when the body mass–abundance scaling exponent should be the same for small and large organisms. This should be the case if the way in which small- and larger bodied species (or individuals) acquire energy is similar and if the way in which body mass and metabolisms scale is the same (as proposed by Brown *et al.*, 2004). For example, if body mass and metabolic rate follow a power-law scaling with a slope of 0.75 on log–log scales and if all species are able to sustain an optimal population growth (because energy availability is not restricted) then the body mass–abundance scaling should have a slope of −0.75 (Brown *et al.*, 2004). Studies carried out on unicellular and multicellular species that overlap in size could test whether they show an indistinguishable body mass–abundance scaling (e.g. data as shown in Petchey *et al.*, 2007), but to our knowledge, so far this has not been tested explicitly. However, small and large organisms should exhibit different slopes if energy was not equally available and/or, if the body mass–metabolic rate scaling differed (as proposed by DeLong *et al.*, 2010).

We compiled a new body mass and abundance data set for protist-, micro-metazoan- and insect species. We obtained these data for 354 protist species [258 species of pelagic and benthic protists (mostly protozoans) compiled by Finlay, 2002 and 96 ciliate species from a benthic data set recently described in Reiss and Schmid-Araya, 2008). Micro-metazoans included were benthic meiofauna (146 species and 63 species described in Reiss and Schmid-Araya, 2008 and Schmid *et al.*, 2000, respectively). Finally, we added data on 146 species of aquatic insects (macrofauna described in Schmid *et al.*, 2000). For all data, body mass was the average size for a species and abundance was the average abundance of that species in the natural habitat. For the body mass–abundance relationship, we regressed (a) protists and metazoans separately and (b) protists, meiofauna and insects separately. We calculated the ordinary least regression (OLS; the most commonly used type of regression) and used analysis of covariance (ANCOVA) to compare the slopes statistically (ANCOVA will give an OLS slope as the output). We used the statistical software R 2.5.1 (R Development Core Team, 2007).

We found that the slope for the body mass–abundance scaling was significantly ($P=0.0026$) steeper for protists (OLS slope of -0.86) than for metazoans (OLS slope of -0.65; Figure 4). This finding is in agreement with the conclusions of DeLong *et al.* (2010) who find a steeper body mass–metabolic rate scaling for protist species compared to metazoan species. However, the results of our meta-analysis might also have different reasons, such as compiling data for different trophic groups and environments or not computing size bins for individuals, which could introduce considerable error. Nonetheless, the large number of species and taxa (even phyla) in our data set makes the findings of general relevance. In any case, they highlight that the abundance of unicellular and multicellular species might not be governed by the same factors (e.g. because body mass and metabolism scale differently), but may be by the same laws (i.e. metabolism is the underlying cause of abundance patterns). In a wider context, this suggests that when testing general

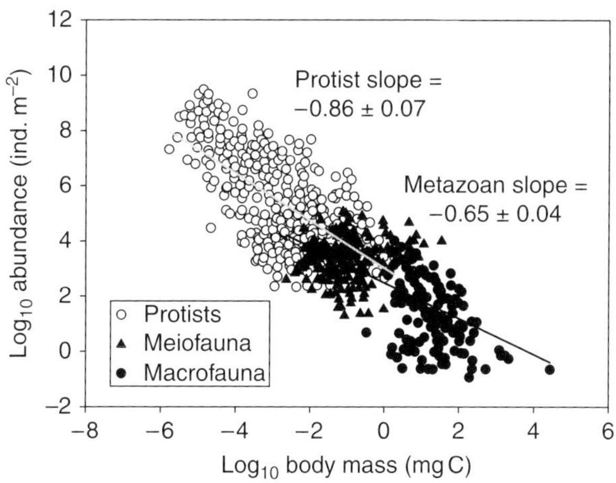

Figure 4 Body mass–abundance scaling of protists, meiofauna and macrofauna in aquatic habitats. Fits are OLS regression slopes ± SE as estimated by ANCOVA. The two slopes (one for protists and one for metazoans) are significantly different (ANCOVA, $P<0.0001$). The regression equations are $y=-0.85x+3.03$, $r^2=0.41$, $F=257.80$, $P<0.0001$ and $y=-0.65x+2.49$, $r^2=0.42$, $F=257.80$, $P<0.0001$ for protists and metazoans, respectively.

ecological theory by using small species as analogues for larger species, these differences have to be taken into account.

Although there might be fundamental differences in the ecology of unicellular and multicellular organisms, small multicellular animals may provide useful analogues of larger multicellular animals. In our analysis, we used data on ectotherm micro-metazoans and insect larvae. Surprisingly, when meiofauna and insect larvae are plotted separately from each other, the meiofauna shows no significant scaling relationship between body mass and abundance (slope of 0.10 ± 0.15 SE; regression line not shown, see Appendix for statistics) while the insect larvae still show the significant negative slope (-0.68 ± 0.12 SE; regression line not shown, see Appendix for statistics). This result was to be expected by using data from Schmid *et al.* (2002) who found scale invariance for the meiofaunal body mass–abundance data. However, exploring the data by Reiss and Schmid-Araya (2008) separately produces an identical result (Appendix). This indicates that within this size range metazoans might conform to somewhat different rules to those set out by the MTE and/or that other factors override the "predictive power of metabolism". Meiofauna and other micrometazoans are often used in (primarily laboratory-based) experiments addressing general ecological theory, such as food web dynamics. Given our results caution should be exercised when extrapolating to natural systems or to larger metazoans or when modelling community size structure (see Yvon-Durocher, in press).

As pointed out by Petchey and Belgrano (2010), body-size distributions and size spectra are used to assess the state of ecosystems, that is, slopes of the mass–abundance relationship are linked to anthropogenic influences such as exploitation. While the authors propose that taxonomic information should be integrated when size bins are used, our results further highlight that there might be differences between groups of organisms that are not only taxonomy-based (i.e. there are difference between protists, micro-metazoans and larger metazoans).

C. Biodiversity and Ecosystem Functioning

1. The Theories and Hypotheses

The high rates at which species are being lost from ecosystems on a global scale have stimulated interest in determining how biodiversity loss alters ecological processes that are vital to the functioning of ecosystems. Several hypotheses have been proposed to explain biodiversity–ecosystem functioning (B–EF) relationships (see Loreau *et al.*, 2002b). For instance, according to the rivet hypothesis, species loss has a slight negative effect on ecosystem functions down to a critical diversity level, beyond which further losses seriously compromise ecosystem functioning (Ehrlich and Ehrlich, 1981; Loreau *et al.*, 2002b). Similarly, the redundancy hypothesis predicts an asymptotic relationship, whereby species loss is of little consequence as long as all functional groups are well represented and if species within each group can compensate for the loss of others (Loreau *et al.*, 2002b; Walker, 1992). B–EF relationships can also be idiosyncratic, that is, the relationship between species richness and ecosystem functioning does not follow a consistent pattern, but depends crucially on species identity (e.g. Straub and Snyder, 2006), the order in which they are lost (Jonsson *et al.*, 2002), the environmental context (Cardinale *et al.*, 2000; McKie *et al.*, 2009) and/or the available resources (Bärlocher and Corkum, 2003).

To date, there is no clear consensus as to which of these particular hypotheses (and other B–EF hypotheses; see Loreau *et al.*, 2002b) best describes B–EF relationships, but generally species richness has been positively associated with primary productivity and plant biomass (Spehn *et al.*, 2005), or other ecosystem processes such as decomposition (Gessner *et al.*, 2010). Meta-analyses have also pointed to mostly positive biodiversity effects (Balvanera *et al.*, 2006; Cardinale *et al.*, 2007), but these are far from ubiquitous and negligible or negative effects have also been reported (e.g. Hooper and Dukes, 2004; McKie *et al.*, 2009; Perkins *et al.*, 2010; Reiss *et al.*, 2010; Weiss *et al.*, 2007), especially in aquatic ecosystems (see Stachowicz *et al.*, 2007 and Woodward, 2009 and references therein).

Positive B–EF relationships have been explained by a range of mechanisms, for example the complementarity effect or sampling effect. The former can occur due to facilitative interactions between species or niche differentiation, and the latter when diverse communities contain particular species that dominate in terms of performance (Loreau and Hector, 2001). Recently, there has been a shift towards measuring and evaluating B–EF relationships in a more complex way, for example by assessing functional diversity or by measuring multiple ecosystem processes (see Gamfeldt *et al.*, 2008; Hillebrand and Matthiessen, 2009; Ptacnik *et al.*, 2010; Reiss *et al.* 2009, 2010).

2. *Is It Different for Small Organisms?*

In the following we assess how small fungi (e.g. aquatic hyphomycetes) have been used to test B–EF relationships (see also Table 2 for examples on other groups and references given in Ptacnik *et al.*, 2010 for examples from planktonic communities). Positive relationships between aquatic fungal species richness and ecosystem functioning (leaf litter decomposition) were suggested in five of six microcosm studies (Bärlocher and Corkum, 2003; Duarte *et al.*, 2006; Pascoal *et al.*, 2010; Raviraja *et al.*, 2006; Treton *et al.*, 2004a, but not Dang *et al.*, 2005). Complementarity effects among aquatic fungal species have been demonstrated for biomass production (Duarte *et al.*, 2006; Pascoal *et al.*, 2010) and litter decomposition (Treton *et al.*, 2004b). Further, diversity effects of aquatic fungi (Bärlocher and Corkum, 2003; Duarte *et al.*, 2006; Pascoal *et al.*, 2010; Raviraja *et al.*, 2006) have also been attributed to species identity, indicating that traits of certain species may have a greater impact on ecosystem processes than species diversity *per se* (Pascoal and Cássio, 2008). These findings confirm those of other B–EF studies ranging from other microscopically small (Table 2) to large organisms and meta-analyses (Balvanera *et al.*, 2006) that have found richness to lead to elevated functioning (e.g. resource capture, biomass production or respiration).

Several studies on microscopic species have, however, also found that species richness might have negligible effects on ecosystem processes, as has also been reported in studies involving other organisms. For example, communities with low microbial richness (six soil fungal taxa) were as effective as the most diverse community (43 soil fungal taxa) in maintaining ecosystem functioning (Setala and McLean, 2004), indicating that biodiversity effects saturate at low levels. This might be interpreted as a signal of functional redundancy, and has been observed in other systems (Kinzig *et al.*, 2001). Dang *et al.* (2005) found no effects of fungal species richness on average rates of litter decomposition and spore production and this redundancy has also been found for aquatic invertebrates and leaf decomposition (Perkins *et al.*, 2010; Reiss *et al.*, 2010). An explanation for these findings might be that leaf decomposition by both fungi and invertebrates could mainly be driven by their metabolic demands (Reiss *et al.*, 2010).

One important aspect of B–EF studies using small organisms has been that they have highlighted the context dependency of species richness effects and redundancy, and that biodiversity might provide insurance or buffering against environmental changes (Naeem and Li, 1997), as proposed by the BIH (Yachi and Loreau, 1999). This means that even in a system where redundancy seems to be high within the assemblage and/or where there are no interactions between species, high species richness can reduce variability of ecosystem processes such as CO_2 flux, decomposition or community biomass or abundance (e.g. Doak *et al.*, 1998; Pascoal *et al.*, 2010). These concepts have been confirmed by studies on aquatic fungi: for instance, the variability of leaf decomposition decreased with increasing fungal species number under zinc stress (Pascoal *et al.*, 2010), that is, species were not redundant when the system was stressed. Similarly, Dang *et al.* (2005) concluded that although diversity of aquatic fungi did not affect average process rates (i.e. the magnitude of processes) in their laboratory experiment, there was a consistent tendency towards greater variability across fungal communities at lower diversity levels. The context dependency of species richness was highlighted in an experiment by Bärlocher and Corkum (2003) who showed that process rates depended on species richness and nutrient levels. These context dependency of fungal species richness effects on ecosystem processes confirms theoretical models (Doak *et al.*, 1998; Tilman *et al.*, 1998), findings for other microscopic organisms (e.g. McGrady-Steed and Morin, 2000; Morin and McGrady-Steed, 2004; see Table 2) and also larger organisms (Cardinale *et al.*, 2000; Jonsson *et al.*, 2002; McKie *et al.*, 2009).

Laboratory experiments with fungi and other microscopic species have informed the field of B–EF and are increasing in number (see examples in Table 2). The above examples show that the multitude of B–EF relationships found for a wide variety of systems has been demonstrated for microbial species. Some ecologists have suggested that B–EF relationships might have little meaning for small organisms in the real world, based on the idea that microbes are ubiquitous in nature, that is, "everything is everywhere", due to their perceived enormous dispersal capability and abundance (Fenchel and Finlay, 2004; Finlay, 2002). However, two arguments have been put forward which show that local and global diversity of microbial species is not random and can be high. They highlight the relevance of studying ecosystem responses to changes in microbial biodiversity. Firstly, there is the assumption that the environment selects; therefore, environmental tolerance would determine the occurrence of different microbial assemblages (Martiny *et al.*, 2006). Secondly, small organisms can be dispersal limited and drive local ecosystem dynamics and processes (Ptacnik *et al.*, 2010).

Small organisms have been proposed to drive the bulk of ecosystem processes (Loreau *et al.*, 2002a,b) and studies on microbial species show that their diversity can indeed affect ecosystem functioning. Many B–EF findings for small organisms also apply to larger ones, for example, the importance of species identity and traits within biodiversity. Advances in molecular techniques are helping to reveal much of the still hidden microbial biodiversity in natural systems (Purdy *et al.*, 2010), but to date we still have an insufficient knowledge of the full range of species interactions that could be operating in natural assemblages. In general, it seems that small organisms are useful models for studying B–EF relationships—and given that they drive most ecosystem processes they are probably more relevant study species than the more commonly used larger organisms in this context.

IV. CONCLUSIONS

Can we use knowledge and theories based on small organisms to formulate and test general ecological theories? Our review and case studies have highlighted studies addressing general ecological theories and hypotheses that have benefited from the use of microscopic organisms. Importantly, they have also informed or inspired the formulation of mathematical models and their empirical tests (e.g. Fox, 2007; Gause, 1934). Generally, set-ups with small organisms can provide useful models for testing many theories, that is, these simplified subsets of the "real world" can be used to extract rules that operate in nature.

The importance of such model systems has been discussed elsewhere (Benton *et al.*, 2007; Bonsall and Hassell, 2005; Bonsall *et al.*, 2005; Cadotte *et al.*, 2005; Jessup *et al.*, 2004; Lawton, 1995; Petchey *et al.*, 2002) and we propose that there are three particularly important reasons for using species with a small body size. Firstly, small organisms used in laboratory experiments or studied in the field can exhibit responses or patterns that are also found for larger species (e.g. in the case of the insurance hypothesis). Secondly, there might be the implicit, but not widely discussed, assumption that nature is self-similar (i.e. subsets resemble each other and the whole). In this sense, small organisms represent a snapshot in a continuum of complexity and are a subset of a self-similar, larger whole, that is, they resemble the whole. Finally, microscopic species play a central role in ecosystems because of their species richness, their high primary and secondary production, their position at the base of the food web and because they regulate and maintain many ecosystem processes.

Despite these appealing attributes, in our three case studies, we have seen not only generalities but also evidence of some key and fundamental differences between small and large organisms, and in certain cases, it is clearly not appropriate to generalise across all organisms. It is therefore vital to consider the level of observation and to test whether general rules are likely to be valid across a wide range of organisms and systems. For example, both single-celled and multicellular organisms seem to follow the TSR, but the mechanisms that govern the phenomena (species grow larger when reared at low temperature) might be very different for subsets of organisms (e.g. protozoans and metazoans). Our second case study addressed allometric scaling relationships that link population-level patterns to the metabolic needs of individual organisms. The universal rules here are that these population-level patterns seem to be governed by power laws connected to metabolism and in this respect different organism groups are similar. However, the "nuts and bolts" of metabolism appear to vary between evolutionary groups (i.e. prokaryotes, protist and metazoans; DeLong *et al.*, 2010) and assuming one law for all may produce erroneous predictions and interpretations.

Some of these fundamental concepts are unlikely to be the main focus of applied ecology, where the questions are more likely to be related to whether small organisms can provide answers to pressing issues such as species loss or global warming. In our case study on B–EF relationships, we have highlighted the use of small organisms for predicting the consequences of species loss in general. It is vital to understand their role in ecosystems to forecast the effects of species loss in response to environmental stress. For example, fungi are important drivers of decomposition in aquatic habitats (Reiss *et al.*, 2010) although the importance of their biodiversity is still not well understood.

In conclusion, we have highlighted examples of where it appears reasonable to extrapolate and also where we need to be cautious in terms of extracting general rules from small to large organisms or between different groups (especially protozoans and metazoans). Nonetheless, the importance of microscopic organisms for testing and advancing general ecological theory is clear. Moreover, many pressing ecological questions, such as how species loss will affect ecosystem functioning, will have to be informed by small organisms: for extracting both fundamental principles and also for assessing their importance in their own right.

ACKNOWLEDGEMENTS

We would like to thank two anonymous referees for their valuable comments that improved an earlier version of this manuscript. J. F. was supported by a NERC grant awarded to A. G. H. (NE/G523655/1).

APPENDIX

In case study 2 (allometric scaling of abundance), we regressed body mass and abundance of meiofauna and insects. Meiofauna and insect larvae were plotted separately from each other and the meiofauna showed no significant scaling relationship between body mass and abundance (slope of 0.10 ± 0.15 SE; regression line not shown, $y = 0.10x + 3.43$, $r^2 = 0.01$, $F = 2.27$, $P = 0.133$) while the insect larvae still showed the significant negative slope (0.68 ± 0.12 SE; regression line not shown, $y = -0.68x + 2.27$, $r^2 = 0.17$, $F = 30.82$, $P < 0.0001$). This result was to be expected by using data from Schmid *et al.* (2002) who found scale invariance for the meiofaunal body mass–abundance data, and exploring the data by Reiss and Schmid-Araya (2008) separately produces an identical result ($y = 0.15x + 3.72$, $r^2 = 0.54$, $F = 8.15$, $P = 0.0049$).

REFERENCES

Angilletta, M.J., and Dunham, A.E. (2003). The temperature-size rule in ectotherms: Simple evolutionary explanations may not be general. *Am. Nat.* **162**, 332–342.

Atkinson, D. (1994). Temperature and organism size—A biological law for ectotherms. *Adv. Ecol. Res.* **25**, 158.

Atkinson, D., Ciotti, B.J., and Montagnes, D.J.S. (2003). Protists decrease in size linearly with temperature: Ca. 2.5% degrees C-1. *Proc. R. Soc. Lond. B Biol.* **270**, 2605–2611.
Balciunas, D., and Lawler, S.P. (1995). Effects of basal resources, predation, and alternative prey in microcosm food-chains. *Ecology* **76**, 1327–1336.
Baldauf, S.L. (2003). The deep roots of eukaryotes. *Science* **300**, 1703–1706.
Balvanera, P., Pfisterer, A.B., Buchmann, N., He, J.S., Nakashizuka, T., Raffaelli, D., and Schmid, B. (2006). Quantifying the evidence for biodiversity effects on ecosystem functioning and services. *Ecol. Lett.* **9**, 1146–1156.
Bärlocher, F., and Corkum, M. (2003). Nutrient enrichment overwhelms diversity effects in leaf decomposition by stream fungi. *Oikos* **101**, 247–252.
Becks, L., Hilker, F.M., Malchow, H., Jürgens, K., and Arndt, H. (2005). Experimental demonstration of chaos in a microbial food web. *Nature* **435**, 1226–1229.
Bell, G., and Gonzalez, A. (2009). Evolutionary rescue can prevent extinction following environmental change. *Ecol. Lett.* **12**, 942–948.
Bell, T., Newman, J.A., Silverman, B.W., Turner, S.L., and Lilley, A.K. (2005). The contribution of species richness and composition to bacterial services. *Nature* **436**, 1157–1160.
Benton, T.G., Solan, M., Travis, J.M.J., and Sait, S.M. (2007). Microcosm experiments can inform global ecological problems. *Trends Ecol. Evol.* **22**, 516–521.
Beveridge, O.S., Humphries, S., and Petchey, O.L. (2010). The interacting effects of temperature and food chain length on trophic abundance and ecosystem function. *J. Anim. Ecol.* **79**, 693–700.
Blanco, J.M., Echevarria, F., and Garcia, C.M. (1994). Dealing with size spectra: Some conceptual and mathematical problems. *Sci. Mar.* **58**, 17–29.
Bohannan, B.J.M., and Lenski, R.E. (1999). Effect of prey heterogeneity on the response of a model food chain to resource enrichment. *Am. Nat.* **153**, 73–82.
Bonsall, M.B., and Hassell, M.P. (2005). Understanding ecological concepts: The role of laboratory systems. *Adv. Ecol. Res.* **37**, 1–36. Population Dynamics and Laboratory *Ecology*.
Bonsall, M.B., Bull, J.C., Pickup, N.J., and Hassell, M.P. (2005). Indirect effects and spatial scaling affect the persistence of multispecies metapopulations. *Proc. R. Soc. Lond. B Biol.* **272**, 1465–1471.
Boxshall, G.A., and Defaye, D. (2008). Global diversity of copepods (Crustacea: Copepoda) in freshwater. *Hydrobiologia* **595**, 195–207.
Brown, J.H., Gillooly, J.F., Allen, A.P., Savage, V.M., and West, G.B. (2004). Toward a metabolic theory of ecology. *Ecology* **85**, 1771–1789.
Buckling, A., Kassen, R., Bell, G., and Rainey, P.B. (2000). Disturbance and diversity in experimental microcosms. *Nature* **408**, 961–964.
Bunker, A.J., and Hirst, A.G. (2004). Fecundity of marine planktonic copepods: Global rates and patterns in relation to chlorophyll a, temperature and body weight. *Mar. Ecol. Prog. Ser.* **279**, 161–181.
Burns, C.W., and Schallenberg, M. (1996). Relative impacts of copepods, cladocerans and nutrients on the microbial food web of a mesotrophic lake. *J. Plankton Res.* **18**, 683–714.
Cadotte, M.W., Drake, J.A., and Fukami, T. (2005). Constructing *Nature*: Laboratory models as necessary tools for investigating complex ecological communities. *Adv. Ecol. Res.* **37**, 333–353.
Cardinale, B.J., Nelson, K., and Palmer, M.A. (2000). Linking species diversity to the functioning ecosystems: On the importance of environmental context. *Oikos* **91**, 175–183.

Cardinale, B.J., Wrigh, J.P., Cadotte, M.W., Carroll, I.T., Hector, A., Srivastava, D.S., Loreau, M., and Weis, J.J. (2007). Impacts of plant diversity on biomass production increase through time because of species complementarity. *Proc. Natl. Acad. Sci. USA* **104**, 18123–18128.

Carpenter, S.R. (1996). Microcosm experiments have limited relevance for community and ecosystem ecology. *Ecology* **77**, 677–680.

Carpenter, S.R., Kitchell, J.F., and Hodgson, J.R. (1985). Cascading trophic interactions and lake productivity: Fish predation and herbivory can regulate lake ecosystems. *Bioscience* **35**, 634–639.

Cavender-Bares, K.K., Rinaldo, A., and Chisholm, S.W. (2001). Microbial size spectra from natural and nutrient enriched ecosystems. *Limnol. Oceanogr.* **46**, 778–789.

Curtis, T.P., Sloan, W.T., and Scannell, J.W. (2002). Estimating prokaryotic diversity and its limits. *Proc. Natl. Acad. Sci. USA* **99**, 10494–10499.

Damuth, J. (1981). Population density and body size in mammals. *Nature* **290**, 699–700.

Damuth, J. (1987). Interspecific allometry of population density in mammals and other animals: The independence of body mass and population energy-use. *Biol. J. Linn. Soc.* **31**, 193–246.

Dang, C.K., Chauvet, E., and Gessner, M.O. (2005). Magnitude and variability of process rates in fungal diversity-litter decomposition relationships. *Ecol. Lett.* **8**, 1129–1137.

Daufresne, M., Lengfellner, K., and Sommer, U. (2009). Global warming benefits the small in aquatic ecosystems. *Proc. Natl. Acad. Sci. USA* **106**, 12788–12793.

De Jong, G. (2010). A biophysical interpretation of temperature-dependent body size in *Drosophila aldrichi* and *D. buzzatii*. *J. Therm. Biol* **35**, 85–99.

Delaney, M.P. (2003). Effects of temperature and turbulence on the predator-prey interactions between a heterotrophic flagellate and a marine bacterium. *Microb. Ecol.* **45**, 218–225.

DeLong, J.P., Okie, J.G., Moses, M.E., Sibly, R.M., and Brown, J.H. (2010). Shifts in metabolic scaling, production, and efficiency across major evolutionary transitions of life. *Proc. Natl. Acad. Sci. USA*. www.pnas.org/cgi/doi/10.1073/pnas.1007783107.

Diehl, S., and Feissel, M. (2000). Effects of enrichment on three-level food chains with omnivory. *Am. Nat.* **155**, 200–218.

Diehl, S., and Feissel, M. (2001). Intraguild prey suffer from enrichment of their resources: A microcosm experiment with ciliates. *Ecology* **82**, 2977–2983.

Dinmore, T.A., and Jennings, S. (2004). Predicting abundance-body mass relationships in benthic infaunal communities. *Mar. Ecol. Prog. Ser.* **276**, 289–292.

Doak, D.F., Bigger, D., Harding, E.K., Marvier, M.A., O'Malley, R.E., and Thomson, D. (1998). The statistical inevitability of stability-diversity relationships in community Ecology. *Am. Nat.* **151**, 264–276.

Duarte, S., Pascoal, C., Cássio, F., and Bärlocher, F. (2006). Aquatic hyphomycete diversity and identity affect leaf litter decomposition in microcosms. *Oecologia* **147**, 658–666.

Ehrlich, P.R., and Ehrlich, A.H. (1981). Extinction: The Causes and Consequences of the Disappearance of Species. Random House, New York.

Enquist, B.J. (2002). Universal scaling in tree and vascular plant allometry: Toward a general quantitative theory linking plant form and function from cells to ecosystems. *Tree Physiol.* **22**, 1045–1064.

Fenchel, T., and Finlay, B.J. (2004). The ubiquity of small species: Patterns of local and global diversity. *Bioscience* **54**, 777–784.

Finlay, B.J. (2002). Global dispersal of free-living microbial eukaryote species. *Science* **296**, 1061–1063.
Foissner, W., and Berger, H. (1996). A user-friendly guide to the ciliates (Protozoa, Ciliophora) commonly used by hydrobiologists as bioindicators in rivers, lakes, and waste waters, with notes on their ecology. *Freshwater Biol.* **35**, 375–482.
Forster, J., Hirst, A.G., Atkinson, D. (unpublished manuscript) How do organisms change size with changing temperature? The importance of reproductive method.
Fox, J.W. (2007). Testing the mechanisms by which source-sink dynamics alter competitive outcomes in a model system. *Am. Nat.* **170**, 396–408.
Fox, J.W. (2008). Testing whether productivity mediates the occurrence of alternate stable states and assembly cycles in a model microcosm system. *Oikos* **117**, 1153–1164.
Fox, J.W., and McGrady-Steed, J. (2002). Stability and complexity in microcosm communities. *J. Anim. Ecol.* **71**, 749–756.
Fox, J.W., McGrady-Steed, J., and Petchey, O.L. (2000). Testing for local species saturation with nonindependent regional species pools. *Ecol. Lett.* **3**, 198–206.
Fukami, T. (2004). Assembly history interacts with ecosystem size to influence species diversity. *Ecology* **85**, 3234–3242.
Gamfeldt, L., and Hillebrand, H. (2005). Species richness changes across two trophic levels simultaneously affect prey and consumer biomass. *Ecol. Lett.* **8**, 696–703.
Gamfeldt, L., Hillebrand, H., and Jonsson, P.R. (2008). Multiple functions increase the importance of biodiversity for overall ecosystem functioning. *Ecology* **89**, 1223–1231.
Garstecki, T., and Wickham, S.A. (2003). The response of benthic rhizopods to sediment disturbance does not support the intermediate disturbance hypothesis. *Oikos* **103**, 528–536.
Gasol, J.M., Guerrero, R., and Pedrosalio, C. (1991). Seasonal variations in size structure and prokaryotic dominance in sulphurous Lake Ciso. *Limnol. Oceanogr.* **36**, 860–872.
Gause, G.F. (1934). The Struggle for Existence. Williams & Wilkins Co., Baltimore.
Gessner, M.O., Swan, C.M., Dang, C.K., McKie, B.G., Bardgett, R.D., Wall, D.H., and Hättenschwiler, S. (2010). Diversity meets decomposition. *Trends Ecol. Evol.* **25**, 372–380.
Gillooly, J.F. (2000). Effect of body size and temperature on generation time in zooplankton. *J. Plankton Res.* **22**, 241–251.
Gillooly, J.F., Brown, J.H., West, G.B., Savage, V.M., and Charnov, E.L. (2001). Effects of size and temperature on metabolic rate. *Science* **293**, 2248–2251.
Glazier, D.S. (2010). A unifying explanation for diverse metabolic scaling in animals and plants. *Biol. Rev.* **85**, 111–138.
Griffiths, D. (1998). Sampling effort, regression method, and the shape and slope of size-abundance relations. *J. Anim. Ecol.* **67**, 795–804.
Haeckel, E. (1904). Kunstformen der Natur. Bibliographisches Institut, Leipzig.
Hausmann, K., Hülsmann, N., and Radek, R. (2003). Protistology. 3rd edition. E. Schweizerbart'sche Verlagsbuchhandlung, Stuttgart.
Hillebrand, H., and Matthiessen, B. (2009). Biodiversity in a complex world: Consolidation and progress in functional biodiversity research. *Ecol. Lett.* **12**, 1405–1419.
Hillebrand, H., Watermann, F., Karez, R., and Berninger, U.G. (2001). Differences in species richness patterns between unicellular and multicellular organisms. *Oecologia* **126**, 114–124.

Holyoak, M. (2000). Effects of nutrient enrichment on predator-prey metapopulation dynamics. *J. Anim. Ecol.* **69**, 985–997.
Holyoak, M., and Lawler, S.P. (1996). The role of dispersal in predator-prey metapopulation dynamics. *J. Anim. Ecol.* **65**, 640–652.
Holyoak, M., and Lawler, S.P. (2005). The contribution of laboratory experiments on protists to understanding population and metapopulation dynamics. *Adv. Ecol. Res.* **37**, 245–271.
Hooper, D.U., and Dukes, J.S. (2004). Overyielding among plant functional groups in a long-term experiment. *Ecol. Lett.* **7**, 95–105.
Hudson, C.T., and Gosse, P.H. (1886). The Rotifers or Wheel-Animalcules, Both British and Foreign. Longmans Green, London.
Hulot, F.D., Lacroix, G., Lescher-Moutoue, F.O., and Loreau, M. (2000). Functional diversity governs ecosystem response to nutrient enrichment. *Nature* **405**, 340–344.
Jennings, S., Pinnegar, J.K., Polunin, N.V.C., and Warr, K.J. (2002). Linking size-based and trophic analyses of benthic community structure. *Mar. Ecol. Prog. Ser.* **226**, 77–85.
Jessup, C.M., Kassen, R., Forde, S.E., Kerr, B., Buckling, A., Rainey, P.B., and Bohannan, B.J.M. (2004). Big questions, small worlds: Microbial model systems in Ecology. *Trends Ecol. Evol.* **19**, 189–197.
Jiang, L., and Morin, P.J. (2004). Temperature-dependent interactions explain unexpected responses to environmental warming in communities of competitors. *J. Anim. Ecol.* **73**, 569–576.
Jonsson, M., Dangles, O., Malmqvist, B., and Guerold, F. (2002). Simulating species loss following perturbation: Assessing the effects on process rates. *Proc. R. Soc. Lond. B Biol.* **269**, 1047–1052.
Jost, J.L., Drake, J.F., Fredrickson, A.G., and Tsuchiya, H.M. (1973). Interaction of *Tetrahymena pyriformis*, *Escherichia coli*, *Azobacter vinelandii*, and glucose in a minimal medium. *J. Bacteriol.* **113**, 834–840.
Jürgens, K. (1994). Impact of Daphnia on planktonic microbial food webs—A review. *Mar. Microb. Food Webs* **8**, 295.
Jürgens, K., Wickham, S.A., Rothhaupt, K.O., and Santer, B. (1996). Feeding rates of macro- and microzooplankton on heterotrophic nanoflagellates. *Limnol. Oceanogr.* **41**, 1833–1839.
Kahl, A. (1935). Urtiere Oder Protozoa I: Wimpertiere oder Ciliata (Infusoria). Gustav Fischer, Jena.
Kassen, R., Buckling, A., Bell, G., and Rainey, P.B. (2000). Diversity peaks at intermediate productivity in a laboratory microcosm. *Nature* **406**, 508–512.
Kaunzinger, C.M.K., and Morin, P.J. (1998). Productivity controls food-chain properties in microbial communities. *Nature* **395**, 495–497.
Kingsolver, J.G., and Huey, R.B. (2008). Size, temperature, and fitness: Three rules. *Evol. Ecol. Res.* **10**, 251–268.
Kinzig, A.P., Pacala, S.W., and Tilman, D. (2001). The Functional Consequences of Biodiversity. Empirical Progress and Theoretical Extensions. Princeton University Press, Princeton, Oxford.
Langenheder, S., and Ragnarsson, H. (2007). The role of environmental and spatial factors for the composition of aquatic bacterial communities. *Ecology* **88**, 2154–2161.
Law, R., Weatherby, A.J., and Warren, P.H. (2000). On the invasibility of persistent protist communities. *Oikos* **88**, 319–326.

Lawler, S.P., and Morin, P.J. (1993). Food web architecture and population dynamics in laboratory microcosms of protists. *Am. Nat.* **141**, 675–686.

Lawton, J.H. (1995). Ecological experiments with model systems. *Science* **269**, 328–331.

Leary, D.J., and Petchey, O.L. (2009). Testing a biological mechanism of the insurance hypothesis in aquatic microcosms. *J. Anim. Ecol.* **78**, 1143–1151.

Li, W.K.W. (2002). Macroecological patterns of phytoplankton in the northwestern North Atlantic Ocean. *Nature* **419**, 154–157.

Loreau, M., and Hector, A. (2001). Partitioning selection and complementarity in biodiversity experiments. *Nature* **412**, 72–76.

Loreau, M., Downing, A., Emmerson, M., Gonzalez, A., Hughes, J., Inchausti, P., Joshi, J., Norberg, J., and Sala, O. (2002a). A new look at the relationship between diversity and stability. In: *Biodiversity and Ecosystem Functioning. Synthesis and Perspectives* (Ed. by M. Loreau, S. Naeem and P. Inchausti), pp. 79–91. Oxford University Press, New York.

Loreau, M., Naeem, S., and Inchausti, P. (2002b). Biodiversity and Ecosystem Functioning. Synthesis and Perspectives. Oxford University Press, Oxford.

Martiny, J.B.H., Bohannan, B.J.M., Brown, J.H., Colwell, R.K., Fuhrman, J.A., Green, J.L., Horner-Devine, M.C., Kane, M., Krumins, J.A., Kuske, C.R., Morin, P.J., Naeem, S., *et al.* (2006). Microbial biogeography: Putting microorganisms on the map. *Nat. Rev. Microbiol.* **4**, 102–112.

May, R.M. (1986). The search for patterns in the balance of nature: Advances and retreats. *Ecology* **67**, 1115–1126.

May, R. (1999). Unanswered questions in ecology. *Philos. Trans. R. Soc. B* **354**, 1951–1959.

McGrady-Steed, J., and Morin, P.J. (2000). Biodiversity, density compensation, and the dynamics of populations and functional groups. *Ecology* **81**, 361–373.

McGrady-Steed, J., Harris, P.M., and Morin, P. (1997). Biodiversity regulates ecosystem predictability. *Nature* **390**, 162–165.

McKie, B.G., Schindler, M., Gessner, M.O., and Malmqvist, B. (2009). Placing biodiversity and ecosystem functioning in context: Environmental perturbations and the effects of species richness in a stream field experiment. *Oecologia* **160**, 757–770.

Mohr, C.O. (1940). Comparative populations of game, fur and other mammals. *Am. Midl. Nat.* **24**, 581–584.

Morin, P. (1999). Productivity, intraguild predation, and population dynamics in experimental food webs. *Ecology* **80**, 752–760.

Morin, P.J., and McGrady-Steed, J. (2004). Biodiversity and ecosystem functioning in aquatic microbial systems: A new analysis of temporal variation and species richness-predictability relations. *Oikos* **104**, 458–466.

Moss, B., McKee, D., Atkinson, D., Collings, S.E., Eaton, J.W., Gill, A.B., Harvey, I., Hatton, K., Heyes, T., and Wilson, D. (2003). How important is climate? Effects of warming, nutrient addition and fish on phytoplankton in shallow lake microcosms. *J. Appl. Ecol.* **40**, 782–792.

Naeem, S., and Li, S.B. (1997). Biodiversity enhances ecosystem reliability. *Nature* **390**, 507–509.

Naeem, S., Hahn, D.R., and Schuurman, G. (2000). Producer-decomposer co-dependency influences biodiversity effects. *Nature* **403**, 762–764.

Nee, S., Read, A.F., Greenwood, J.J.D., and Harvey, P.H. (1991). The relationship between abundance and body size in British birds. *Nature* **351**, 312–313.

Neill, W.E. (1975). Experimental studies of microcrustacean competition, community composition and efficiency of resource utilization. *Ecology* **56**, 809–826.

Newsham, K.K., and Garstecki, T. (2007). Interactive effects of warming and species loss on model Antarctic microbial food webs. *Funct. Ecol.* **21**, 577–584.

Pascoal, C., and Cássio, F. (2008). Linking fungal diversity to the functioning of freshwater ecosystems. In: *Novel Techniques and Ideas in Mycology* (Ed. by K.R. Sridhar, F. Bärlocher and K.D. Hyde), pp. 1–15. Fungal Diversity Research Series, Fungal Diversity Press, Hong Kong.

Pascoal, C., Cássio, F., Nikolcheva, L., and Bärlocher, F. (2010). Realized fungal diversity increases functional stability of leaf litter decomposition under zinc stress. *Microb. Ecol.* **59**, 84–93.

Perkins, D.M., McKie, B.G., Malmqvist, B., Gilmour, S.G., Reiss, J., and Woodward, G. (2010). Environmental warming and biodiversity-ecosystem functioning in freshwater microcosms: partitioning the effects of species identity, richness and metabolism. *Adv. Ecol. Res.* **43,** 177–209.

Petchey, O.L. (2000). Environmental colour affects aspects of single-species population dynamics. *Proc. R. Soc. Lond. B Biol.* **267**, 747–754.

Petchey, O., and Belgrano, A. (2010). Body-size distributions and size-spectra: universal indicators of ecological status? *Biol. Lett.* **6**, 434–437.

Petchey, O.L., McPhearson, P.T., Casey, T.M., and Morin, P.J. (1999). Environmental warming alters food-web structure and ecosystem function. *Nature* **402**, 69–72.

Petchey, O.L., Morin, P.J., Hulot, F.D., Loreau, M., McGrady-Steed, J., and Naeem, S. (2002). Contributions of aquatic model systems to our understanding of biodiversity and ecosystem functioning. In: *Biodiversity and Ecosystem Functioning: Synthesis and Perspectives* (Ed. by M. Loreau, S. Naeem and P. Inchausti), pp. 127–138. Oxford University Press, Oxford.

Petchey, O.L., Long, Z.T., and Morin, P.J. (2007). The consequences of body size in model microbial ecosystems. In: *Body Size: The Structure and Function of Ecosystems* (Ed. by A.G. Hildrew, D. Raffaelli and R. Edmonds-Brown). Cambridge University Press, Cambridge.

Peters, R.H. (1983). The Ecological Implications of Body Size. Cambridge University Press, Cambridge.

Ptacnik, R., Moorthi, S.D., and Hillebrand, H. (2010). Hutchinson reversed, or why there need to be so many species. *Adv. Ecol. Res.* **43**, 1–43.

Purdy, K.J., Hurd, P.J., Moya-Laraño, J., Trimmer, M., and Woodward, G. (2010). Systems biology for ecology. *Adv. Ecol. Res.* **43**, 87–149.

R Development Core Team (2007). R: A Language and Environment for Statistical Computing. R Foundation for Statistical Computing, Vienna.

Raviraja, N.S., Nikolcheva, L.G., and Bärlocher, F. (2006). Fungal growth and leaf decompostion are affected by amount and type of inoculum and by external nutrients. *Sydowia* **58**, 98–104.

Reiss, J., and Schmid-Araya, J.M. (2008). Existing in plenty: Abundance, biomass and diversity of ciliates and meiofauna in small streams. *Freshwater Biol.* **53**, 652–668.

Reiss, J., Bridle, J.R., Montoya, J.M., and Woodward, G. (2009). Emerging horizons in biodiversity and ecosystem functioning research. *Trends Ecol. Evol.* **24**, 505–514.

Reiss, J., Bailey, R.A., Cássio, F., Woodward, G., and Pascoal, C. (2010). Assessing the contribution of micro-organisms and macrofauna to biodiversity-ecosystem functioning relationships in freshwater microcosms. *Adv. Ecol. Res.* **43**, 151–176.

Replansky, T., and Bell, G. (2009). The relationship between environmental complexity, species diversity and productivity in a natural reconstructed yeast community. *Oikos* **118**, 233–239.

Replansky, T., Koufopanou, V., Greig, D., and Bell, G. (2008). *Saccharomyces sensu stricto* as a model system for evolution and ecology. *Trends Ecol. Evol.* **23**, 494–501.
Rundle, S., Robertson, A.L., and Schmid-Araya, J.M. (2002). Freshwater meiofauna: Biology and Ecology. Blackhuys.
Ruttner-Kolisko, A. (1977). Suggestions for biomass calculations of plankton rotifers. *Arch. Hydrobiol. Beihe. Ergebn. Limnol.* **8**, 71–76.
Salles, J.F., Poly, F., Schmid, B., and Le Roux, X. (2009). Community niche predicts the functioning of denitrifying bacterial assemblages. *Ecology* **90**, 3324–3332.
Schmid, P.E., Tokeshi, M., and Schmid-Araya, J.M. (2000). Relation between population density and body size in stream communities. *Science* **289**, 1557–1560.
Schmid, P.E., Tokeshi, M., and Schmid-Araya, J.M. (2002). Scaling in stream communities. *Proc. R. Soc. Lond. B Biol.* **269**, 2587–2594.
Schmid-Araya, J.M., Hildrew, A.G., Robertson, A., Schmid, P.E., and Winterbottom, J. (2002). The importance of meiofauna in food webs: Evidence from an acid stream. *Ecology* **83**, 1271–1285.
Schrag, S.J., and Mittler, J.E. (1996). Host-parasite coexistence: The role of spatial refuges in stabilizing bacteria-phage interactions. *Am. Nat.* **148**, 348–377.
Setala, H., and McLean, M.A. (2004). Decomposition rate of organic substrates in relation to the species diversity of soil saprophytic fungi. *Oecologia* **139**, 98–107.
Sieburth, J.M., Smetacek, V., and Lenz, J. (1978). Pelagic ecosystem structure: Heterotrophic compartments of plankton and their relationship to plankton size fractions—Comment. *Limnol. Oceanogr.* **23**, 1256–1263.
Smith, D.W., and Cooper, S.D. (1982). Competition among Cladocera. *Ecology* **63**, 1004–1015.
Spehn, E.M., Hector, A., Joshi, J., Scherer-Lorenzen, M., Schmid, B., Bazeley-White, E., Beierkuhnlein, C., Caldeira, M.C., Diemer, M., Dimitrakopoulos, P.G., Finn, J.A., Freitas, H., *et al.* (2005). Ecosystem effects of biodiversity manipulations in European grasslands. *Ecol. Monogr.* **75**, 37–63.
Stachowicz, J.J., Bruno, J.F., and Duffy, J.E. (2007). Marine biodiversity and ecosystem function. *Annu. Rev. Ecol. Evol. Syst.* **38**, 739–766.
Stead, T.K., Schmid-Araya, J.M., Schmid, P.E., and Hildrew, A.G. (2005). The distribution of body size in a stream community: One system, many patterns. *J. Anim. Ecol.* **74**, 475–487.
Stelzer, C.P. (2002). Phenotypic plasticity of body size at different temperatures in a planktonic rotifer: Mechanisms and adaptive significance. *Funct. Ecol.* **16**, 835–841.
Straub, C.S., and Snyder, W.E. (2006). Species identity dominates the relationship between predator biodiversity and herbivore suppression. *Ecology* **87**, 277–282.
Streble, H., and Krauter, D. (1988). Das Leben im Wassertropfen: Mikroflora und Mikrofauna des Süßwassers. Ein Bestimmungsbuch. Franckh-Kosmos Verlag GmbH, Stuttgart.
Tilman, D., Lehman, C.L., and Bristow, C.E. (1998). Diversity–stability relationships: Statistical inevitability or ecological consequence? *Am. Nat.* **151**, 277–282.
Traunspurger, W. (2000). The biology and ecology of lotic nematodes. *Freshwater Biol.* **43**, 1–17.
Treton, C., Chauvet, E., and Charcosset, J.-Y. (2004a). Competitive interaction between two aquatic hyphomycete species and increase in leaf litter breakdown. *Microb. Ecol.* **48**, 439–446.
Treton, C., Chauvet, E., and Charcosset, J.Y. (2004b). Competitive interaction between two aquatic hyphomycete species and increase in leaf litter breakdown. *Microb. Ecol.* **48**, 439–446.

van der Have, T.M., and de Jong, G. (1996). Adult size in ectotherms: Temperature effects on growth and differentiation. *J. Theor. Biol.* **183**, 329–340.
Walker, B.H. (1992). Biodiversity and ecological redundancy. *Conserv. Biol.* **6**, 18–23.
Warren, P.H., and Weatherby, A.J. (2006). Energy input and species diversity patterns in microcosms. *Oikos* **113**, 314–324.
Warren, P.H., Law, R., and Weatherby, A. (2003). Mapping the assembly of protist communities in microcosms. *Ecology* **84**, 1001–1011.
Weatherby, A.J., Warren, P.H., and Law, R. (1998). Coexistence and collapse: an experimental investigation of the persistent communities of a protist species pool. *J. Anim. Ecol.* **67**, 554–566.
Weiss, J.J., Cardinale, B.J., Forshay, K.J., and Ives, A.R. (2007). Effects of species diversity on community biomass production change over the course of succession. *Ecology* **88**, 929–939.
Wickham, S.A. (1995). Trophic relations between cyclopoid copepods and ciliated protists: Complex interactions link the microbial and classic food webs. *Limnol. Oceanogr.* **40**, 1173–1181.
Wittebolle, L., Marzorati, M., Clement, L., Balloi, A., Daffonchio, D., Heylen, K., De Vos, P., Verstraete, W., and Boon, N. (2009). Initial community evenness favours functionality under selective stress. *Nature* **458**, 623–626.
Woodward, G. (2009). Biodiversity, ecosystem functioning and food webs in freshwaters: Assembling the jigsaw puzzle. *Freshwater Biol.* **54**, 2171–2187.
Woodward, G., Benstead, J.P., Beveridge, O.S., Blanchard, J., Brey, T., and Brown, L. (2010). Ecological networks in a changing climate. *Adv. Ecol. Res.* **42**, 71–138.
Worsfold, N.T., Warren, P.H., and Petchey, O.L. (2009). Context-dependent effects of predator removal from experimental microcosm communities. *Oikos* **118**, 1319–1326.
Yachi, S., and Loreau, M. (1999). Biodiversity and ecosystem productivity in a fluctuating environment: The insurance hypothesis. *Proc. Natl. Acad. Sci. USA* **96**, 1463–1468.
Yoshida, T. (2005). Toward the understanding of complex population dynamics: Planktonic community as a model system. *Ecol. Res.* **20**, 511–518.
Yvon-Durocher, G., Jones, J.I., Montoya, J.M., Trimmer, M., and Woodward, G. (2010a). Warming increases the proportion of primary production emitted as methane from freshwater mesocosms. *Glob. Change Biol.* 10.1111/j.1365-2486.2010.02289.x
Yvon-Durocher, G., Jones, J.I., Woodward, G., Trimmer, J.M., and Montoya, J.M. (2010b). Warming alters the metabolic balance of ecosystems. *Philos. Trans. R. Soc. B* **365**, 2117–2126.
Yvon-Durocher, G., Montoya, J.M., Trimmer, M., and Woodward, G. (2010c). Warming alters the size spectrum and shifts the distribution of biomass in freshwater ecosystems. *Glob. Change Biol.* 10.1111/j.1365-2486.2010.02321.x
Yvon-Durocher, G., Reiss, J., Blanchard, J., Ebenman, B., Perkins, D.M., Reuman, D., Thierry, A., Woodward, G., Petchey, O. (in press) Across ecosystem comparison of size structure. *Oikos*

Systems Biology for Ecology: From Molecules to Ecosystems

KEVIN J. PURDY, PAUL J. HURD, JORDI MOYA-LARAÑO,
MARK TRIMMER, BRIAN B. OAKLEY AND GUY WOODWARD

ADVANCES IN ECOLOGICAL RESEARCH VOL. 43 0065-2504/10 $35.00
© 2010 Elsevier Ltd. All rights reserved
DOI: 10.1016/S0065-2504(10)43003-2

SUMMARY

Ecology stands on the edge of a true paradigm shift, fuelled by a recent technological revolution in our ability to measure both taxonomic and functional biodiversity via the application of metagenomics and transcriptomics. The advent of 'next generation sequencing' (NGS) in molecular biology is rapidly opening the black box of microbial ecology, providing us with some of the first glimpses of a previously hidden world. This is now enabling microbial ecology to become firmly embedded as a core subdiscipline within ecology, and to test general theories about biodiversity, biogeography and ecosystem functioning using a combination of molecular and more traditional techniques. In addition, NGS offers a means of not only measuring the abundance and diversity of the main drivers of many of the planet's key biogeochemical processes, but also of linking the microscopic and macroscopic worlds that have, until now, been largely studied in isolation. We provide a detailed review of the rise of NGS, as well as highlighting areas that offer special promise for addressing general ecological questions across a range of levels of organisation, from individuals to ecosystems: essentially, how a 'systems biology for ecology' might be developed. We consider the current limitations and future prospects for NGS, and also how it offers potential economic benefits, for instance via bioprospecting the environment for commercially valuable genes and their products within the metagenome of natural ecosystems.

I. INTRODUCTION

A. Towards a Systems Biology for Ecology?

All organisms interact with the physical and biological world that surrounds them, from the human gut microbiome to tropical forests and the open ocean. Any study aimed at understanding the development and existence of a multi-organism system falls within the broad remit of ecology, although most of us are more familiar with far a narrower definition. Despite this, ecology has, often out of necessity, been studied in relative isolation from many other fields of biology and has itself been calved into numerous discrete subdisciplines that deal with different levels of organisation (e.g. molecular, organismal, community and ecosystem ecology; Figure 1). In reality, of course, these are rather arbitrary and subjective delimitations, and there are numerous links and feedbacks across organisational levels (Woodward *et al.*, 2010b; Yvon-Durocher *et al.*, 2010b). In recent years ecology has become an increasingly holistic discipline that now draws on, for instance, molecular biology at one extreme and socioeconomics at the other. Here we

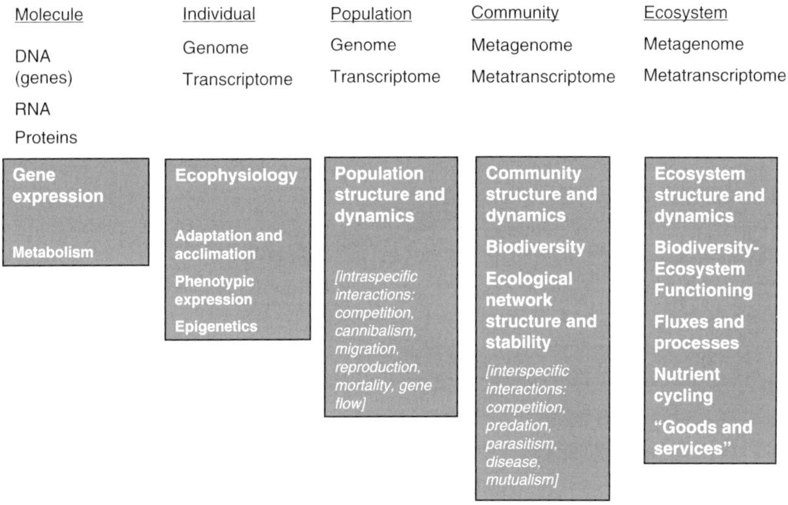

Figure 1 The hierarchy of biological organisation, in increasing order of complexity from molecules to ecosystems. Beneath each heading the relevant genetic and ecological (shaded boxes) measures and fields of study are highlighted.

consider the future of ecology in the context of understanding a system from the chemicals and molecules to the individual organisms that comprise the community and, ultimately, the processes they perform that contribute to the functioning of the ecosystem as a whole. We propose that an inclusive and integrative ecology should be able to encompass all these aspects: essentially, a 'Systems Biology for Ecology' (Figure 2). The opening section of the paper focuses on the dramatic advances that have been opened up in microbial ecology by the advent of new molecular techniques, as a case study, with subsequent consideration of the potential application of these technologies to more general ecology, particularly at the higher levels of organisation.

A major challenge facing ecology is the need to determine how whole ecosystems operate and, hence to be able to predict their responses to future environmental change. The ecological responses at these higher levels of organisation and larger spatial-temporal scales (Figure 1), however, cannot be predicted by simply scaling up from studies of single species in isolation: it is the diversity of species, their ecological roles, and the interactions between them that are key to understanding ecosystem functioning, and we need to understand these links between different levels (Woodward *et al.*, 2010a; Yvon-Durocher *et al.*, 2010b). Achieving this understanding of complex multi-species systems requires far more information than can be obtained by simply measuring bulk stocks and fluxes in the classic Lindeman (1942) tradition of ecosystem ecology: the community and the ecosystem are inextricably interlinked, as emphasised by the plethora of biodiversity–ecosystem functioning (B–EF)

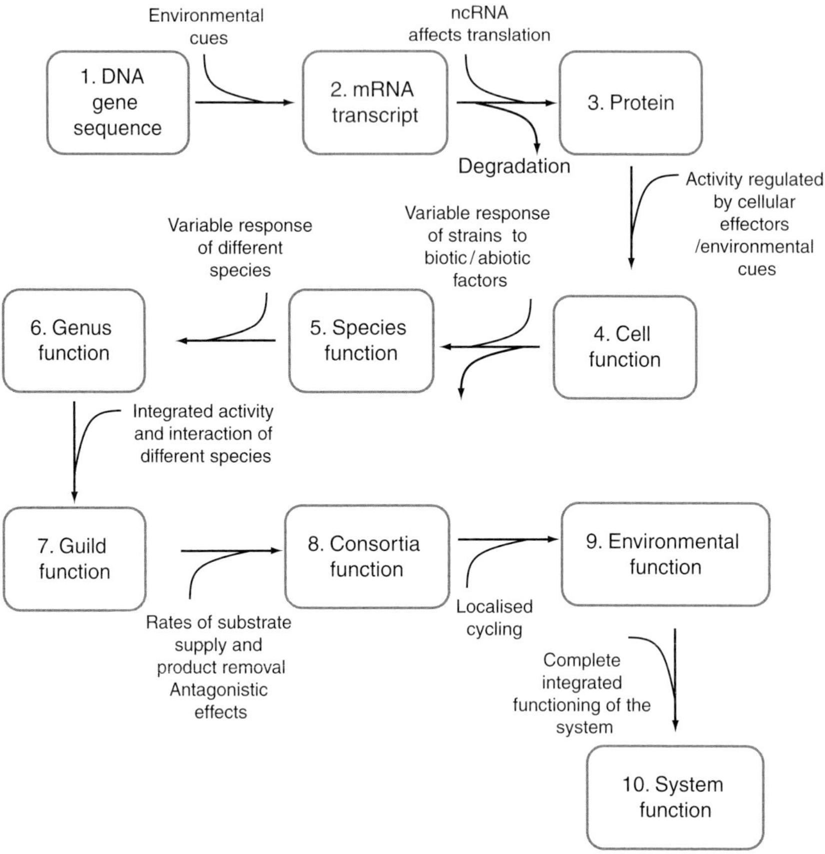

Figure 2 From molecules to ecosystems: a systems biology perspective highlighting a range of potential biotic and abiotic drivers and responses, and their interactions.

studies that have appeared in recent years (Woodward, 2009; Ptacnik *et al.*, 2010 and references therein). The resilience and stability of an ecosystem ultimately depends on the contributions made by its constituent individual organisms, from the smallest viruses and microbes to the largest plants and animals.

Unfortunately, the long-standing dichotomy between community and ecosystem ecology has hindered progress in linking structure to function, with most studies to date being compartmentalised into one or another approach. Much of community ecology, for instance, has focused on the primacy of species identity and diversity as the key variables at this level of organisation, leading some to suggest that it is in danger of being enslaved by the 'the curse of the Latin binomial' (Raffaelli, 2007; Woodward *et al.*, 2010c). Conversely, ecosystems ecology has tended to apply a more broad-brush approach to taxonomy by focussing on stocks and fluxes of biomass, nutrients, or energy among often coarse aggregations of functional units, which often contain unrelated groups of taxa (e.g. cyanobacteria, diatoms and plants lumped as primary producers). The recent rise of B–EF research has, however, started to address structure–function links explicitly, but primarily from an experimental and macroscopic perspective, rather than in the natural environment (e.g. Perkins *et al.*, 2010). Also most B–EF studies have focused on higher plants or metazoans as the focal taxa of interest, whilst largely ignoring the critical role that microscopic organisms in general, and bacteria and archaea in particular, play in driving ecosystem processes in the real world (but see Reiss *et al.*, 2010a).

Much of the reason for the disparity between microbial community and ecosystem ecology can be ascribed to a methodological bottleneck: until relatively recently it has simply not been possible to characterise microbial diversity accurately, or to link it to specific ecosystem processes, in natural systems. For instance, taxonomic identification of the vast majority of bacterial and archaeal species by microscopy is impossible because of a lack of defining morphological characteristics, while microscopic identification of microbial eukaryotes is extremely time consuming, expensive, and also reliant on the assumption that, once detected, individual cells can be ascribed to species by visual inspection. Genetic techniques have assisted with the latter problem in particular and now underpin most of the taxonomy applied in microbial systematics and ecology. These molecular approaches have also been limited until very recently by the rate of data generation and also potential biases resulting from, for instance, the patchy availability of markers and the lack of quantitative resolution beyond simple presence–absence information on community composition. The use of molecular biology and whole community analysis approaches, such as metagenomics, offer exciting new ways to integrate structure and function, and ecology and evolution (DeLong, 2009; Falkowski *et al.*, 2008; He *et al.*, 2010b). Clearly, however, enormous challenges still remain when attempting to link the physical and biological parameters that characterise the functioning of ecosystems across multiple levels of organisation, but we are now at least able to move considerably faster in this direction than was possible even just a few years ago (Figure 2).

B. The Microbial Black Box

While understanding the functional role of plants and animals in ecosystems is itself a challenge, the microbial components of ecosystems, whether bacteria, archaea or microscopic eukaryotes, have for the most part been confined to a functional 'black box'. Material goes in and comes out altered, but with no comprehension of the processes or identity of the organisms that perform these functions (Vandenkoornhuyse *et al.*, 2010). Given that all biogeochemical cycles depend on the organisms within these black boxes (Falkowski *et al.*,

2008; Zak *et al.*, 2006), the fact that most studies have not explicitly identified the critical drivers of ecosystem functioning represents a fundamental lacuna in our current understanding (Woodward, 2009). In this context, it is widely perceived that the sum of an organism's functional traits defines their contribution to ecosystem functioning (Fitter, 2005). Which traits are important, or how they interact with one another or respond to environmental change, is still poorly defined even for plants and animals (Reiss *et al.*, 2009) and often entirely undefined for the microbial components of an ecosystem.

The primary reason for treating the microbes as a black box is simply that these organisms have not been amenable to study using traditional ecological methods: just 20 years ago, studying natural microbial communities in any real depth and consistency was almost impossible (Fenchel, 1992). The application of molecular tools has revolutionised microbial ecology, such that it is now possible to investigate these communities in a scientifically robust and ecologically meaningful way (Hugenholtz *et al.*, 1998). The immense taxonomic and functional diversity that is still being unearthed in microbial communities (Falkowski *et al.*, 2008; Hugenholtz *et al.*, 1998; Oren, 2004) implies that to truly understand ecosystem functioning it is necessary to view a system across the full range of levels of organisation that affect it. This starts at the molecules that an individual consumes and produces, the genes and proteins that facilitate that usage, and the species populations within the community that carry these capabilities through to the manifestation of processes at the ecosystem level (Figures 1 and 2). There is a need for an analysis of the systems that can integrate across all levels: a systems biology for ecology that works from molecules to ecosystems. The incorporation of molecular and microbial ecology into the more established fields of community and ecosystem ecology is a therefore critical prerequisite for developing such a conceptual framework. At this juncture it is necessary to evaluate briefly how microbial ecology, and the application of molecular techniques within this particular field, has developed over the last 30 years or so, and how these different disciplines have started to converge more recently, before returning to discuss the new research opportunities that are now emerging on the horizon.

II. A BRIEF HISTORY OF MOLECULAR MICROBIAL ECOLOGY

A. Characterising Diversity, Abundance and Functional Traits

Microbes play a key role in the functioning of all ecosystems, but the lack of clear defining characteristics meant microbial ecology was, for many years, perceived as being effectively impossible to study (Stanier and Smith, 1960;

Stanier and van Niel, 1962): even the three most basic questions in ecology (who is there? how many are there? and what do they do?) could not be addressed. However, the field was revolutionised in the late 1970s by the recognition that Zuckerkandl and Pauling's (1965) idea of using molecules as markers of evolutionary history could overcome the lack of defining morphological characteristics that hindered microbial taxonomy, and hence ecology. The application of molecular biology to microbial taxonomy, using the small-subunit ribosomal RNA gene (16S rRNA in bacteria and archaea and 18S rRNA in eukaryotes) revolutionised not just microbiology but altered our view of the phylogeny of all organisms with the proposal of the Three Domains Tree of Life (Fox *et al.*, 1980; Woese, 1983; Woese and Fox, 1977). This replaced Whittaker's Five Kingdoms tree (Whittaker, 1969), dismissed the eukaryotic/prokaryotic dichotomy (Chatton, 1938; Stanier and van Niel, 1962) as a natural division in taxonomy, and showed that the vast majority of biological diversity on the planet was microbial and not in the macroscopic animalia and planta. The so-called Tree of Life is still under constant refinement and revision, via, for instance, the application of phylogenomics and other advanced molecular techniques (Dunn *et al.*, 2008).

The initial sequencing of 16S rRNA sequences led to the development of short 16S rRNA-targeted oligonucleotide probes that could be used to detect microbes of specific phylogenetic lineages (DeLong *et al.*, 1994; Giovannoni *et al.*, 1988). These phylogenetic probes, whether radio- or fluorescently labelled (Amann and Fuchs, 2008), proved that previously unknown and uncultured bacteria and archaea dominated bacterioplankton communities in the global oceans (DeLong *et al.*, 1994; Giovannoni *et al.*, 1996; Morris *et al.*, 2002).

The advent of the polymerase chain reaction (PCR, Saiki *et al.*, 1986, 1988) made previously inaccessible, uncultured natural microbial communities available for analysis for the first time by the amplification of specific gene fragments from either whole communities or specific organisms (Britschgi and Giovannoni, 1991; Giovannoni *et al.*, 1990; Pace *et al.*, 1986). Since then the recognised diversity of the microbial world has increased enormously (Hugenholtz *et al.*, 1998), with many new, previously unrecognised clades of microbes being detected in a host of unexpected and often seemingly very hostile environments, from the rocks of the deep subsurface to high in the atmosphere (Hugenholtz *et al.*, 1998; Oren, 2004). In many cases, however, all that exists of most of these new 'organisms' are short sequences of phylogenetic marker genes, such as the 16S/18S rRNA gene, which give little or no insight into the physiology of the organism detected. PCR analysis of functional genes associated with specific respiratory processes, such as sulphate reduction (*dsrAB*; Joulian *et al.*, 2001; Wagner *et al.*, 1998) or denitrification (*nirS* and *nirK*; Braker *et al.*, 1998; Throbäck *et al.*, 2004) or metabolic processes, such as ammonia oxidation (*amoA*;

Sinigalliano *et al.*, 1995, Ward and O'Mullan, 2005) allowed the diversity of organisms involved in specific processes to be investigated: at last links could start to be forged between microbial community structure and specific ecosystem processes. Further links between community structure and functioning could be made by analysing rRNA or functional gene mRNA using reverse transcriptase PCR (RT-PCR) as the detection of either rRNA or mRNA from an organism is an indication that it is active *in situ* (Bernhard *et al.*, 2007; Chin *et al.*, 2008; Dar *et al.*, 2007; Kandeler *et al.*, 2006; Smith *et al.*, 2007). There is good evidence that the level of rRNA in a bacterial cell is correlated with the level of its activity (Kemp *et al.*, 1993; Rosset *et al.*, 1966), although such a link is less clear for the presence of mRNA and activity (Feder and Walser, 2005).

The more recent development of quantitative PCR has made it possible to address the issues of abundance more directly, by counting the number of specific genes or mRNA transcripts in a sample (Bernhard *et al.*, 2007; Smith *et al.*, 2007). Even so, qPCR counts the number of copies of specific target genes, not organisms themselves. As gene copy number per cell can vary, especially for 16S rRNA genes and even for protein-coding genes (VanGuilder *et al.*, 2008), such data needs to be interpreted with a modicum of caution.

The enormous growth in the number of sequences in the global databases has made it possible to target a large number of organisms using array and microarray technologies. Here short pieces of DNA (probes), either from a specific organism or derived from database analyses, are attached to a slide and then hybridised to DNA and/or RNA extracted from a sample, to produce a fluorescent signal proportional to the amount of binding of the applied DNA/RNA to the probes on the array. Microarray technologies have been used to study microbial communities, as these allow very large numbers of potential organisms or genes to be analysed in a single experiment. Both phylogenetic arrays (DeSantis *et al.*, 2007) and functional gene arrays (He *et al.*, 2007, 2010a) have been used with some success. The major limitations of microarrays are that only those organisms targeted by the probes put on the slide can be detected and arrays tend to have relatively poor dynamic detection ranges and poor signal-to-noise ratios (Bloom *et al.*, 2009; van Vliet, 2010).

Despite having access to markers for the presence and activity of microbes it is still difficult to make links between the structure of a community (who is present, and in what numbers) and the processes taking place *in situ* (what ecological roles are being performed): taking microbial ecology out of the laboratory and into the natural environment is still a challenging step. Specific analytical methods have been used to make such links, including the uptake of stable and radio isotopes of tracer molecules or by the experimental manipulation of communities to determine the effects of added substrates (Boschker *et al.*, 1998; Lee *et al.*, 1999; Manefield *et al.*, 2002; Ouverney and Fuhrman, 1999; Purdy *et al.*, 2002, 2003b; Radajewski *et al.*, 2000) with some notable successes. However, these are still inevitably focused on specific processes and organisms and involve perturbations and manipulations of communities, rather than assessing structural–functional linkages in natural ecosystems.

B. Genomics and Post Genomics in Microbial Ecology

A further major step forward in molecular microbial ecology came with the advent of genome sequencing, initially of single strains of organisms (Fleischmann *et al.*, 1995) and then of genomic DNA fragments extracted directly from the environment (Handelsman, 2004). These new environmental analyses, initially called environmental, community or functional genomics, or ecogenomics, but now generally known as metagenomics (Handelsman, 2004; Tringe and Rubin, 2005), largely avoids the selective problems associated with PCR-based methods and could conceivably allow access to the whole gene pool within an environment. This has opened up a vast new range of applications and research vistas and has finally placed molecular and microbial biology firmly within the core of general ecology, much of which has been dominated for decades by community-level studies of macro-organisms. The potential to completely sequence all the genes in an environment suggested that it was also possible to define all the metabolic pathways that could occur within a system (Buckley, 2004). Metagenomics was rapidly followed by the application of post-genomics technologies, transcriptomics, proteomics and metabolomics (DeLong, 2009 and references therein) and the potential to define not just what the microbes in a system could potentially do but also what they were actually doing. Thus, it appeared that all of the barriers that had stood in the way of microbial ecology were finally falling, due to the application of large scale analytical methods that simply sequenced everything to expose the capacity and functions of all of the organisms within a sample (DeLong, 2009). Arguably, the challenge now is not so much the difficulty of extracting sufficient data, but rather the need for advanced bioinformatics approaches to handle the vast amounts of information that can now be generated rapidly from even a small number of samples.

Although they have provided important breakthroughs in their own right, these technical advances and sophisticated methods do not in themselves create the understanding and integration across different ecological levels that is needed to build a true systems-based approach to ecology (Vandenkoornhuyse *et al.*, 2010). This could be provided, however, by the range of next generation sequencing (NGS) technologies that have appeared very recently and which seem destined to make a major contribution to ecology. Given their potential to generate significant paradigm shifts in all spheres of ecology, but particularly in integrating community and ecosystems approaches, we will consider NGS technologies in more detail in the context of what they are, what they offer, and the prospects for future advances.

III. NEXT-GENERATION SEQUENCING TECHNOLOGIES

A. A New Paradigm

Occasionally a new technology is developed that revolutionises biology. Between 1975 and 1977 Sanger and colleagues devised a method of DNA sequencing based on dideoxynucleotide chain-termination (Sanger and Coulson, 1975; Sanger *et al.*, 1977a,b). With subsequent technological developments (Hunkapiller *et al.*, 1991; Smith *et al.*, 1986), this method resulted in the successful sequencing of the ~3 billion base pairs of the human genome (Collins *et al.*, 2004; Lander *et al.*, 2001) after 13 years of work by hundreds of researchers and hundreds of DNA sequencing machines running in parallel, at an estimated cost of $3 billion. For comparison, in 2009, a human genome was sequenced for just $48,000 by three people using only four DNA sequencing machine runs (Pushkarev *et al.*, 2009); an indication of the exponential increase in sequencing power and declining costs over that period (Figure 3). This monumental shift in both cost and efficiency is based on new chemical techniques, miniaturization, parallelization, high-throughput and technological advances in computing and image detection, which in combination are commonly referred to as NGS. The paradigm shift associated with NGS is rapidly changing the field of biology at all levels of organization, from molecules to ecosystems, and promises to alter the way ecologists view the natural world in many unexpected ways.

There are currently four commercially available NGS platforms, and one near-commercial platform, which differ in both protocols and in the nature of the data generated. All have particular strengths and weaknesses, with some more suited to specific biological questions than others. Here we present a brief introduction and technical review, to highlight these new technologies, and discuss the relative merits and potential applications of NGS in an ecological context.

NGS requires only a few microgram quantities of DNA, and therefore the time-consuming cloning into bacterial hosts followed by multiple rounds of PCR amplification can be bypassed. Instead, the template is isolated directly from the source, fragmented, amplified in a limited way (or not at all), and

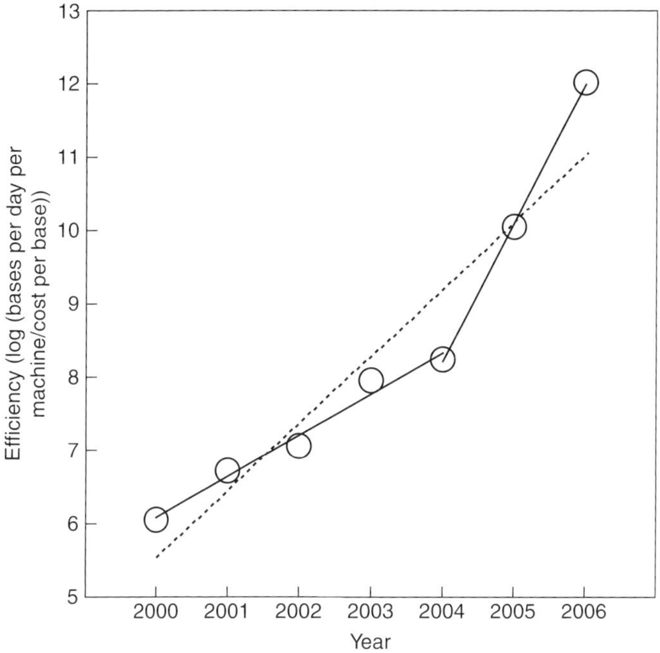

Figure 3 More bang for your buck: the rapid growth of sequencing power and declining costs over the last decade, expressed as an exponential increase in efficiency over time (data extracted from Hudson (2008)). On average there is close to an order of magnitude ($\beta = 0.9$) increase in efficiency per year over the time series (dashed line; $r^2 = 0.906$). However, the relationship is better described by two separate regressions, shown as the solid lines ($r^2 = 0.978$ and $r^2 = 0.999$, respectively), as a step change occurs from 2004 onwards as NGS technologies emerge and improvements in efficiency increase dramatically, with the scaling coefficient increasing from $\beta = 0.6$ to 2 orders of magnitude per year.

attached to a solid surface or support (see Figures 4–8). Either clonally amplified templates from single DNA molecules or single DNA-molecule (non-amplified) templates can be used here, with the former being derived from either emulsion PCR (emPCR) (Dressman *et al.*, 2003) or solid-phase amplification (Fedurco *et al.*, 2006). NGS platforms use three types of mechanism for sequencing the templates: (1) single-nucleotide addition (454 pyrosequencing); (2) cyclic reversible termination (CRT) (Illumina/Solexa and Helicos) and (3) real-time sequencing (Pacific Biosciences). Sequencing by ligation (ABI SOLiD) is also discussed. The key feature of all these methodologies is sequencing by synthesis: that is, the DNA templates are sequenced at the same time as DNA polymerization takes place. A description of these platforms is given below and a brief outline of some of their features and how they have been used is given in Table 1.

Table 1 Comparison of next-generation sequencing platforms and their advantages and disadvantages in the context of ecological research

Platform	454 Pyrosequencer GS FLX Titanium (Roche)	Solexa GA_{II} (Illumina)	HeliScope (Helicos BioSciences)	SOLiD 4 (Life/Applied Biosystems)	SMRT (Pacific Biosciences)
Library/template preparation	Fragment Mate-pair emPCR	Fragment Mate-pair Solid-phase	Fragment Mate-pair (Single molecule)	Fragment Mate-pair emPCR	Fragment (Single-molecule)
Chemistry	Single-nucleotide addition	Cyclic-reversible termination	Cyclic-reversible termination	Sequencing by ligation	Real-time
Read length (basepairs)	~400	≤150 (average 75)	≤55 (average 35)	≤50	964 (average)
Gbp per run	~0.4	Up to 18	Up to 35	Up to 100	N/A
Pros	• Fast run times • Longer reads (improve mapping)	• Most widely used platform • Fast run times	• Single-molecule sequencing, therefore non-biased representation of templates	• Two-base encoding provides inbuilt error correction	• Longest read-length • Single-molecule sequencing, therefore non-biased representation of templates
Cons	• High error rates in homopolymer repeats.	• Short reads	• High error rates • Short reads	• Long run times • Short reads	• High error rates
Applications	• *De novo* genome assembly • Metagenome sequencing • PCR amplicon sequencing	• Variant discovery by whole-genome resequencing • Seq-bases methods (RNA-Seq and ChIP-Seq)	• Seq-based methods (RNA-Seq and ChIP-Seq)	• Variant discovery by whole-genome resequencing	• *De novo* genome assemblies • Full-length transcriptome sequencing

Microbiome studies (within-organism ecosystems)	• Analysis of obesity associated gut microbiome (Turnbaugh *et al.*, 2006) • Oral microbiome (Zaura *et al.*, 2009)	• Human gut microbial gene catalogue (Qin *et al.*, 2010)	• None at present	• Used to confirm the role of a proposed causative agent of a disease in citrus trees (Tyler *et al.*, 2009)	• None at present
Environmental ecosystems	• Functional profiling of nine biomes (Dinsdale *et al.*, 2008) • Analysis of honeybee colony collapse disorder (Cox-Foster *et al.*, 2007)	• Genome sequencing from enrichment culture of novel microbe (Ettwig *et al.*, 2010)	• None at present	• None at present	• None at present

B. 454 Pyrosequencing (Roche)

The 454 system was the first commercially available NGS platform (Margulies *et al.*, 2005). Genomic DNA is randomly fragmented and universal DNA adaptors of known sequence are ligated to the free ends (Figure 4A). The resultant DNA fragments are denatured into single strands and captured on micro-beads coated with complementary adaptors under conditions that favour the addition of only one DNA molecule per bead (Figure 4B). Each DNA template is clonally amplified using emPCR. This reaction takes place within single aqueous droplets, each containing a single micro-bead with a bound DNA template (Figure 4C). Following amplification, each micro-bead, containing millions of identical copies of the DNA template, is enriched and deposited onto an array (Figure 4D). It is here that the DNA sequencing takes place via single-nucleotide addition, in a method known as pyrosequencing (Ronaghi *et al.*, 1998). Additional beads carrying the enzymes sulphurylase and luciferase are added to each well followed by the introduction of a single-nucleotide species, which flows across the wells (Figure 4E). Where this results in the incorporation of nucleotide, pyrophosphate is released which, in an enzymatic-catalyzed reaction, is converted into a burst of light (Figure 4F) whose intensity is proportional to the number of incorporated nucleotides and recorded as a series of peaks called a flowgram, from which a sequence can then be determined (Figure 4G). The sequencing cycle is repeated with each of the four nucleotides added sequentially until the desired read length is attained.

Using the GS FLX Titanium platform, up to 1 million beads, each coated with a clonally amplified DNA molecule, are pyrosequenced in parallel. With individual sequence read lengths of up to 400 bases, a single run can generate 400 Mb of sequence in 10 h. Using an improved sequencing chemistry, Roche currently report an average read length of 657 bases, which would boost coverage to over 650 Mb per run. The long reads (400+ bps) now being produced by 454 sequencing are sufficient to allow robust phylogenetic analysis of rRNA or functional genes. The capacity to acquire data that can be analysed robustly without the need to assemble contigs greatly reduces analysis time and increases the level of confidence with which such data can be interpreted. These are important advantages in ecological studies that may involve analyzing many samples, and are particularly well-suited to quantifying biodiversity in natural or experimental communities.

C. Illumina Genome Analyzer

Commonly referred to as ‘Solexa’, this platform currently dominates the market (Bentley *et al.*, 2008). DNA libraries are constructed from fragmented genomic DNA and clonal amplification is achieved via solid phase.

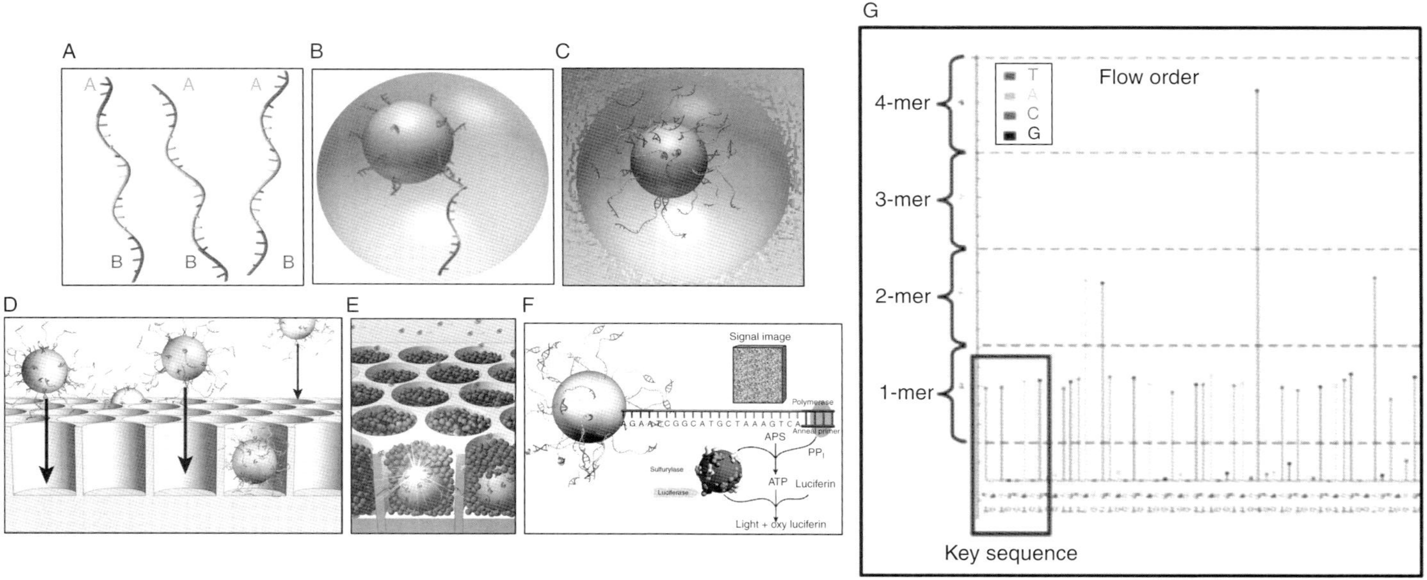

Figure 4 Outline of the GS FLX titanium 454 sequencing workflow. (A) Randomly fragment genomic DNA and ligate to 454-specific adapters (indicated as A and B). (B) The DNA is separated into single strands and captured onto beads through hybridisation of adapter sequences with primer sequences coating the bead. Conditions are used that favour one DNA molecule per bead. (C) An oil-aqueous emulsion is created to encapsulate single bead-DNA complexes within a single aqueous drop. emPCR is performed within these drops to create beads containing thousands of copies of the same DNA template. (D) Beads are then loaded into individual PicoTitre wells. (E) Additional beads coupled with sulphurylase and luciferase are added to each well, followed by a single type of nucleotide (in this case adenine). Incorporation of the nucleotide generates light, which is recorded by a CCD camera located underneath the PicoTitre plate. The cycle is then repeated with the other three nucleotides in a step-wise, pre-determined order until the desired read-length is achieved. (F) The pyrosequencing reaction. (G) The light generated by the enzymatic cascade is recorded as a series of peaks called a flowgram. Images copyright of 454 Life Sciences/Roche (2010). All rights reserved. Used with permission of 454 Life Sciences/Roche.

Single-stranded DNA fragments are ligated to universal adapters of known sequence (Figure 5, Step 1) and randomly arrayed on a glass slide containing adapters which also act as primers for the sequencing reaction (Figure 5, Step 2). Each DNA fragment is clonally amplified by bridge amplification to

A

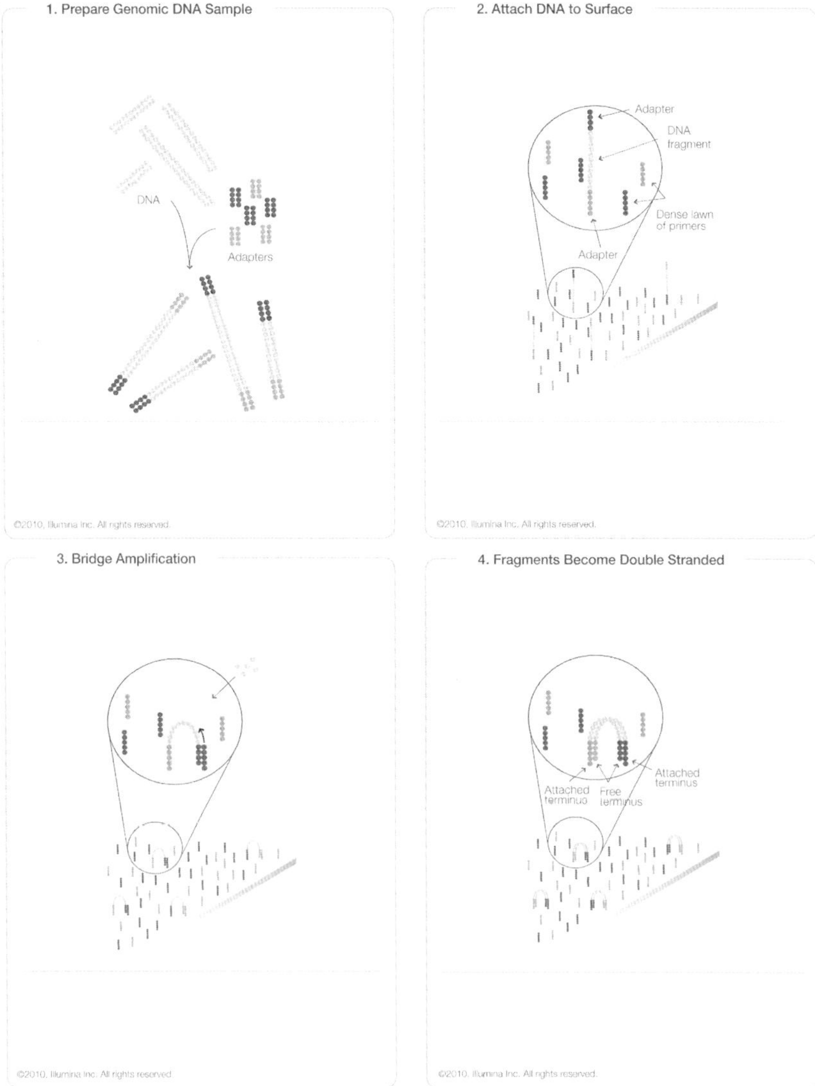

Figure 5 (Continued)

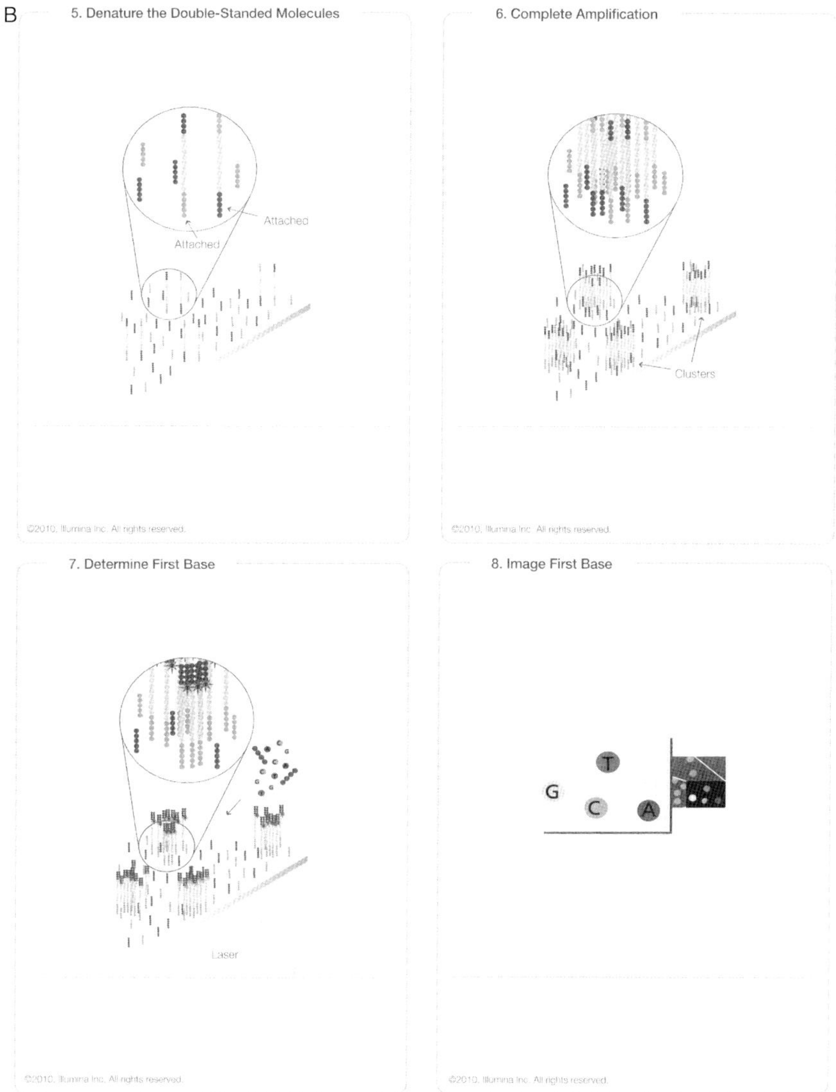

Figure 5 (Continued)

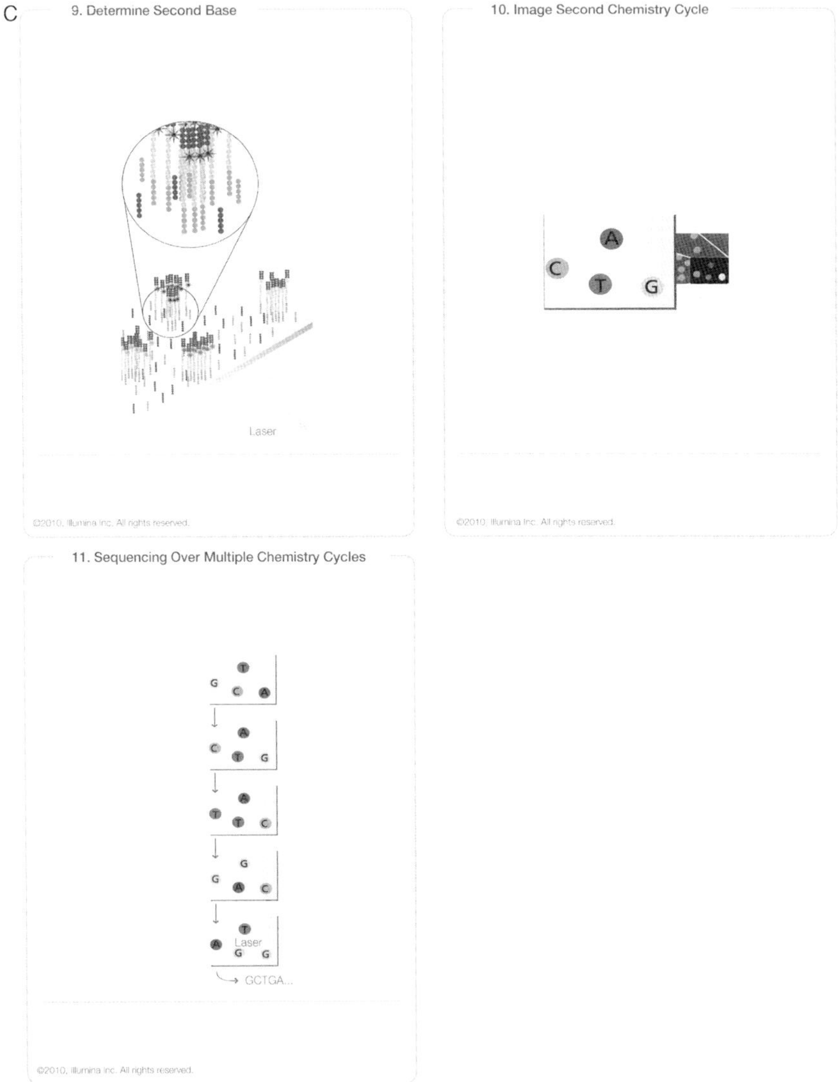

Figure 5 Outline of the Illumina Genome Analyzer workflow. (1) Randomly fragment genomic DNA and ligate adapters to both ends of the fragments. (2) Bind single-stranded fragments randomly to the inside surface of the flow cell channels. (3) Add unlabelled nucleotides and enzyme to initiate solid-phase bridge amplification. (4) The enzyme incorporates nucleotides to build double-stranded bridges on the solid-phase substrate. (5) Denaturation leaves single-stranded templates anchored to the substrate. (6) Several million dense clusters of double-stranded DNA are

produce clusters (Figure 5, Steps 3–6). Eight independent lanes are contained on a single flow-cell, meaning eight independent libraries can be sequenced in parallel during the same instrument run, thus reducing time and reagent costs, whilst increasing throughput.

After cluster generation, the single-stranded amplicons are sequenced by CRT (Figure 5, Step 7). All four dye-labeled modified nucleotides containing reversible blocking groups are added in a cyclical process that includes nucleotide incorporation, fluorescence imaging and cleavage (Bentley *et al.*, 2008). Due to the presence of the blocking group on the modified nucleotide, DNA polymerase adds just one nucleotide, complementary to the template strand, before synthesis terminates. Unincorporated nucleotides are washed away and the incorporated nucleotide is detected by laser imaging (Figure 5, Step 8). A cleavage step then removes the blocking group from the incorporated nucleotide, allowing another modified nucleotide to be added and the cycle is repeated (Figure 5, Steps 9 and 10). In the Illumina/Solexa method, each of the four nucleotides carries a different fluorophore, allowing discrimination of the incorporated base. By repeated cycles of this method with imaging after cycle, the DNA sequence of a particular cluster of templates is determined by computer analysis of that cluster on each sequential image (Figure 5, Step 11).

Read lengths of up to 150 bases are already routinely possible but, to date, generally average about 75 bases, with the number predicted to rise rapidly. The recently introduced Genome Analyzer$_{IIx}$ can generate up to 320 million reads (640 million paired-end reads), with a total throughput of up to 6.5 Gb per day, the equivalent of two complete human genome a day. The massive sequencing capacity and reasonable read-length delivered by Solexa makes deep sequencing of many samples possible, as shown by Qin *et al.* (2010) in an analysis of the human gut microbiome of over 100 people. It is also extremely useful in analyzing small molecules such as non-coding RNA (ncRNA (Perkins *et al.*, 2009)).

generated in each channel of the flow cell. (7) The first sequencing cycle begins by adding four differently labelled reversible terminators, primers and DNA polymerase. (8) After laser excitation, the emitted fluorescence from each cluster is captured and the first base is identified. (9) The next cycle repeats the incorporation of four labelled reversible terminators, primers and DNA polymerase. (10) After laser excitation, the image is captured as before, and the identity of the second base is recorded. (11) The sequencing cycles are repeated to determine the sequence of bases in a fragment, one base at a time until the desired read-length is achieved. The data are subsequently aligned and compared to a reference, and sequencing differences are identified. Images copyright of Illumina, Inc. (2010). All rights reserved. Used with permission of Illumina, Inc.

D. HeliScope

A more recent advancement by Helicos Bioscience Corporation has seen the introduction of NGS based on single-molecule templates (Harris *et al.*, 2008). Here, no amplification is required and < 1 μg of starting material can be analysed. The key advantage of single-molecule sequencing is that no PCR-

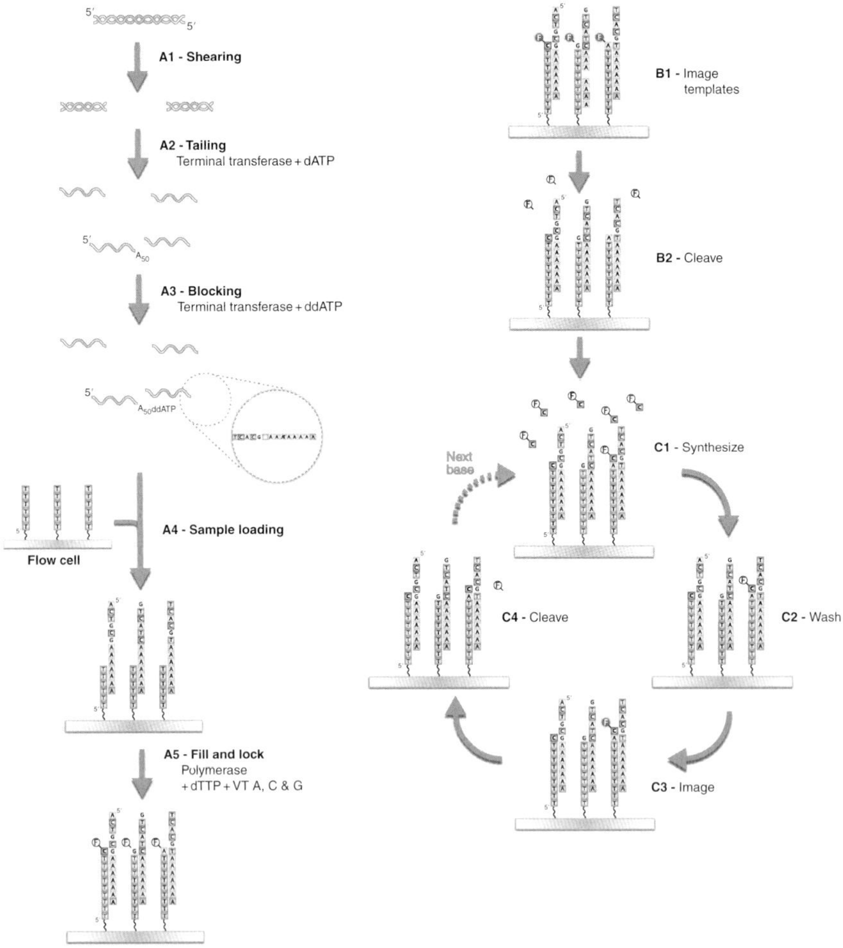

Figure 6 Outline of the HeliScope workflow. (A) 1. Genomic DNA is randomly fragmented. 2. DNA fragments are separated into single strands and a polyA universal priming sequence is added to the 3′-end using Terminal Transferase. 3. Each strand is then blocked by addition of a modified adenosine nucleotide. 4. These labelled DNA strands are then hybridised to the Helicos flow cell, which contains millions of oligo-dT

based errors or amplification biases are introduced into the sequencing template. Genomic DNA is randomly fragmented and single-stranded DNA molecules are generated, to which polyA tails are added in an enzyme-catalysed reaction (Figure 6, Steps A1–A3). DNA templates are then immobilised on a solid support attached to the flow cell, which contains spatially distributed primers (oligo-dT), complementary to the polyA tails (Figure 6, Step A4).

The CRT method of sequencing differs from that provided by Illumina/Solexa since each dye-labeled nucleotide is added sequentially. Following incorporation of the first dye-labeled nucleotide, a laser illuminates the surface of the flow cell, which determines the location of each fluorescently labeled single-molecule template. A camera then produces a map of the flow cell surface by taking sequential images across it. A cleavage step then removes the dye and the blocking groups to permit the addition of the next dye-labeled nucleotide in the following cycle (Figure 6, Steps B2–C4). This sequencing cycle is repeated with the other three dye-labeled nucleotides, with the number of cycles determining read length. The DNA sequence of a particular single molecule is determined by analysis of the images taken after each sequencing round. Read lengths of up to 55 bases (average read length 36 bases) have been reported with up to 35 Gb per run and as much as 1 Gb per hour.

This technology can produce 10 times as much data per run as the Solexa sequencer but the very short reads are best suited to analyses of small molecules and transcriptomes of previously sequenced organisms. While it is still difficult, at present, to see obvious broad ecological applications, it has been used to accurately quantify a yeast transcriptome over 4 orders of

capture sites that are immobilised on a solid-support. 5. The extra adenosine nucleotides on the template strands are filled in by addition of thymine nucleotides, and the template strand is locked in place at the first non-adenosine base by the addition of Virtual Terminator (VT) nucleotides. This also serves to label the templates. (B) 1. A laser then illuminates the surface of the flow cell showing the location of each fluorescently labelled template. A CCD camera then produces a map of the templates on the flow cell surface by taking a sequence of images across the flow cell in a step-wise pattern. 2. After the templates have been imaged the template label is cleaved and washed away. (C) 1. The sequencing reaction begins by the introduction of DNA polymerase and a fluorescently labelled nucleotide (in this case cytosine). The labelled nucleotide is incorporated in a template-dependent manner. 2. A wash step removes the DNA polymerase and any unincorporated labelled nucleotide. 3. The templates that have incorporated a cytosine are then visualised by illuminating and imaging the entire flow cell surface. 4. After imaging a cleavage step removes the fluorescent labels and the process (C1–C4) continues by addition of each of the other three labelled nucleotides in a step-wise, pre-determined manner and repeats until the desired read-length is achieved. Images copyright 2010. Helicos BioSciences Corporation. All rights reserved. Used with permission of Helicos BioSciences Corporation.

magnitude (Lipson *et al.*, 2009), so it could be especially useful for characterising the functioning and gene expression of certain key microbial taxa.

E. SOLiD (Life/Applied Biosystems)

In the Support Oligonucleotide Ligation Detection (SOLiD) system, sequencing is achieved by using a DNA ligase rather than a polymerase (Valouev *et al.*, 2008). DNA libraries are prepared and amplified on microbeads by emPCR (as in 454 pyrosequencing, see Figure 4A–C). Sequential rounds of ligation to a collection of fluorescently labeled probes (octamers) are performed, with imaging used to determine their identity after every round of probe hybridization (Figure 7A, Steps 1–4). The octamers are structured such that the identity of specific positions within the probes correlate with the identity of the fluorescent label (see Figure 7B). After ligation, the octamer is chemically cleaved to remove the fluorescent label, and the cycle is repeated nine more times to give 10 colour calls (Figure 7A, Steps 1–4). The number of ligation cycles determines the eventual read length. The primer is then removed and a second round of ligation is performed with an '$n-1$' primer, which shifts the interrogation bases and the corresponding colour calls one position to the left (Figure 7A, Steps 5 and 6). Nine more ligation cycles follow, followed by three more rounds of ligation cycles with 'base-shifted' primers ($n-2$, $n-3$ and $n-4$). Colour calls from the five ligation rounds are then arranged into a linear sequence (Figure 7A, Step 7). Using the SOLiD 4 system, read lengths of up to 50 bases are currently possible, with an expected throughput of up to 100 Gb per run.

As with the HelioScope the short reads of the SOLiD system make widespread ecological applications more difficult to envisage than with 454 pyrosequencing, but the technology has been used, along with Solexa sequencing, to analyse ncRNA in grapevines (Picardi *et al.*, 2010) and to identify the causative agent of a disease in citrus trees (Tyler *et al.*, 2009).

F. SMRT Technology (Pacific Biosciences)

The near-commercial SMRT technology from Pacific Biosciences offers a further revolution in DNA sequencing (Eid *et al.*, 2009). Like the HelioScope, single DNA molecules are sequenced but in real-time, such that nucleotides do not halt the DNA polymerase and synthesis. Incorporated dye-labeled nucleotides are imaged continuously during the latter process. In this approach, genomic DNA is randomly fragmented and hairpin structures

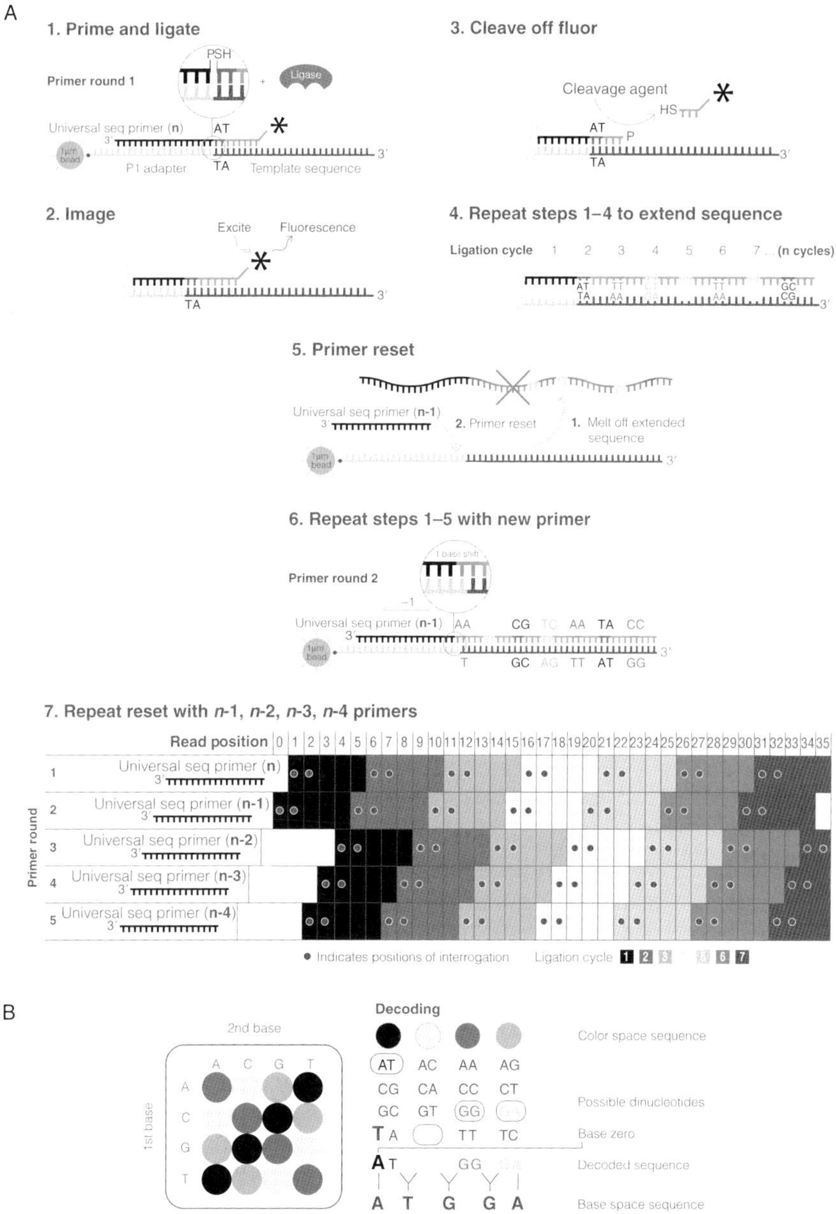

Figure 7 Outline of the SOLiD 4 workflow. DNA libraries are prepared and amplified on beads by emPCR (see Figure 4A–C). (A) Step 1: Universal sequencing primers hybridise to the P1 adapters. Libraries of fluorescently labelled 1,2-nucleotide probes compete for ligation to the template strand. The 1,2-nucleotide probes are designed to

are ligated to both ends of the DNA (Figure 8A). Single molecules of a strand-displacing bacteriophage DNA polymerase are attached to a solid support and template molecules are added to the immobilised polymerase (Figure 8B) within individual nanostructures (zero-mode waveguide [ZMW]; Levene *et al.*, 2003) with an observation volume of just 10^{-21} l. At this volume, the activity of a single molecule can be detected (Figure 8C) and the real-time incorporation of one of four differently dye-labeled nucleotides can be observed by monitoring the fluorescence pulse (Figure 8D).

With reported average read lengths of 964 bases (Metzker, 2010), and with each ZMW capable of producing up to 400 kb of sequence per day, an array of only 14,000 ZMWs would be required to produce the equivalent of one human diploid genome per day (Eid *et al.*, 2009). Currently SMRT DNA sequencing is performed on proprietary SMRT Cells, each having an array of approximately 75,000 ZMWs: that is, offering the potential detection of approximately 75,000 single-molecule sequencing reactions in parallel. As this technology has not yet been released onto the market there are currently no published reports of its use in any area of ecology, however, the very long reads could enable deep sequencing of samples that produce robust data of environmental samples. As with all new technologies, the real test is not in the claims of what it can do but what it actually does when faced with real samples.

G. Improving Analysis of Multiple Samples and Increasing Read Length in NGS Runs

Various strategies exist in order to utilise the sequencing capacity of NGS platforms fully. Although application-dependent, for many studies, including organisms with small genomes or for those involving the genome-wide

---

interrogate the first and second positions adjacent to the sequencing primer (shown here as TA), such that the 16 possible di-nucleotides are encoded by four dyes (see also Figure 7B). Step 2: Following ligation, fluorescence is detected by laser excitation during four-colour imaging. Step 3: The ligated 1,2-dinucleotide probes are chemically cleaved between the fifth and sixth nucleotides with silver ions to generate a free 5′-phosphate group. Step 4: The SOLiD cycle (Steps 1–3) is then repeated nine more times. Step 5 and 6: The extended primer is then stripped off and a second round of 10 ligations is then performed with a universal sequencing primer one base shorter than the previous one ('$n-1$'). Step 7: Three more rounds of ligation cycles are performed with $n-2$, $n-3$ and $n-4$ primers. Colour calls from the five ligation rounds are then arranged into a linear sequence. Through the primer reset process, virtually every base is interrogated in two independent ligation reactions by two different primers. For example, the base at read position 5 is assayed by primer number 2 in ligation cycle 2 and by primer number 3 in ligation cycle 1. (B) A two-base encoding scheme in which four di-nucleotide sequences are associated with one colour. The decoding matrix allows a sequence of di-nucleotides to be converted to a base sequence, as long as one of the two bases is known. Images copyright 2010. Life/Applied Biosystems. All rights reserved. Used with permission of Life/Applied Biosystems.

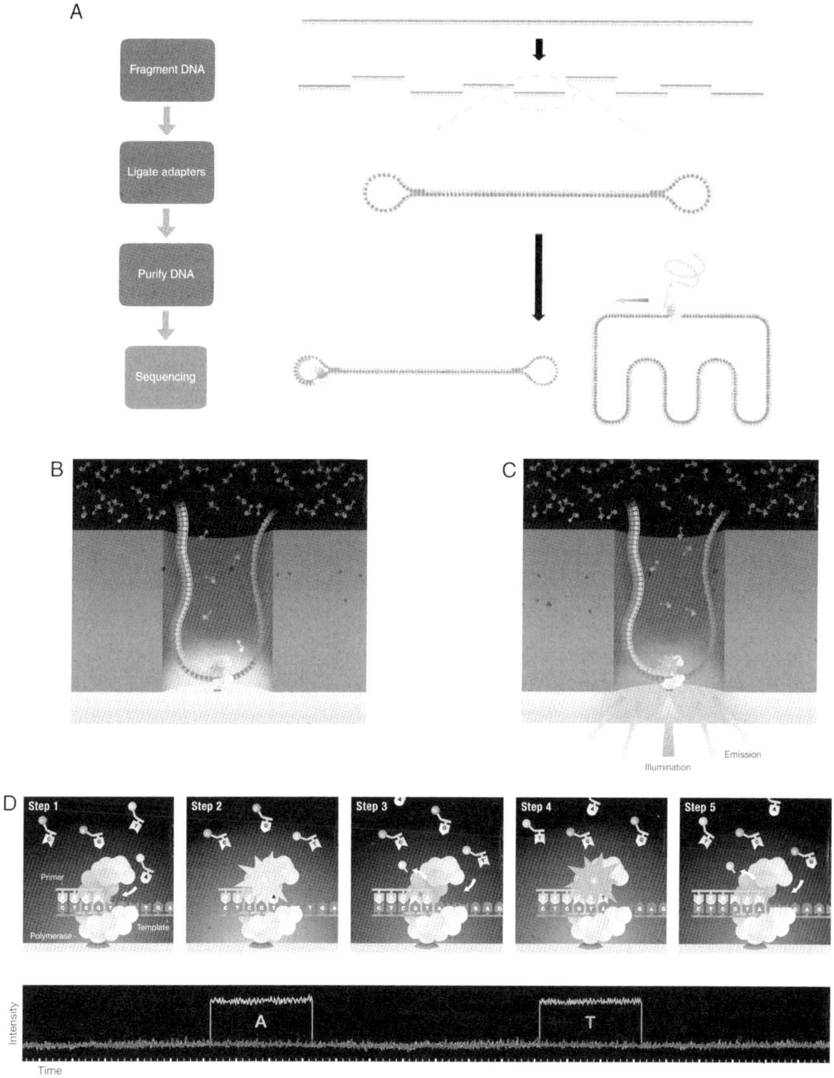

Figure 8 Outline of the SMRT workflow. (A) Genomic DNA is randomly fragmented and hairpin adapters are ligated to both ends. The templates can then go through the sequencing reaction. A strand displacing bacteriophage DNA polymerase opens up the DNA structure into a circular template and can generate independent reads, both forward and reverse of the same DNA molecule. (B) The bacteriophage DNA polymerase is immobilised at the bottom of each ZMW, nucleotides diffuse into the ZMW chamber and are processively incorporated in real-time to produce a long continuous template-directed DNA molecule. (C) Only the bottom 30 nm of the ZMW is illuminated, therefore only those nucleotides near the bottom fluoresce. Both excitation and detection occur without interruption through the transparent

mapping of DNA-associated factors where only a fraction of a genome is sequenced (Hurd and Nelson, 2009), the full capacity of NGS platforms is not required. Furthermore, there are also limitations to the number of different samples that can be processed in parallel at any one time. Two recently reported adaptations to NGS protocols that employ a multiplex (or bar-coding) strategy combined with either the 454 pyrosequencing platform (Meyer *et al.*, 2007, 2008) or the Solexa/Illumina platform (Lefrancois *et al.*, 2009) offer a solution to both these issues. Each uses a bar-coding system, whereby unique 'tags' are included in the oligonucleotide adapters ligated to DNA library fragments. After individual library generation, libraries can be pooled and sequenced in parallel and the tags used subsequently to identify the sample origin of individual reads. Bar-coding of multiple samples reduces the depth of sequencing (number of sequences reads per sample) for each sample but this can be extremely useful when using an NGS technology to analyse an amplified gene, where several thousand sequences will probably prove to be sufficient for a comprehensive analysis. Such an approach has been used recently in a temporal analysis of a lake protist community Nolte *et al.* (2010).

Mate-paired reads also offer increased sequencing reads per machine run. This strategy is particularly useful for the *de novo* assemblies of genomes and in the analysis of structural variants (SV) between related genomes (see Section V and references therein). Here, DNA libraries are prepared with fragments of a known size, with sequencing directed from both ends of the DNA fragments, rather than from just one, thus doubling the amount of sequence obtained. Since the sequenced ends are of a known distance apart, reads can be aligned to a reference genome and distances compared (as with SV analysis), or overlapping mate-pair libraries can be used to build larger contiguous sequences of DNA for *de novo* assembly.

These methods can be useful in an ecological context, where for instance we might wish to split a previously pooled sample to identify, for instance, replicates within an experimental treatment, dates within a seasonal study or quadrats positions within a spatial survey. The great advantage of this is that the efficiency of the yield-to-cost ratio can be tweaked to acquire the optimum output, to keep running costs down whilst also minimizing machine time, by running multiple samples simultaneously (e.g. replicates can be marked, pooled, sequenced, and then subsequently disentangled).

glass bottom of the SMRT Cell. (D) Step 1: Fluorescent phospholinked labeled nucleotides are introduced to the DNA polymerase. Step 2: The base being incorporated is held in the detection volume for tens of milliseconds, producing a bright flash of light, which is recorded as a pulse. Step 3: The phosphate chain is cleaved, releasing the attached dye molecule. Steps 4 and 5: The process repeats. Images copyright 2010. Pacific Biosciences. All rights reserved. Used with permission of Pacific Biosciences.

IV. GENOME SEQUENCING: FUNCTIONAL DIVERSITY IN ECOLOGY

Once DNA sequences have been generated, they can be aligned to a known reference sequence or assembled *de novo*. The alignment method is useful for identifying genetic variation, SV, or single-nucleotide variants (SNV) in highly related genomes. For example, genome variation has been analyzed in this way for various species of bacteria (Moran *et al.*, 2009; Nusbaum *et al.*, 2009; Srivatsan *et al.*, 2008) and eukaryotes (Hillier *et al.*, 2008; Ossowski *et al.*, 2008; Sarin *et al.*, 2008; Shen *et al.*, 2008), including humans (Ahn *et al.*, 2009; Bentley *et al.*, 2008; Kim *et al.*, 2009; Wang *et al.*, 2008). The SOLiD and HeliScope platforms have also been used to sequence the human genome (McKernan *et al.*, 2009; Pushkarev *et al.*, 2009). Within these new technologies there exists the capacity to analyse a sample to detect the real extent of diversity within specific organismal groups. This is particularly relevant for microbial populations where small differences between strains can be critical to their functional role in the environment (see references above).

Alignment of sequencing reads to complex genomes can be complicated by the presence of repetitive regions. Mate-pair reads are commonly employed to overcome this particular issue, where one read of the pair is unique to the genome, and in this way reads can be anchored to the reference genome. This approach also allows for the identification of SVs: larger regions of genomic variation such as substitutions, insertions, deletions, inversions, duplications and copy-number variations (Korbel *et al.*, 2007).

Existing NGS platforms generate relatively short reads, and therefore *de novo* assemblies have so far been limited to relatively small genome sizes, mostly among bacteria and archaea (Chaisson and Pevzner, 2008; Hernandez *et al.*, 2008; Margulies *et al.*, 2005). More recently Li *et al.* (2010b) published the first true *de novo* assembly of a large eukaryotic genome (giant panda) based entirely on short-reads generated by NGS, and the complete genomes of humans have also now been assembled *de novo* (Li *et al.*, 2010c).

Since relatively small amounts of starting material are required for NGS approaches, the ability to sequence genomes from ancient organisms has been revolutionised, enabling ecologists and evolutionary biologists to look back through deep time. Palaeogenomics has previously focused on sequencing the relatively abundant mitochondrial DNA that routinely requires many rounds of PCR amplification prior to traditional DNA sequencing. Owing to the low amount of genomic material from fossilised or preserved materials, however, the nuclear genome has thus far proved elusive. NGS allows not only an unbiased approach, but by aligning sequencing reads to phylogenetically related species, this allows contaminating DNA from microbial species to be removed from the analyses, or studied in parallel. These types of studies facilitate the exploration of evolutionary pathways and phylogenetic relationships, which could, for instance, be mapped onto previous climates or fossil communities within long-extinct ecosystems: NGS has resulted in the partial sequencing of the nuclear genomes of the cave bear, *Ursus spelaeus* from 40,000-year-old bones (Noonan *et al.*, 2005), the woolly mammoth, *Mammuthus primigenius*, from 20,000-year-old hair shafts (Miller *et al.*, 2008), a 4000 year-old male Palaeo-Eskimo from permafrost-preserved hair (Rasmussen *et al.*, 2010) and from two 38,000-year-old fossilised Neanderthals (Green *et al.*, 2006; Noonan *et al.*, 2006).

This ability of NGS to deal with tiny sample sizes also holds great promise for its potential applications to contemporary as well as palaeo-ecology. For instance, it could be particularly useful for 'remote sensing' of the diets and gut flora of rare or protected species via faecal analysis in food web studies. Similarly suctorial feeders (e.g. aphid herbivores and many predatory beetles) have no visually recognizable gut contents and are therefore often omitted from traditional food web studies (e.g. Woodward *et al.*, 2010c), despite the fact that they can be very abundant and may perform key functional roles. Sufficient DNA could potentially be extracted from their guts and characterised using molecular techniques, rather than the less efficient and more time-consuming immunological responses (Carreon-Martinez and Heath, 2010). As previously noted, NGS can also be used to detect interactions between eukaryotes and potential pathogens within food webs, which have also been largely ignored in traditional ecology, for instance, via characterizing viromes in the faeces of vectors or intermediate hosts (e.g. Li *et al.*, 2010a).

V. TRANSCRIPTOMICS: FUNCTIONAL EXPRESSION

The transcriptome is the complete set of RNA transcripts in a cell, including mRNA, ncRNAs, and small RNAs. Such information allows the elucidation and determination of the transcriptional structure of genes, in particular enabling changes in expression levels and patterns of transcripts to be quantified in response to different cellular conditions/stimuli or changes in whole communities within environments (metatranscriptomics).

NGS allows the mapping and quantification of transcriptomes (RNA-Seq), via the isolation of total RNA (or enrichment of a particular RNA species) and enzymatically converting it to a library of complementary DNAs (cDNA) using reverse transcriptase. Universal adapters of known sequences are then ligated to the ends of the cDNAs, and NGS is then performed. Transcripts can be analysed from both 5′ and 3′ ends to determine both the start and the end of a particular transcript, and a mate-pair strategy (see Section IV.G) allows sequencing of both ends simultaneously. Transcriptome studies have already been carried out in a small number of eukaryote species (Cloonan *et al.*, 2008; Lister *et al.*, 2008; Morin *et al.*, 2008; Mortazavi *et al.*, 2008; Nagalakshmi *et al.*, 2008;

Wilhelm *et al.*, 2008). Using these methods it has been shown that up to 90% of the genome in yeast appears to be transcribed, with many transcripts being non-coding. Such results indicate the complexity of eukaryotic transcriptomes and that their analysis, even solely in terms of the data produced, is challenging even before the data is interpreted in a functional sense.

Transcriptome analyses in bacteria and archaea are further complicated by the fact that unlike eukaryotic mRNAs, most bacterial and archaeal mRNAs do not have a polyA-tail. Furthermore, most bacterial and archaeal RNA preparations contain up to 80% rRNA and tRNA. Thus far, transcriptome studies in bacteria have required the enrichment of mRNA, usually by the selective depletion of tRNA and/or rRNA (Liu *et al.*, 2009; Passalacqua *et al.*, 2009; Perkins *et al.*, 2009; Yoder-Himes *et al.*, 2009). In the study by Yoder-Himes *et al.* (2009) two strains of *Burkholderia cenocepacia* were compared, one from soil and another from a cystic fibrosis patient, at the transcriptome level under differing conditions and identified a subset of RNAs that may explain their different habitats and pathogenic potential.

Metatranscriptomics have been applied recently in a few studies that have attempted to characterise gene expression across multiple species, such as the recent study of microscopic eukaryotes (including protists and small metazoans) within soil communities (Bailly *et al.*, 2007). Recent reports suggest, however, that metatranscriptomes are still subject to very poor coverage and are not reliably reproducible (Gifford *et al.*, 2010; Stewart *et al.*, 2010). However, the development of single-molecule technologies could directly address these issues and make metatranscriptomics a more viable methodology in the near future.

VI. APPLICATION OF SEQUENCING IN ECOLOGY

A. Unveiling the Planet's Hidden Biodiversity

The advent of DNA sequencing has revolutionised molecular, microbial and general ecology, by enabling scientists to identify which and how many species are present in the system under study, and to consider both the microscopic and macroscopic world together. Most fundamentally the Tree of Life, now changed beyond recognition, and the discovery of lateral gene transfer as a factor in the evolution of all organisms has altered our understanding of both the ecological and evolutionary relationships between organisms. In the context of building a systems biology for ecology, probably the most dramatic recent change is the application of NGS to analyse communities, whether via metagenomics or by the analysis of amplified PCR products. We will highlight some metagenomics studies and critique the use of this, and other post-genomics approaches in ecology.

Most published metagenomics studies predate NGS, and were based on Sanger sequencing of cloned vectors, often plasmids. Such studies can be characterised as ‘discovery’ science, as there were generally no specific *a priori* hypotheses to be tested. Even now, many studies are primarily ‘fishing expeditions’, although a few have started to move into more sophisticated empirical end experimental cause-and-effect studies as the field has begun to mature as a scientific discipline (e.g. He *et al.*, 2010b). To illustrate the types of studies, and the data and insight they have tended to produce, four case studies are discussed briefly below.

B. Four Ecological Metagenomic Case Studies

There are many genomics and post-genomics studies in microbial ecology, but here we have selected four different case studies to highlight some general points. These are an analysis of an extreme, low diversity acid mine drainage (AMD) biofilm system by the Banfield group (Baker *et al.*, 2006; Lo *et al.*, 2007; Ram *et al.*, 2005; Tyson and Banfield, 2005; Tyson *et al.*, 2004); Venter *et al.*’s ambitious Global Ocean Sampling (GOS) expedition (Rusch *et al.*, 2007; Venter *et al.*, 2004); the application of Solexa sequencing to the human gut microbiome (Qin *et al.*, 2010); and the use of single species metagenomics by Rodriguez-Valera *et al.* (2009) to investigate the role of viral predation on biodiversity.

The Banfield group’s work on the environmentally extreme ($pH < 1$) and low diversity AMD biofilm system produced two almost complete genomes, from a Group II *Leptospirillum* and a type II *Ferroplasma*. Analysis of these genomes showed that, while both organisms had the potential? or did they measure rates? aerobically oxidised ferrous iron to ferric iron (Fe^{2+} to Fe^{3+}), the bacterial *Leptospirillum* fixed carbon which the archaeal *Ferroplasma* utilised. Interestingly, the *Leptospirillum* genome showed very little nucleotide polymorphism, suggesting either strong selection for this genotype or a very strong founder effect following the colonisation of the mine. In contrast the *Ferroplasma* genome appeared to be a mosaic of three ancestral genomes, with as many as 400 individual recombination events creating the contemporary genome. This suggests significant localised diversity in the *Ferroplasma* population, which could increase species resilience to perturbation. Several other partial *Leptospirillum* and *Ferroplasma* genomes were also constructed showing distinct divergence from the two complete genomes. In the almost complete genomes about 20% of the genes detected could not be assigned a cellular function (Tyson *et al.*, 2004). Within this metagenome of 76.2 Mbp just a single complete nitrogen-fixation operon was found and further work led to the isolation of the bacterium *Leptospirillum ferrodiazotrophum*, the putative sole nitrogen-fixing organism in the system (Tyson *et al.*, 2005).

Subsequent metaproteomics analysis showed high levels of expression of genes associated with oxidative stress and iron oxidation, matching the harsh conditions and predominant oxidative process in the biofilm (Ram *et al.*, 2005). This study still represents one of the most complete and successful of all genomics and post-genomics analyses of a defined ecosystem, and points the way for a host of future studies into real-world ecotoxicology and B–EF relationships in response to environmental stressors.

The GOS *Sorceror II* expedition was an extremely large and ambitious shotgun sequencing project, led by Craig Venter. Results from the survey within the Sargasso Sea (Venter *et al.*, 2004) were surprising in that, despite collecting over 1 Gbp of data, the study could not reconstruct even a single genome from these oligotrophic waters. An important finding in the study was the linking of a gene for aerobic ammonia oxidation to an archaeal scaffold which, along with a similar finding in soil (Treusch *et al.*, 2005) and the isolation of marine crenarchaeaon capable of aerobic ammonia oxidation (Könneke *et al.*, 2005), contributed to a shift in our understanding of this part of the global nitrogen cycle (Erguder *et al.*, 2009). The authors concluded that their samples contained at least 1800 distinct genomic species and identified a surprising diversity of genes encoding for bacteriorhodopsin (now called proteorhodopsin), a light-harvesting protein once thought to be limited to halophilic archaea but that had been previously identified in bacterioplankton (Beja *et al.*, 2000). The detection of such a natural diversity of proteorhodopsin genes indicated this gene was widespread in bacterioplankton, suggesting that it plays an important role in carbon cycling in the oceans. Of the 1.2 million genes identified in the study, ~65% could not be assigned a cellular function nor aligned with any known organism. This study highlighted a major issue with metagenomics: that the diversity within even what was perceived as a relatively simple system simply overwhelmed the enormous sequencing effort applied to it.

NGS have been used to characterise an 'internal ecosystem' – the human gut microbiome in over 100 people. Qin *et al.* (2010) exploited the massively paralleled sequencing capacity of the Solexa technology to analyse this ecosystem: over 500 Gbp of data were collected, encoding 3.3 million genes from 1000 to 1150 microbial species. These data showed that each host had a bacterial population of ~160 species. Analysis of a subset of the data showed that there were clear differences in the gene complements of healthy and diseased subjects, suggesting a role for gut microbiota in chronic conditions such as Crohn's disease and ulcerative colitis. The data has expanded the depth of knowledge of the potential microbial metabolism within the human gut by highlighting the potential for gut microbes to use carbohydrates, such as pectin and sorbitol, which are poorly absorbed by humans. Presumably these sugars are used by the microbes as an energy source that the human host does not require. However, even though nearly 200 gut microbes have been genome-sequenced, only 12% of the sequenced genes in this study could be associated with existing human gut microbial genome sequences and >80% could not be assigned a cellular function at all.

Rodriguez-Valera *et al.* (2009) provide an excellent example for how metagenomics can be used to understand microbial diversity driven by phage predation. Using single-species metagenomics these authors identified that natural populations of bacteria and archaea are relatively rich in metagenomic islands (MGIs), regions of the bacterial genome that are under-represented in the population, but that include an over-representation of genes related to phage resistance. These genes are therefore among the most variable in the population. They then used computer simulations to propose that Constant Diversity (CD) dynamics, in which density-dependent phage predation prevents competitively superior strains from dominating the community, could be behind the maintenance of relatively high strain richness, with each strain being maintained with a sub-optimal and nearly equal fitness. They further proposed that the maintenance of such highly diverse populations can increase ecosystem efficiency (i.e. the decomposition of organic material), by each strain specializing in exploiting a different micro-resource instead of an over-abundant one taking most of them. These findings have direct parallels with classic ecology: first, they show how (biotic) heterogeneity can explain the maintenance of high diversity (the 'paradox of the plankton' – Hutchinson, 1961); and second, they show how predation may contribute to maintain strains within populations by suppressing inter-strain competition, in a manner similar to that shown in classic experiments with metazoans (Paine, 1966).

Over the 6 years since publication of the Tyson *et al.* (2004) paper the amount of data that can be collected via NGS has increased by nearly four orders of magnitude, yet proportionately fewer of the genes sequenced can be unambiguously assigned to a cellular function. The initial issues in metagenomics of poor depth of sampling has thus given way to a less conclusive analysis of more data, as well as concerns about systematic errors in metagenome libraries (Gomez-Alvarez *et al.*, 2009). Clearly, there are still major issues to be resolved in the use of NGS technologies, some of which may be remedied by more sophisticated bioinformatics approaches when analysing these vast datasets. It should be noted that the majority of metagenomic studies still tend to conclude with rather generic statements that the majority of the genes detected are associated with core metabolic processes, carbohydrate utilisation, protein metabolism and lipid biosynthesis. It is perhaps difficult to be excited by the unsurprising finding that cells use carbon and make/degrade proteins and lipids, and these rather bland conclusions highlight the embryonic state of the field at present; the potential for truly novel insights has barely been tapped and the field is still wide open for exploration and discovery.

C. Moving Beyond Traditional Metagenomics

There have been some important and potentially powerful developments in the field of metagenomics field, including its use in comparative studies (Gianoulis *et al.*, 2009; Tringe *et al.*, 2005) to determine real differences among systems from the complex mass of data. These studies have also highlighted the clear need for proper replication in order to draw statistically significant conclusions: in many instances statistical rigour has been largely ignored, undermining the validity of interpretations based on the data collected (Gilbert *et al.*, 2008, 2009). Other developments include using metagenomics to build genomes of important uncultured organisms, either from low diversity environments or from enrichment cultures (Chivian *et al.*, 2008; Erkel *et al.*, 2006; Strous *et al.*, 2006) and as an analysis tool for hypothesis-driven experimental studies, including those using stable isotope probing (Chen and Murrell, 2010 and references therein) and some simple, but elegant, microcosm or even field experiments (He *et al.*, 2010b; Mou *et al.*, 2008).

The problems and limitations that constrain metagenomics also affect metatranscriptomics, the analysis of all the transcripts from a system. At present, transcriptomes often suffer from poor rates of assignment for detected putative genes (Gilbert *et al.*, 2008), poor coverage (Gifford *et al.*, 2010) and technical replication such that it has been reported that only 17% of transcripts were found in common in duplicate runs of the same templates (Stewart *et al.*, 2010). These issues are exacerbated by the problems associated with studying microbial mRNA, including rapid degradation and poor separation from ncRNA species (Feder and Walser, 2005; van Vliet, 2010). There is clearly much that still needs to be done before metatranscriptomics becomes a readily available ecological tool for molecular analysis of communities or ecosystems. Even so, successful studies that have focused on clearly defined subsets of transcripts have been reported, with a large number of presumably regulatory small non-coding RNA (sRNA) molecules detected in ocean waters (Shi *et al.*, 2009), and a small scale metatranscriptomics analysis was used to characterise eukaryotic functional diversity within soil ecosystems (Bailly *et al.*, 2007). Given the rapid advances made using NGS within the space of just a few years, it seems likely that metatranscriptomics also has the potential to become an integral part of the next generation of molecular ecology techniques.

D. Next-Generation Sequencing and Biodiversity

NGS technologies have also been used in non-genomics studies, as the very large numbers of sequences reads make it possible to sample microbial communities completely. Using 454 sequencing ($\sim$400 bp reads) PCR-amplified functional gene fragments (*dsrB*) from a specific bacterial genus

(*Desulfobulbus*) was sampled completely sampled from 10 sites along an estuary in the United Kingdom (Figure 9A, unpublished data). As few as 1000 sequence reads were required to effectively analyse the complete diversity of this genus, with nearly 100 operational taxonomic units (OTUs, which are analogous to 'morphospecies') detected along the estuary. Furthermore, the data from the 454 sequences matched the differential distributions of four specific *Desulfobulbus dsrB* genotypes detected independently by molecular fingerprinting analysis and genotype-targeted qPCR (Figure 10, unpublished data). Therefore, the perennial problem in microbial ecology of

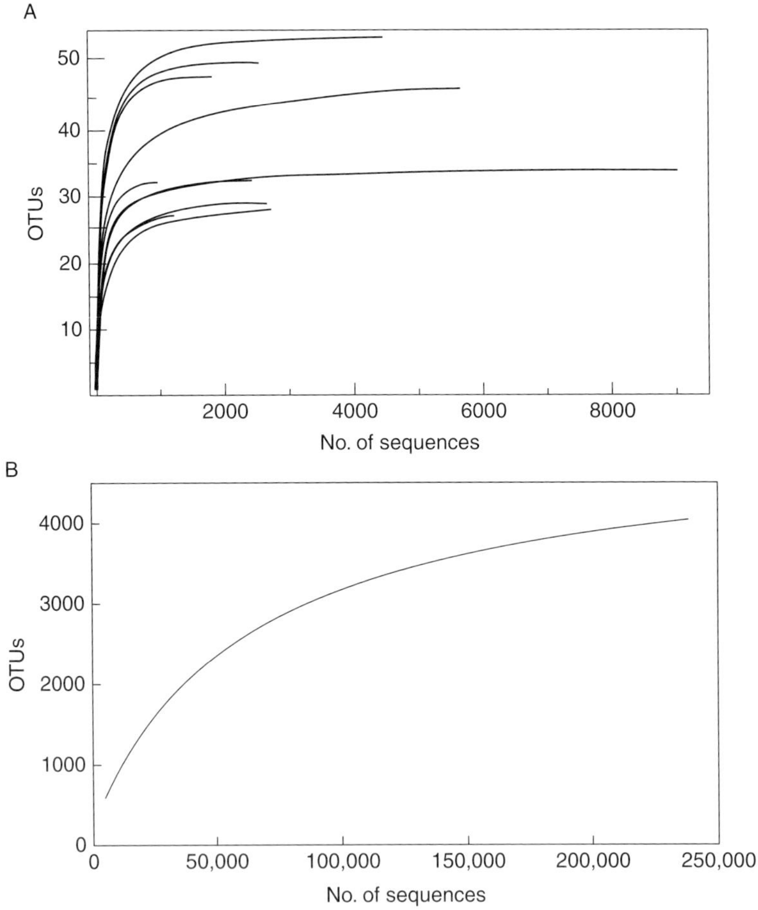

Figure 9 Rarefaction analysis of (A) 454 data (~400 bp) of amplified *Desulfobulbus dsrB* gene fragments from 10 sites along the Colne estuary, UK. All samples approach an asymptote; Coverage is >98% (Good, 1953). (B) Protist diversity in a lake ecosystem, as revealed from 454 sequencing, redrawn after Nolte *et al.* (2010).

undersampling may be solved using NGS (Figure 9). However, these methods are not necessarily the magic bullet that will solve all diversity problems in microbial ecology. The first major 454 diversity study, by Sogin *et al.* (2006), was unable to completely sample their targeted community, leading to the proposal of the idea of the rare biosphere, in which there exists a very long tail of low abundance microbial species. While this was initially seen as a major breakthrough in understanding microbial biodiversity, the validity of Sogin *et al.*'s analysis has been vigorously challenged (Kunin *et al.*, 2009; Quince *et al.*, 2009). Criticisms include the use of very short reads, which make accurate phylogenetic assignments very difficult, poor quality control, and a lack of real support for the general idea of the rare biosphere.

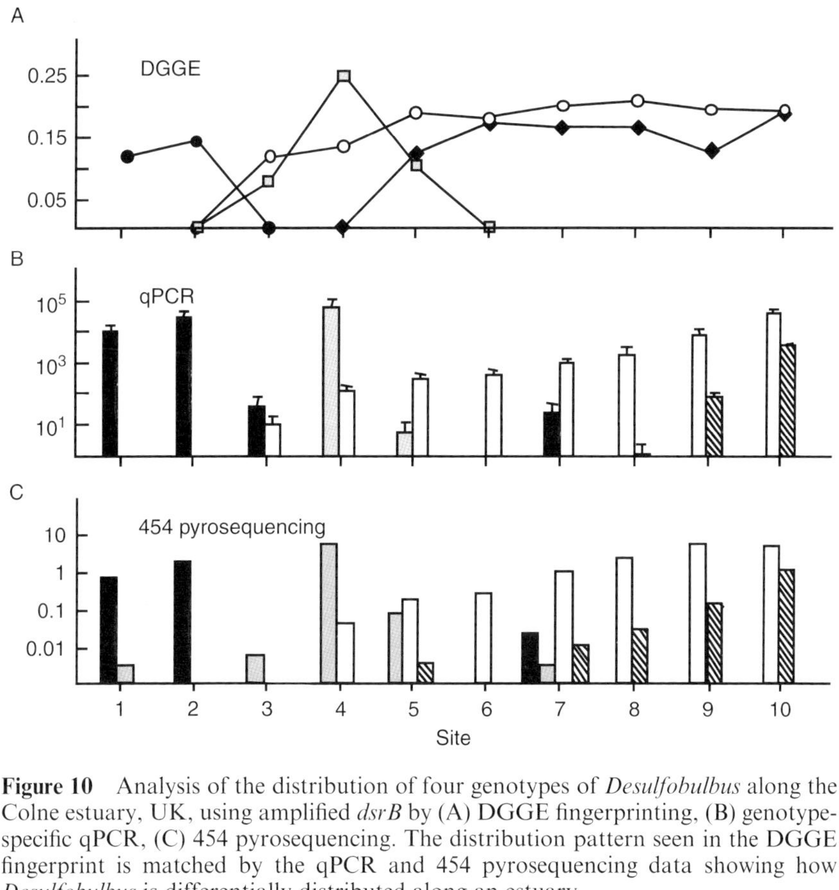

Figure 10 Analysis of the distribution of four genotypes of *Desulfobulbus* along the Colne estuary, UK, using amplified *dsrB* by (A) DGGE fingerprinting, (B) genotype-specific qPCR, (C) 454 pyrosequencing. The distribution pattern seen in the DGGE fingerprint is matched by the qPCR and 454 pyrosequencing data showing how *Desulfobulbus* is differentially distributed along an estuary.

VII. LINKING ACROSS MULTIPLE LEVELS OF ORGANISATION: THE KEY TO UNDERSTANDING THE SYSTEM

A. Scaling from Molecules to Ecosystems

A systems approach to ecology ultimately requires the ability to scale from molecules to ecosystems, but making these connections is an extremely challenging task (DeLong, 2009). Figure 2 shows a simple diagram defining the flow of information from a gene sequence to an ecosystem process. While this simplistic figure is useful it fails to highlight the full range of feedback interactions that can and do occur in the environment. All aspects of organismal functioning are related to the external environmental conditions and so, by extension, ecosystem functioning will also be dependent on those factors (Yvon-Durocher *et al.*, 2010c). For example, the extent to which sulphate reduction or methanogenesis dominates terminal oxidation of organic matter is dependent on many factors, including sulphate and organic carbon availability, temperature, and the presence of alternative respiratory compounds, such as nitrate or oxygen (Purdy *et al.*, 2003b; Raskin *et al.*, 1996). A systems approach to ecology will require the incorporation of the effects of such environmental feedback.

The importance of environmental feedback can be seen by considering at what level it might affect organisms. Microbes are very directly affected by their immediate surrounding and control their metabolism as a consequence of this. This can be represented as regulation during gene expression (Figure 2), by mRNA half-life and the effects of ncRNA, translational control and of course protein functional regulation, all of which are basic biological concepts (Nelson and Cox, 2009). These regulation controls will be dependent on the environment a cell finds itself in at that particular time and are further complicated by inter-cellular functions. Consequently there may be substantial disconnects between the genes present in a system and the activity of their products (Feder and Walser, 2005), and we therefore need more robust empirical data to gauge how strong (or weak) these feedbacks might be. Furthermore, ecologists commonly assume a substantial phenotypic coherence within defined taxonomic groups, but this is not a valid assumption for many microbes, which are much less clearly defined as species (Cohan, 2001): recent studies on microbial genomes suggest that bacteria and archaea have a flexible genome structure that includes a core genome, comprising the essential and defining genes of a particular clade, and also an accessory genome, which can vary substantially between individual strains. This has led to the definition of the pangenome, the total potential genome of a microbial species group (Medini *et al.*, 2005; Tettelin *et al.*, 2008). Therefore, it would appear that no matter how well defined the species concept becomes for microbes, it will never define the exact phenotype of any particular strain within that species. Despite these potential caveats and limitations, which add complexity, but also reality, to a system, the study of microbial members of communities in terms of ecosystem functioning is becoming increasingly feasible in natural or experimental systems (Figure 11).

B. Linking Microbial Community Structure and Ecosystem Functioning

There are many reports that make coherent and well-supported links between microbial community structure and ecosystem functioning, including experimental manipulations of microbial communities that have linked specific organisms to the degradation of specific substrates (He *et al.*, 2010b; Purdy *et al.*, 1997, 2002, 2003a; Figure 11). Although these studies have made some very definitive links between organisms and substrates, they have also highlighted the complexity and variation of microbial responses to environmental stimuli. For example, in fatty acid amended sediment slurries from an

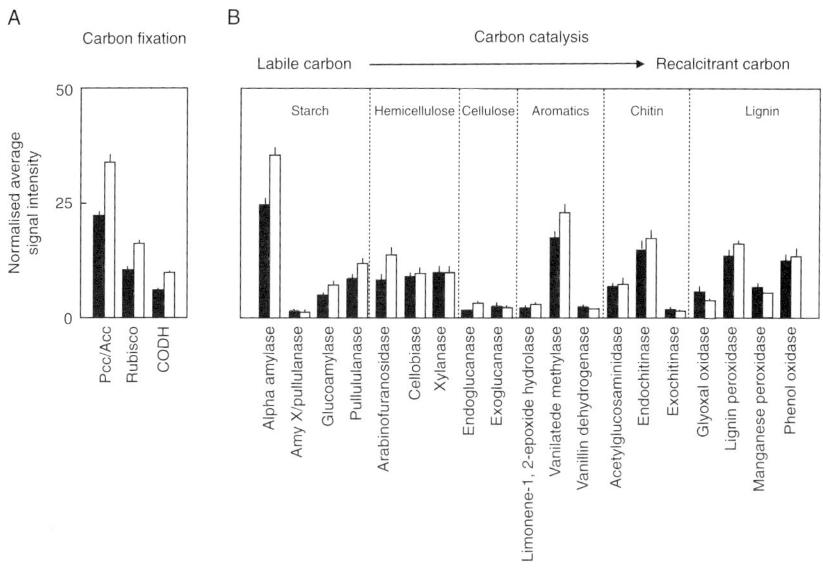

Figure 11 Linking microbial diversity to ecosystem functioning in a long-term field experiment. The metagenomic expression of metabolic processes associated with the carbon cycle differed markedly between soils that had been exposed to a decade of elevated CO_2 concentrations relative to ambient controls. Redrawn after He *et al.* (2010b).

estuary the addition of acetate, or its *in situ* production from lactate, significantly enriched the sulphate-reducing bacteria *Desulfobacter*, whereas the production of acetate from butyrate did not (Purdy *et al.*, 1997). This suggests that the immediate biological history of a system can affect the way in which it responds to a perturbation. The development of the use of stable isotope analysis has further advanced the range of methods that directly link organisms to activities (Boschker *et al.*, 1998; Chen and Murrell, 2010; Radajewski *et al.*, 2000). It is also important to recognise that the activity of many microbes is intimately linked to the activity of other, often taxonomically unrelated, organisms (Dolfing, 2001; Jackson and McInerney, 2002) making the assignment of a role to a specific microbial group often dependent upon at least some knowledge of the other organisms in the system. Madsen (2005) has detailed the very real issues associated with defining an ecological role to specific microbes.

The task of linking the different organisational levels in an ecosystem that incorporates microbial activity is clearly challenging (Figure 12). This is where systems approaches can be very powerful, as a discovery-based, wide-ranging approach to investigation is likely to throw up new and unexpected drivers and processes, such as light-harvesting by heterotrophs (via proteorhodopsin, Beja *et al.*, 2000). Such structural-functional interactions

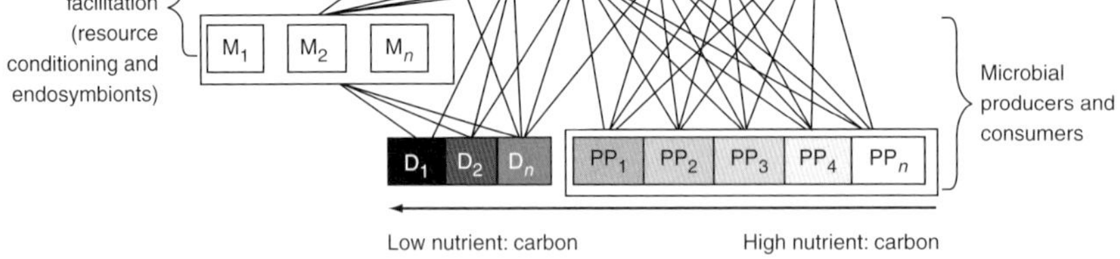

Key:
C, carnivorous predator; O, omnivorous predator; PC, primary consumer; M, microbial decomposer; D, detritus (e.g. leaf-litter); PP, microbial/protist primary producers (e.g. diatoms, cyanobacteria). Redrawn after Woodward (2009).

Figure 12 The microbial 'black box' within a schematic community food web. The vast majority of community and ecosystem studies have ignored the diversity and functional roles of the taxa at the base of the web, even though these are the ultimate drivers of bottom-up processes and the supply of energy and nutrients to the higher trophic levels.

will then need to be proven to have a real biological effect on a system (Fuhrman and Steele, 2007), which requires carefully designed and rigorous controlled experiments and manipulations. On the ecosystem side, manipulating environmental conditions in reasonably sized and replicated meso- or microcosms (e.g., Yvon-Durocher *et al.*, 2010a,c) will allow important questions related to the links between structure and functioning at the higher levels of organisation to be addressed (Ptacnik *et al.*, 2010; Reiss *et al.*, 2010a). Large-scale mesocosm experiments, which are needed to identify potential mechanistic relationships in the field, will facilitate the application of molecular microbial ecological studies within a wider framework and so should facilitate the development of links across different scales of organisation within an ecosystem.

C. New Questions in Ecology: Opening the Microbial Black Box

The aim of a systems approach is to define an ecosystem in terms of its components to predict how it will respond to change. If it is possible to open the microbial black box, and to understand what we find inside, then comparative analyses of different ecosystems will improve, as will our ability to predict responses to future change.

The strength of integrating NGS technologies within general ecology includes the ability to analyse a community without the need to amplify the DNA, to achieve complete coverage of a community and to handle multiple samples with limited sample manipulation. A 'complete' metagenome would, at least in theory, allow the mapping of all possible metabolisms, even those that are presently unknown (but which could be revisited repeatedly as bioinformatic databases mature), that were encoded within a system, and hence the full range of processes an ecosystem could potentially sustain. Such data could raise a vast new range of questions that could be asked of ecosystems, and importantly to test whether many of the existing theories of community ecology, which have developed primarily from macro-organismal studies, can also be applied to the microscopic world (Ptacnik *et al.*, 2010; Reiss *et al.*, 2010a,b). This leads to the challenge of marrying traditional ecology with microbial ecology.

There is an ever-increasing body of data that supports the idea that many aspects of the ecology of large organisms are not necessarily fundamentally different to that of microbes (Brown *et al.*, 2004; Ptacnik *et al.*, 2010). Nonetheless, there are in fact many obvious differences (e.g. reproduction via binary fission, reliance on nutrient uptake via diffusion, etc.) that could be critically important, depending on the question being addressed. The sheer diversity of microbes and the difficulty of defining an all-encompassing species concept pose real challenges (Reiss *et al.*, 2010b). Delong *et al.* (2010), for example, have recently reported that allometric mass-metabolism scaling coefficients in bacteria and archaea differ from that in microscopic and large eukaryotes, suggesting that not all factors are independent of body size, although differences in metabolic scaling are far less than differences in body mass *per se* (Makarieva *et al.*, 2008). Notwithstanding some of these more obvious differences, there are also many areas of solid ground that link microbial and general ecology: for example, there is now strong evidence that microbes exhibit biogeographic patterns of distribution (Cho and Tiedje, 2000; Oakley *et al.*, 2010; Whitaker *et al.*, 2003) and that, similar to larger organisms, the number of species increases in a power-law relationship with increases in area or volume (Green and Bohannan, 2006; van der Gast *et al.*, 2005). Latitudinal diversity gradients (with higher species richness occurring at lower latitudes; Fuhrman *et al.*, 2008) and spatial heterogeneity explaining the structure of local microbial communities (Östman *et al.*, 2010) have also been reported recently.

NGS offers an excellent opportunity to test hypotheses for the processes involved in determining these diversity patterns, such as the suggestion that latitudinal diversity gradients are formed in part due to the enhancement that higher temperatures, and in terrestrial environments water availability, have on encounter rates among organisms and thus on the rate of biotic interactions (Moya-Laraño, 2010). One prediction of this hypothesis is that in aquatic ecosystems, and across individuals within a single species, functional diversity should increase with temperature (up to a certain threshold). NGS, metagenomics, metaproteomics and metabolomics can be used to test the hypothesis that once controlled for community size (cell number), species richness and phylogenetic distance, communities at higher temperatures (up to 30 °C) hold a higher number of functional units. In aquatic bacterial and archaeal cells, it has been hypothesised that temperature enhances cell-to-cell encounter rates and the diversity of interactions is even more likely because cell and metabolite mobility is higher in environments of low viscosity (Kummerli *et al.*, 2009) and viscosity decreases with increasing temperature (Woodward *et al.*, 2010a).

The advent of NGS may resolve the issue of undersampling of microbes but differences in sampling and analysis between the two fields will also need to be bridged. Perhaps the most relevant area where new questions can be raised is in the contribution microbial studies could make to ecology. Ecology has a strong history of theoretical developments, but rarely directly testing the validity of these theories (Raffaelli and Moller, 2000) due to the difficulties of performing controlled and replicated experiments at suitable spatiotemporal scales in field ecology (Woodward *et al.*, 2010a,b). Microbial communities, however, because of their large population sizes and rapid generation times, can be manipulated and experimented upon under clearly defined and controlled conditions (Petchey *et al.*, 1999). Hence, a microbial ecologist should be able to test the wide variety of mechanisms that have been put forward to explain macroecological patterns among the larger taxa (McGeogh and Gaston, 2002), in addition to testing microbial ecology theories in the real world. For instance, the apparent stability in terms of the species composition and metabolic potential of viral and microbial assemblages observed in natural ecosystems appears to contradict theoretical models and viral–microbial dynamics observed in simple systems, where cycling of the dominant taxa is more typical (Rodriguez-Brito *et al.*, 2010).

D. Functional Redundancy: Do Species Matter?

A common assumption in general ecology has been that microbes can be treated as a black box because of massive functional redundancy (Vandenkoornhuyse *et al.*, 2010), yet there is little compelling evidence to support this view. In fact, data on reconstructed microbial communities suggests that more diverse microbial communities are also more productive (Bell *et al.*, 2005), in line with the positive B–EF relationships reported for many macro-organismal studies (Cardinale *et al.*, 2006). In contrast, rapid changes at the fine-grained level of viral genotypes and microbial strains overlain on the apparent community and ecosystem stability reported by Rodriguez-Brito *et al.* (2010) suggest that functionally redundant taxa might cycle at the level of viral genotypes and virus-sensitive microbial strains. In a different study, the cyanobacterium, *Prochlorococcus*, which is the most abundant oxygen fixing phototroph in the world's oceans, with an estimated global population of 10^{27} cells, underpins the pelagic food webs of the nutrient-poor tropical and subtropical gyres – among the largest ecosystems on Earth. Recent metagenomics studies have revealed that uncultured *Prochlorococcus* possess the genetic potential for nitrate assimilation, and metatranscriptomics data from the field have confirmed that these genes are expressed (Martiny *et al.*, 2009); yet earlier work on isolates found no evidence for this metabolic capacity (Rocap *et al.*, 2003). That this functional trait might not be ubiquitous even within a single species has implications for understanding not only the biogeography of this key primary producer and its role within the oceanic carbon and nitrogen cycles, but also the level of organization at which functional redundancy or uniqueness is expressed.

Whether microbial communities are functionally redundant is clearly important for understanding the role of microbes in ecosystem functioning, and this represents one of the great unanswered questions in ecology. A useful starting point would be to determine whether 'poorly-performing' or seemingly unstable ecosystems contain reduced diversity in specific critical ecosystem processes. If ecosystem process rates can be manipulated by increasing or reducing functional capacity, which could be validated via NGS techniques, in controlled small-scale experiments, then this would suggest that redundancy is not as prevalent as is often imagined. This is a non-trivial issue and its resolution is critical to understanding the role of microbial biodiversity in an ecosystem context.

E. Beyond Bacteria: Opening the Eukaryotic Black Box

Although we have focused primarily on bacterial and archaeal examples, as this is where most of the recent advances have been made in this field, metagenomics and NGS techniques also offer promise in the future for studying eukaryotic micro- and macro-organisms, such as host–pathogen and host–symbiont ecology and also in the study of eukaryotic assemblages *per se*. Woyke *et al.* (2006) for instance revealed that bacteria can provide their eukaryote host with multiple sources of nutrition, and a recent study by Coolon *et al.* (2010) has characterised metagenomic shifts in the microbial assemblages in the environment and also in vertebrate guts across a pollution gradient. Widespread distributions of associations between bacteria and sponges have also been revealed in a range of marine ecosystems on a biogeographical scale (Lafi *et al.*, 2009). Similarly, in terrestrial systems, Jumpponen *et al.* (2010) have discovered strong seasonality in the diversity and dynamics of oak–ectomycorrhizae associations in urban versus rural environments.

Piganeau *et al.* (2008) have found unexpected picoeukaryotic sequences within the Sargasso Sea metagenome, and other recent studies have found evidence of surprisingly diverse eukaryotic assemblage in seemingly very harsh environments, including anoxic marine sediments (Stoeck *et al.*, 2010; Table 2). Clearly, as more sequences and datasets are assigned to functions in eukaryotic cells, both from uni- and multi-cellular organisms, we will be better able to use NGS to understand the functions present in an entire system as well as to assign each function to a particular species (or group of species) within communities, thus contributing to solving the structure of ecological networks accurately and to understanding the reasons behind their apparent robustness and stability.

F. Food Webs: Identifying the Missing Links

Microbes play many key roles in food webs, particularly at the lower trophic levels, where they are important conduits for the flux of energy and materials to the consumer assemblage. They can also be significant pathogens, but this has been almost completely ignored in food web research to date (Ings *et al.*, 2009). Much of food web ecology has suffered from patchy taxonomic resolution, which tends to be high at the top of the web but poor at the bottom, where the

Table 2 Molecules to ecosystems: selected examples of recent applications or reviews of genomics, metagenomics and NGS as tools for testing ecological theory

Ecological subdiscipline	Approach	Key findings
Organismal functional ecology		
Renaut *et al.* (2010)	454	Transcriptomics: identification of potential adaptive single-nucleotide polymorphisms in fishes
Wolf *et al.* (2010)	454	Transcriptomics: nucleotide divergence and gene expression differentiation in the carrion crow and hooded crow hybrid zone
Population structure and dynamics		
Van Bers *et al.* (2010)	Solexa	Baseline generation of >2bn basepairs of novel sequence information for an ecological model species, the great tit *Parus major*
Goetz *et al.* (2010)	454	Identifies genetic basis for phenotypic differences within trout populations
Microbiomes (internal ecosystems)		
Woyke *et al.* (2006)	Sanger	Symbiosis: bacteria provide eukaryote host with multiple sources of nutrition
Kloch *et al.* (2010)	454	Characterisation of gut parasite loads in bank voles
Community structure and dynamics		
Jumpponen *et al.* (2010)	454	Seasonality of diversity and composition in oak ectomycorrhiza across ecosystems
Schütte *et al.* (2010)	454	Shifts in bacterial diversity over glacial chronosequences
Stoeck *et al.* (2010)	454	Discovery of complex eukaryotic community in anoxic marine ecosystem
Piganeau *et al.* (2008)	Sanger	Picoeukaryotic sequences detected in the Sargasso Sea metagenome
Gilbert *et al.* (2009)	454	Seasonal dynamics of marine microbial communities
Rodriguez-Brito *et al.* (2010)	454	Viral and microbial community dynamics in four aquatic environments
Biogeography		
Dinsdale *et al.* (2008)	454; SEED	Potential for viromes to influence global evolutionary and metabolic processes
Breitbart and Rohwer (2005)	n/a (review)	Review: potential for viruses to form 'global' diaspora
Cuvelier *et al.* (2008)	Sanger; FISH	Discovery of widespread distribution of a unique marine protistan lineage
Biddle *et al.* (2008)	454	Peru Margin subseafloor biosphere revealed as a genetically distinct environment

(*continued*)

Table 2 *(continued)*

Ecological subdiscipline	Approach	Key findings
Lafi *et al.* (2009)	Sanger	Widespread distribution of associations between bacteria and sponges
Biodiversity & ecosystem functioning		
Pernthaler *et al.* (2008)	Magneto-FISH	Nitrogenase genes within metagenome and incorporation in biomass of methane-oxidizing consortia suggest role in new nitrogen inputs by syntrophic assemblages
Béjà *et al.* (2000)	Sanger	New form of phototrophy discovered in marine ecosystems
Food webs and trophic interactions		
Jones and Lennon (2009)	qPCR	Potential importance of indirect trophic interactions in planktonic food webs
Kant and Baldwin (2007)	n/a (review)	Review: Integration of molecular biology and ecology to characterise herbivory
Phillips *et al.* (2003)	n/a (review)	Review: Model of multitrophic molecular signals in the rhizosphere: effects of molecular 'control points' could propagate through other trophic levels
Proulx *et al.* (2005)	n/a (review)	Review: Identifies need to develop conceptual framework for integrating complex ecological, genetic and metabolic networks
Environmental stressors and biomonitoring		
Debroas *et al.* (2009)	Sanger	Overrepresentation of genes involved in xenobiotic degradation in a polluted lake
Dinsdale *et al.* (2008)	454, SEED	Increase in heterotrophic microbes and potential pathogens in human-impacted sites with widespread coral loss
Coolon *et al.* (2010)	454	High rates of microbial turnover in soils and consumer guts across a gradient of heavy metal pollution
He *et al.* (2010b)	454; GeoChip microarray	Long-term exposure of grassland to elevated CO_2 altered structure and functional potential of soil microbial communities
Ecosystem goods and services		
D'Costa *et al.* (2007) Kennedy *et al.* (2008, 2010) Hofmann and Gaines (2008) Marco (2010)	n/a (reviews)	Reviews: Potential for bioprospecting for antibiotics, pharmaceuticals and other bioproducts
Sharma *et al.* (2010)	Database	Bioinformatics database for bioprospecting for commercially useful enzymes

black box approach is still very much in evidence (Figure 12). Kant and Baldwin (2007) have recently highlighted the increasing convergence between ecology and molecular biology for studying plant–herbivore interactions, which are often extremely difficult to characterise using standard ecological techniques. Many consumers, such as suctorial herbivores or predators, have unrecognisable gut contents, and many resources have insufficient hard tissue to allow reliable identification based on morphological structures, and in such instances dissection of gut contents is largely pointless (Woodward *et al.*, 2010c). In such situations, stable isotope analysis can help to detect assimilated tissue, but it lacks sufficiently high-level taxonomic resolution to identify trophic interactions between species pairs, which may be addressed by immunological or, increasingly, molecular techniques (Figure 13). The use of metagenomics has been particularly useful for resolving otherwise difficult to characterise feeding links and species responses to exploitation, but proteomics and metabolomics

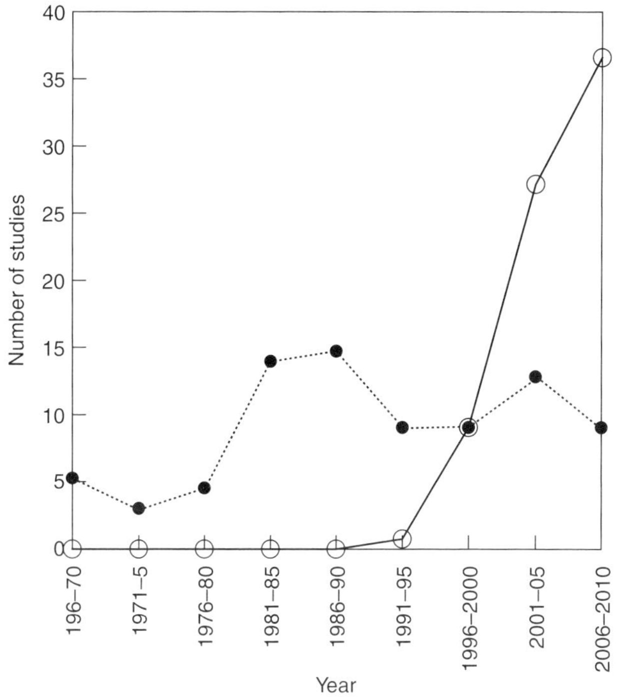

Figure 13 Potential for molecular applications in food web ecology. The dashed line shows the slow increase in studies using immunological techniques to characterise diets, the solid line shows the rapid rise of those based on molecular techniques. Redrawn after Carreon-Martinez and Heath (2010).

have also been used to describe the dramatic metabolic reconfigurations that occur in response to herbivory (Kant and Baldwin, 2007).

Foraging theory (FT), which has been recently used successfully to explain the structure of food webs (Petchey *et al.*, 2008; Woodward *et al.*, 2010c), explains animal foraging decisions, among other things, based on the profitability of food items. Since the microbial community within animal guts may largely determine the digestibility and profitability of the food ingested (Hooper *et al.*, 2002; Stevens and Hume, 1998), microbial-mediated foraging efficiency can contribute to determine predator–prey interactions and food web structure. In other words, the optimality of prey items may change depending on the microbial gut community. In well-resolved food webs, NGS could be used to test the hypothesis that metazoan food web structure is associated with the functional diversity of the microbial gut community. Microbial gut community NGS analysis could then be used to uncover missing links in other food webs.

G. NGS and the Link Between Ecology and Evolution

The link between ecology and evolution is experiencing a renaissance, particularly when it comes to understanding communities and ecosystems (Ellers, 2010; Johnson and Stinchcombe, 2007). Both at micro- and macro-scales, the ecological context may largely drive evolutionary outputs and, conversely, adaptive evolution and phylogenetic structure can determine ecological processes. The need for studying such reciprocal links is also obvious in microbial research, with authors requiring experimental approaches to determine the selective pressures to which microbes are exposed (Fraser *et al.*, 2007) or arguing that to understand adaptive microbe evolution we need to include the ecological context in which evolution is occurring (Kassen and Rainey, 2004). Importantly, adding the spatial context into this ongoing link, and explicitly adding the study of dispersal among communities (i.e. the metacommunity context) may greatly improve our understanding of how species and populations interact in real scenarios and the role of eco-evolutionary interplay in the stability of communities in the long term (Urban *et al.*, 2008).

In addition, the study of ecological networks, that is, networks in which many species interact with each other (e.g. food webs, parasite–host, competitive or pollination networks), is also contributing to the link of ecology and evolution, as both factors may affect network structure, robustness and stability (Bascompte, 2010; Bascompte and Jordano, 2007). The amount of genetic and phenotypic variation (and diversity) of the multifarious phenotypes that are present in populations have been proposed to affect food web structure in a manner that enhance its robustness and stability (Moya-Laraño, in press), for instance by increasing food web connectance (i.e. the proportion of potential links that are realised). However, finding that higher phenotypic diversity correlates with higher connectance in networks would not be conclusive evidence in favour of this hypothesis, because networks may be maintained from phenotypic feedback, in which phenotypic diversity may affect connectance, which may, in turn, facilitate the maintenance of phenotypic diversity, and so on (J. Moya-Laraño, A. Rossberg and J.M. Montoya, manuscript in preparation). Furthermore, these networks of interactions may have complex eco-evolutionary dynamics in which genotypic variation can affect both evolutionary, as well as ecological processes, and these two processes may dynamically affect each other over ecological time scales (Fussmann *et al.*, 2007; Jones *et al.*, 2009; Pelletier *et al.*, 2009; Yoshida *et al.*, 2003). Lastly, it has been suggested that heritability at the level of communities and ecosystems may be substantial if we explicitly consider 'Interspecific Indirect Genetic Effects', by which the genetics of individuals of a species in a community may affect the expression of traits in individuals of a different species, thus allowing the community and its associated ecosystem functional attributes to evolve by natural selection (Shuster *et al.*, 2006).

The above hypotheses could potentially be tested by using NGS across space and time in both natural and experimental communities and by following the fate of sets of species, genes and/or molecules. In addition, by comparing local communities through time, one can potentially track among-community migration rate and include the metacommunity context to understand the ecology and evolution of complex interaction networks. NGS is thus a promising tool that may greatly accelerate our understanding for how ecology and evolution are interconnected via feedback loops.

VIII. SOCIOECONOMIC APPLICATIONS: ECOSYSTEM GOODS AND SERVICES AND BIOPROSPECTING

Ecosystem processes underpin socio-economic 'goods and services' of value to humans, for instance, the supply of clean drinking water, maintenance of a breathable atmosphere, and the provision of medicines. Many of these are supplied by microbes, yet as we have already discussed, their biodiversity and resultant functional roles are still largely unknown: hence their contributions to both relatively directly measurable goods and services (e.g. isolation of commercially exploitable antibiotics) and those that are more difficult to quantify (e.g. maintenance of the global carbon cycle) are also still poorly understood.

The huge untapped potential of biodiversity in natural systems offers great prospects for commercial applications in the future. For instance, bioprospecting for particular enzymes or genes that could offer new medical treatments is a rapidly growing field that is starting to trawl the vast bioinformatics databases that have sprung up in the wake of NGS (D'Costa *et al.*, 2007; Hofmann and Gaines, 2008; Kennedy *et al.*, 2008, 2010; Sharma *et al.*, 2010). To date most of the literature in this field is in the form of reviews that highlight the potential commercial benefits of NGS, rather than concrete examples of metagenomics data being applied in a socioeconomic context *per se*, but this is now changing (Marco, 2010). No doubt these examples will appear at a rapidly increasing rate in the near future, given that the potential is vast: from purely medical applications to the use of bioengineering organisms to deal with pollutants and toxic spills, by identifying genes in the environment associated with metabolising xenobiotics (Marco, 2010). Some of the seemingly more outlandish suggestions that have been put forward include the use of certain bacteria as potential agents for terraforming other planets, based on their unusual metabolism and ability to survive in extreme environments (Ardelean *et al.*, 2008). On a more down-to-earth level, numerous biotechnological applications of metagenomics have been proposed in recent years, including bioremediation of damaged ecosystems and the development of new industrial bioproducts and novel medical treatments (Marco, 2010; Table 2).

IX. CONCLUSIONS

The revolution in ecology that began with the advent of molecular studies has developed increasingly rapidly via the exponential growth in our ability to extract data from a system, as exemplified by the recent emergence of NGS technologies. The application of these methods will undoubtedly lead to a new more inclusive ecology and allow us to begin to make the links required to understand how the different levels of organisation are interconnected, from molecules to whole ecosystems. However, we are not advocating these new technologies as a panacea, as many of the questions NGS data will raise will require testing using the rigorous experimental techniques that have underpinned traditional ecology for decades. It is in this area that a new and more integrative ecology will really alter our understanding: when ecological ideas developed from plants and animals are tested and validated using experimental microbial systems, then a real systems biology for ecology will emerge. When fully integrated within traditional ecology, NGS has the potential to ask many questions that previously could simply not be answered, and to not only forge links across multiple levels of organisation within ecology, but also to bind ecology and evolution more tightly together. We can therefore 'expect the unexpected' in the near future: whilst we have attempted to outline some of the prospects for the development of this field, many of the new and exciting advances that are undoubtedly around the corner will come as surprises to us all.

ACKNOWLEDGEMENTS

K. J. P. and B. B. O. were funded by the EU Commission (MicroComXT a Marie Curie Excellence Team MEXT-CT-2005-024112). The authors thank Andrew R. Leitch and A. Mira for helpful comments, and Murray Thompson for his assistance with the graphics.

REFERENCES

Ahn, S.M., Kim, T.H., Lee, S., Kim, D., Ghang, H., Kim, D.S., Kim, B.C., Kim, S.Y., Kim, W.Y., Kim, C., Park, D., Lee, Y.S., *et al.* (2009). The first Korean genome sequence and analysis: Full genome sequencing for a socio-ethnic group. *Genome Res.* **19**, 1622–1629.

Amann, R., and Fuchs, B.M. (2008). Single-cell identification in microbial communities by improved fluorescence *in situ* hybridization techniques. *Nat. Rev. Microbiol.* **6**, 339–348.

Ardelean, I.I., Moisescu, C., and Popoviciu, D.R. (2008). Magnetotactic bacteria and their potential for terraformation. In: *From Fossils to Astrobiology: Records of Life on Earth and the Search for Extraterrestrial Biosignatures* (Ed. by J. Seckbach and M. Walsh), pp. 335–350. Springer-Verlag, New York.

Bailly, J., Fraissinet-Tachet, L., Verner, M.C., Debaud, J.C., Lemaire, M., Wesolowski-Louvel, M., and Marmeisse, R. (2007). Soil eukaryotic functional diversity, a metatranscriptomic approach. *ISME J.* **1**, 632–642.

Baker, B.J., Tyson, G.W., Webb, R.I., Flanagan, J., Hugenholtz, P., Allen, E.E., and Banfield, J.F. (2006). Lineages of acidophilic archaea revealed by community genomic analysis. *Science* **314**, 1933–1935.

Bascompte, J. (2010). Structure and dynamics of ecological networks. *Science* **329**, 765–766.

Bascompte, J., and Jordano, P. (2007). Plant-animal mutualistic networks: The architecture of biodiversity. *Annu. Rev. Ecol. Evol. Syst.* **38**, 567–593.

Beja, O., Aravind, L., Koonin, E.V., Suzuki, M.T., Hadd, A., Nguyen, L.P., Jovanovich, S., Gates, C.M., Feldman, R.A., Spudich, J.L., Spudich, E.N., and DeLong, E.F. (2000). Bacterial rhodopsin: Evidence for a new type of phototrophy in the sea. *Science* **289**, 1902–1906.

Bell, T., Newman, J.A., Silverman, B.W., Turner, S.L., and Lilley, A.K. (2005). The contribution of species richness and composition to bacterial services. *Nature* **436**, 1157–1160.

Bentley, D.R., Balasubramanian, S., Swerdlow, H.P., Smith, G.P., Milton, J., Brown, C.G., Hall, K.P., Evers, D.J., Barnes, C.L., Bignell, H.R., Boutell, J.M., Bryant, J., *et al.* (2008). Accurate whole human genome sequencing using reversible terminator chemistry. *Nature* **456**, 53–59.

Bernhard, A.E., Tucker, J., Giblin, A.E., and Stahl, D.A. (2007). Functionally distinct communities of ammonia-oxidizing bacteria along an estuarine salinity gradient. *Environ. Microbiol.* **9**, 1439–1447.

Biddle, J.F., Fitz-Gibbon, S., Schuster, S.C., Brenchley, J.E., and House, C.H. (2008). Metagenomic signatures of the Peru Margin subseafloor biosphere show a genetically distinct environment. *Proc. Natl. Acad. Sci. USA* **105**, 10583–10588.

Bloom, J.S., Khan, Z., Kruglyak, L., Singh, M., and Caudy, A.A. (2009). Measuring differential gene expression by short read sequencing: Quantitative comparison to 2-channel gene expression microarrays. *BMC Genomics* **10**, 221.

Boschker, H.T.S., Nold, S.C., Wellsbury, P., Bos, D., de Graaf, W., Pel, R., Parkes, R.J., and Cappenberg, T.E. (1998). Direct linking of microbial populations to specific biogeochemical processes by C^{13}-labelling of biomarkers. *Nature* **392**, 801–805.

Braker, G., Fesefeldt, A., and Witzel, K.P. (1998). Development of PCR primer systems for amplification of nitrite reductase genes (*nirK* and *nirS*) to detect denitrifying bacteria in environmental samples. *Appl. Environ. Microbiol.* **64**, 3769–3775.

Breitbart, M., and Rohwer, F. (2005). Here a virus, there a virus, everywhere the same virus? *Trends Microbiol.* **13**, 278–284.

Britschgi, T.B., and Giovannoni, S.J. (1991). Phylogenetic analysis of a natural marine bacterioplankton population by rRNA gene cloning and sequencing. *Appl. Environ. Microbiol.* **57**, 1707–1713.

Brown, J.H., Gillooly, J.F., Allen, A.P., Savage, V.M., and West, G.B. (2004). Toward a metabolic theory of ecology. *Ecology* **85**, 1771–1789.

Buckley, M.R. (2004). *The Global Genome Question: Microbes as the Key to Understanding Evolution and Ecology*. American Academy of Microbiology, Washington, DC.

Cardinale, B.J., Weis, J.J., Forbes, A.E., Tilmon, K.J., and Ives, A.R. (2006). Biodiversity as both a cause and consequence of resource availability: A study of reciprocal causality in a predator-prey system. *J. Anim. Ecol.* **75**, 497–505.

Carreon-Martinez, L., and Heath, D.D. (2010). Revolution in food web analysis and trophic ecology: Diet analysis by DNA and stable isotope analysis. *Mol. Ecol.* **19**, 25–27.

Chaisson, M.J., and Pevzner, P.A. (2008). Short read fragment assembly of bacterial genomes. *Genome Res.* **18**, 324–330.

Chatton, E. (1938). *Titres et Travaux Scientifiques [Titles and Scientific Works] (1906-1937)*. E. Sottano, Ste, France.

Chen, Y., and Murrell, J.C. (2010). When metagenomics meets stable-isotope probing: Progress and perspectives. *Trends Microbiol.* **18**, 157–163.

Chin, K.-J., Sharma, M., Russell, L., O'Neill, K., and Lovley, D. (2008). Quantifying expression of a dissimilatory (bi)sulfite reductase gene in petroleum-contaminated marine harbor sediments. *Microb. Ecol.* **55**, 489–499.

Chivian, D., Brodie, E.L., Alm, E.J., Culley, D.E., Dehal, P.S., DeSantis, T.Z., Gihring, T.M., Lapidus, A., Lin, L.-H., Lowry, S.R., Moser, D.P., Richardson, P.M., *et al.* (2008). Environmental genomics reveals a single-species ecosystem deep within Earth. *Science* **322**, 275–278.

Cho, J.C., and Tiedje, J.M. (2000). Biogeography and degree of endemicity of fluorescent *Pseudomonas* strains in soil. *Appl. Environ. Microbiol.* **66**, 5448–5456.

Cloonan, N., Forrest, A.R.R., Kolle, G., Gardiner, B.B.A., Faulkner, G.J., Brown, M.K., Taylor, D.F., Steptoe, A.L., Wani, S., Bethel, G., Robertson, A.J., Perkins, A.C., *et al.* (2008). Stem cell transcriptome profiling via massive-scale mRNA sequencing. *Nat. Methods* **5**, 613–619.

Cohan, F.M. (2001). Bacterial species and speciation. *Syst. Biol.* **50**, 513–524.

Collins, F.S., Lander, E.S., Rogers, J., and Waterston, R.H.The International Human Genome Sequencing Consortium (2004). Finishing the euchromatic sequence of the human genome. *Nature* **431**, 931–945.

Coolon, J.D., Jones, K.L., Narayanan, S., and Wisely, S.M. (2010). Microbial ecological response of the intestinal flora of *Peromyscus maniculatus* and *P. leucopus* to heavy metal contamination. *Mol. Ecol.* **19**, 67–80.

Cox-Foster, D.L., Conlan, S., Holmes, E.C., Palacios, G., Evans, J.D., Moran, N.A., Quan, P.L., Briese, T., Hornig, M., Geiser, D.M., Martinson, V., vanEngelsdorp, D., *et al.* (2007). A metagenomic survey of microbes in honey bee colony collapse disorder. *Science* **318**, 283–287.

Cuvelier, M.L., Ortiz, A., Kim, E., Moehlig, H., Richardson, D.E., Heidelberg, J.F., Archibald, J.M., and Worden, A.Z. (2008). Widespread distribution of a unique marine protistan lineage. *Environ. Microbiol.* **10**, 1621–1634.

Dar, S.A., Yao, L., van Dongen, U., Kuenen, J.G., and Muyzer, G. (2007). Analysis of diversity and activity of sulfate-reducing bacterial communities in sulfidogenic bioreactors using 16S rRNA and *dsrB* genes as molecular markers. *Appl. Environ. Microbiol.* **73**, 594–604.

D'Costa, V.M., Griffiths, E., and Wright, G.D. (2007). Expanding the soil antibiotic resistome: Exploring environmental diversity. *Curr. Opin. Microbiol.* **10**, 481–489.

Debroas, D., Humbert, J.F., Enault, F., Bronner, G., Faubladier, M., and Cornillot, E. (2009). Metagenomic approach studying the taxonomic and functional diversity of the bacterial community in a mesotrophic lake (Lac du Bourget – France). *Environ. Microbiol.* **11**, 2412–2424.

DeLong, E.F. (2009). The microbial ocean from genomes to biomes. *Nature* **459**, 200–206.

DeLong, E.F., Wu, K.Y., Prézelin, B.B., and Jovine, R.V.M. (1994). High abundance of Archaea in Antarctic marine picoplankton. *Nature* **371**, 695–697.

DeLong, J.P., Okie, J.G., Moses, M.E., Sibly, R.M., and Brown, J.H. (2010). Shifts in metabolic scaling, production, and efficiency across major evolutionary transitions of life. *Proc. Natl. Acad. Sci. USA* **107**, 12941–12945.

DeSantis, T.Z., Brodie, E.L., Moberg, J.P., Zubieta, I.X., Piceno, Y.M., and Andersen, G.L. (2007). High-density universal 16S rRNA microarray analysis reveals broader diversity than typical clone library when sampling the environment. *Microb. Ecol.* **53**, 371–383.

Dinsdale, E.A., Edwards, R.A., Hall, D., Angly, F., Breitbart, M., Brulc, J.M., Furlan, M., Desnues, C., Haynes, M., Li, L., McDaniel, L., Moran, M.A., *et al.* (2008). Functional metagenomic profiling of nine biomes. *Nature* **452**, 629–632.

Dolfing, J. (2001). The microbial logic behind the prevalence of incomplete oxidation of organic compounds by acetogenic bacteria in methanogenic environments. *Microb. Ecol.* **41**, 83–89.

Dressman, D., Yan, H., Traverso, G., Kinzler, K.W., and Vogelstein, B. (2003). Transforming single DNA molecules into fluorescent magnetic particles for detection and enumeration of genetic variations. *Proc. Natl. Acad. Sci. USA* **100**, 8817–8822.

Dunn, C., Hejnol, A., Matus, D.Q., Pang, K., Browne, W.E., Smith, S.A., *et al.* (2008). Broad phylogenomic sampling improves resolution of the animal tree of life. *Nature* **452**, 745–749.

Eid, J., Fehr, A., Gray, J., Luong, K., Lyle, J., Otto, G., Peluso, P., Rank, D., Baybayan, P., Bettman, B., Bibillo, A., Bjornson, K., *et al.* (2009). Real-time DNA sequencing from single polymerase molecules. *Science* **323**, 133–138.

Ellers, J. (2010). Evolutionary processes in community ecology. In: *Community Ecology: Processes, Models and Applications* (Ed. by H.A. Verhoef and P.J. Morin), pp. 151–162. Oxford University Press, Oxford.

Erguder, T.H., Boon, N., Wittebolle, L., Marzorati, M., and Verstraete, W. (2009). Environmental factors shaping the ecological niches of ammonia-oxidizing archaea. *FEMS Microbiol. Rev.* **33**, 855–869.

Erkel, C., Kube, M., Reinhardt, R., and Liesack, W. (2006). Genome of Rice Cluster I Archaea – The key methane producers in the rice rhizosphere. *Science* **313**, 370–372.

Ettwig, K.F., Butler, M.K., Le Paslier, D., Pelletier, E., Mangenot, S., Kuypers, M.M.M., Schreiber, F., Dutilh, B.E., Zedelius, J., de Beer, D., Gloerich, J., Wessels, H.J.C.T., *et al.* (2010). Nitrite-driven anaerobic methane oxidation by oxygenic bacteria. *Nature* **464**, 543–548.

Falkowski, P.G., Fenchel, T., and Delong, E.F. (2008). The microbial engines that drive Earth's biogeochemical cycles. *Science* **320**, 1034–1039.

Feder, M.E., and Walser, J.C. (2005). The biological limitations of transcriptomics in elucidating stress and stress responses. *J. Evol. Biol.* **18**, 901–910.

Fedurco, M., Romieu, A., Williams, S., Lawrence, I., and Turcatti, G. (2006). BTA, a novel reagent for DNA attachment on glass and efficient generation of solid-phase amplified DNA colonies. *Nucleic Acids Res.* **34**, e22.

Fenchel, T. (1992). What can ecologists learn from microbes: Life beneath a square centimetre of sediment surface. *Funct. Ecol.* **6**, 499–507.

Fitter, A.H. (2005). Darkness visible: Reflections on underground ecology. *J. Ecol.* **93**, 231–243.

Fleischmann, R.D., Adams, M.D., White, O., Clayton, R.A., Kirkness, E.F., Kerlavage, A.R., Bult, C.J., Tomb, J.F., Dougherty, B.A., Merrick, J.M., McKenney, K., Sutton, G., *et al.* (1995). Whole-genome random sequencing and assembly of *Haemophilus influenzae* Rd. *Science* **269**, 496–512.

Fox, G.E., Stackebrandt, E., Hespell, R.B., Gibson, J., Maniloff, J., Dyer, T.A., Wolfe, R.S., Balch, W.E., Tanner, R., Magrun, L., Zabien, L.B., Blakemore, R., *et al.* (1980). The phylogeny of procaryotes. *Science* **209**, 457–463.

Fraser, C., Hanage, W.P., and Spratt, B.G. (2007). Recombination and the nature of bacterial speciation. *Science* **315**, 476–480.

Fuhrman, J.A., and Steele, J.A. (2007). Community structure of marine bacterioplankton: Patterns, networks, and relationships to function. *Aquat. Microb. Ecol.* **53**, 69–81.

Fuhrman, J.A., Steele, J.A., Hewson, I., Schwalbach, M.S., Brown, M.V., Green, J.L., and Brown, J.H. (2008). A latitudinal diversity gradient in planktonic marine bacteria. *Proc. Natl. Acad. Sci. USA* **105**, 7774–7778.

Fussmann, G.F., Loreau, M., and Abrams, P.A. (2007). Eco-evolutionary dynamics of communities and ecosystems. *Funct. Ecol.* **21**, 465–477.

Gianoulis, T.A., Raes, J., Patel, P.V., Bjornson, R., Korbel, J.O., Letunic, I., Yamada, T., Paccanaro, A., Jensen, L.J., Snyder, M., Bork, P., and Gerstein, M.B. (2009). Quantifying environmental adaptation of metabolic pathways in metagenomics. *Proc. Natl. Acad. Sci. USA* **106**, 1374–1379.

Gifford, S., Sharma, S., Rinta-Kanto, J., and Moran, M.A. (2010). Metatranscriptomic analysis of a summer bacterioplankton community in Southeastern U.S. coastal waters. *In13th International Symposium on Microbial Ecology*. ISME, Seattle.

Gilbert, J.A., Field, D., Huang, Y., Edwards, R., Li, W.Z., Gilna, P., and Joint, I. (2008). Detection of large numbers of novel sequences in the metatranscriptomes of complex marine microbial communities. *PLoS ONE* **3**, e3042.

Gilbert, J.A., Field, D., Swift, P., Newbold, L., Oliver, A., Smyth, T., Somerfield, P.J., Huse, S., and Joint, I. (2009). The seasonal structure of microbial communities in the Western English Channel. *Environ. Microbiol.* **11**, 3132–3139.

Giovannoni, S.J., DeLong, E.F., Olsen, G.J., and Pace, N.R. (1988). Phylogenetic group-specific oligodeoxynucleotide probes for identification of single microbial cells. *J. Bacteriol.* **170**, 720–726.

Giovannoni, S.J., Britschgi, T.B., Moyer, C.L., and Field, K.G. (1990). Genetic diversity in Sargasso Sea bacterioplankton. *Nature* **345**, 60–63.

Giovannoni, S.J., Rappé, M.S., Vergin, K.L., and Adair, N.L. (1996). 16S rRNA genes reveal stratified open ocean bacterioplankton populations related to the green non-sulfur bacteria. *Proc. Natl. Acad. Sci. USA* **93**, 7979–7984.

Goetz, F., Rosauer, D., Sitar, S., Goetz, G., Simchick, C., Roberts, S., Johnson, R., Murphy, C., Bronte, C.R., and Mackenzie, S. (2010). A genetic basis for the phenotypic differentiation between siscowet and lean lake trout (*Salvelinus namaycush*). *Mol. Ecol.* **19**, 176–196.

Gomez-Alvarez, V., Teal, T.K., and Schmidt, T.M. (2009). Systematic artifacts in metagenomes from complex microbial communities. *ISME J.* **3**, 1314–1317.

Good, I.J. (1953). The population frequency of species and the estimation of population parameters. *Biometrika* **40**, 237–262.

Green, J., and Bohannan, B.J.M. (2006). Spatial scaling of microbial biodiversity. *Trends Ecol. Evol.* **21**, 501–507.

Green, R.E., Krause, J., Ptak, S.E., Briggs, A.W., Ronan, M.T., Simons, J.F., Du, L., Egholm, M., Rothberg, J.M., Paunovic, M., and Paabo, S. (2006). Analysis of one million base pairs of Neanderthal DNA. *Nature* **444**, 330–336.

Handelsman, J. (2004). Metagenomics: Application of genomics to uncultured microorganisms. *Microbiol. Mol. Biol. Rev.* **68**, 669–685.

Harris, T.D., Buzby, P.R., Babcock, H., Beer, E., Bowers, J., Braslavsky, I., Causey, M., Colonell, J., Dimeo, J., Efcavitch, J.W., Giladi, E., Gill, J., *et al.* (2008). Single-molecule DNA sequencing of a viral genome. *Science* **320**, 106–109.

He, Z., Gentry, T.J., Schadt, C.W., Wu, L., Liebich, J., Chong, S.C., Huang, Z., Wu, W., Gu, B., Jardine, P., Criddle, C., and Zhou, J. (2007). GeoChip: A comprehensive microarray for investigating biogeochemical, ecological and environmental processes. *ISME J.* **1**, 67–77.

He, Z., Deng, Y., Van Nostrand, J.D., Tu, Q., Xu, M., Hemme, C.L., Li, X., Wu, L., Gentry, T.J., Yin, Y., Liebich, J., Hazen, T.C., *et al.* (2010a). GeoChip 3.0 as a high-throughput tool for analyzing microbial community composition, structure and functional activity. *ISME J.* **4**, 1167–1179.

He, Z., Xu, M., Deng, Y., Kang, S., Kellogg, L., Wu, L., Van Nostrand, J.D., Hobbie, S.E., Reich, P.B., and Zhou, J. (2010b). Metagenomic analysis reveals a marked divergence in the structure of belowground microbial communities at elevated CO_2. *Ecol. Lett.* **13**, 564–575.

Hernandez, D., Francois, P., Farinelli, L., Osteras, M., and Schrenzel, J. (2008). De novo bacterial genome sequencing: Millions of very short reads assembled on a desktop computer. *Genome Res.* **18**, 802–809.

Hillier, L.W., Marth, G.T., Quinlan, A.R., Dooling, D., Fewell, G., Barnett, D., Fox, P., Glasscock, J.I., Hickenbotham, M., Huang, W.C., Magrini, V.J., Richt, R. J., *et al.* (2008). Whole-genome sequencing and variant discovery in *C. elegans*. *Nat. Methods* **5**, 183–188.

Hofmann, G.E., and Gaines, S.D. (2008). New tools to meet new challenges: Emerging technologies for managing marine ecosystems for resilience. *Bioscience* **58**, 43–52.

Hooper, L.V., Midtvedt, T., and Gordon, J.I. (2002). How host-microbial interactions shape the nutrient environment of the mammalian intestine. *Annu. Rev. Nutr.* **22**, 283–307.

Hudson, M.E. (2008). Sequencing breakthroughs for genomic ecology and evolutionary biology. *Mol. Ecol. Resour.* **8**, 3–17.

Hugenholtz, P., Goebel, B.M., and Pace, N.R. (1998). Impact of culture-independent studies on the emerging phylogenetic view of bacterial diversity. *J. Bacteriol.* **180**, 4765–4774.

Hunkapiller, T., Kaiser, R.J., Koop, B.F., and Hood, L. (1991). Large-scale and automated DNA-sequence determination. *Science* **254**, 59–67.

Hurd, P.J., and Nelson, C.J. (2009). Advantages of next-generation sequencing versus the microarray in epigenetic research. *Brief. Funct. Genomic Proteomic* **8**, 174–183.

Hutchinson, G.E. (1961). The paradox of the plankton. *Am. Nat.* **95**, 137–145.

Ings, T.C., Montoya, J.M., Bascompte, J., Bluthgen, N., Brown, L., Dormann, C.F., Edwards, F., Figueroa, D., Jacob, U., Jones, J.I., Lauridsen, R.B., Ledger, M.E., *et al.* (2009). Ecological networks – Beyond food webs. *J. Anim. Ecol.* **78**, 253–269.

Jackson, B.E., and McInerney, M.J. (2002). Anaerobic microbial metabolism can proceed close to thermodynamic limits. *Nature* **415**, 454–456.

Johnson, M.T.J., and Stinchcombe, J.R. (2007). An emerging synthesis between community ecology and evolutionary biology. *Trends Ecol. Evol.* **22**, 250–257.

Jones, S.E., and Lennon, J.T. (2009). Evidence for limited microbial transfer of methane in a planktonic food web. *Aquat. Microb. Ecol.* **58**, 45–53.

Jones, L.E., Becks, L., Ellner, S.P., Hairston, N.G., Yoshida, T., and Fussmann, G.F. (2009). Rapid contemporary evolution and clonal food web dynamics. *Philos. Trans. R. Soc. B-Biol. Sci.* **364**, 1579–1591.

Joulian, C., Ramsing, N.B., and Ingvorsen, K. (2001). Congruent phylogenies of most common small-subunit rRNA and dissimilatory sulfite reductase gene sequences retrieved from estuarine sediments. *Appl. Environ. Microbiol.* **67**, 3314–3318.

Jumpponen, A., Jones, K.L., Mattox, D., and Yaege, C. (2010). Massively parallel 454-sequencing of fungal communities in *Quercus* spp. ectomycorrhizas indicates seasonal dynamics in urban and rural sites. *Mol. Ecol.* **19**, 41–53.

Kandeler, E., Deiglmayr, K., Tscherko, D., Bru, D., and Philippot, L. (2006). Abundance of *narG, nirS, nirK*, and *nosZ* genes of denitrifying bacteria during primary successions of a Glacier Foreland. *Appl. Environ. Microbiol.* **72**, 5957–5962.

Kant, M.R., and Baldwin, I.T. (2007). The ecogenetics and ecogenomics of plant-herbivore interactions: Rapid progress on a slippery road. *Curr. Opin. Genet. Dev.* **17**, 519–524.

Kassen, R., and Rainey, P.B. (2004). The ecology and genetics of microbial diversity. *Annu. Rev. Microbiol.* **58**, 207–231.

Kemp, P.F., Lee, S.-H., and LaRoche, J. (1993). Estimating the growth rate of slowly growing marine bacteria from RNA content. *Appl. Environ. Microbiol.* **59**, 2594–2601.

Kennedy, J., Marchesi, J.R., and Dobson, A.D.W. (2008). Marine metagenomics: Strategies for the discovery of novel enzymes with biotechnological applications from marine environments. *Microb. Cell Fact.* **7**, 27.

Kennedy, J., Flemer, B., Jackson, S.A., Lejon, D.P.H., Morrissey, J.P., O'Gara, F., and Dobson, A.D.W. (2010). Marine metagenomics: New tools for the study and exploitation of marine microbial metabolism. *Mar. Drugs* **8**, 608–628.

Kim, J.I., Ju, Y.S., Park, H., Kim, S., Lee, S., Yi, J.H., Mudge, J., Miller, N.A., Hong, D., Bell, C.J., Kim, H.S., Chung, I.S., *et al.* (2009). A highly annotated whole-genome sequence of a Korean individual. *Nature* **460**, 1011–1015.

Kloch, A., Babik, W., Bajer, A., Sinski, E., and Radwan, J. (2010). Effects of an MHC-DRB genotype and allele number on the load of gut parasites in the bank vole *Myodes glareolus*. *Mol. Ecol.* **19**, 255–265.

Könneke, M., Bernhard, A.E., de la Torre, J.R., Walker, C.B., Waterbury, J.B., and Stahl, D.A. (2005). Isolation of an autotrophic ammonia-oxidizing marine archaeon. *Nature* **437**, 543–546.

Korbel, J.O., Urban, A.E., Affourtit, J.P., Godwin, B., Grubert, F., Simons, J.F., Kim, P.M., Palejev, D., Carriero, N.J., Du, L., Taillon, B.E., Chen, Z.T., *et al.* (2007). Paired-end mapping reveals extensive structural variation in the human genome. *Science* **318**, 420–426.

Kummerli, R., Griffin, A.S., West, S.A., Buckling, A., and Harrison, F. (2009). Viscous medium promotes cooperation in the pathogenic bacterium *Pseudomonas aeruginosa*. *Proc. R. Soc. B-Biol. Sci.* **276**, 3531–3538.

Kunin, V., Engelbrektson, A., Ochman, H., and Hugenholtz, P. (2009). Wrinkles in the rare biosphere: Pyrosequencing errors can lead to artificial inflation of diversity estimates. *Environ. Microbiol.* **12**, 118–123.

Lafi, F.F., Fuerst, J.A., Fieseler, L., Engels, C., Goh, W.W.L., and Hentschel, U. (2009). Widespread distribution of Poribacteria in Demospongiae. *Appl. Environ. Microbiol.* **75**, 5695–5699.

Lander, E.S., Linton, L.M., Birren, B., Nusbaum, C., Zody, M.C., Baldwin, J., Devon, K., Dewar, K., Doyle, M., FitzHugh, W., Funke, R., Gage, D., *et al.* (2001). Initial sequencing and analysis of the human genome. *Nature* **409**, 860–921.

Lee, N., Nielsen, P.H., Andreasen, K.H., Juretschko, S., Nielsen, J.L., Schleifer, K.H., and Wagner, M. (1999). Combination of fluorescent in situ hybridization and microautoradiography – A new tool for structure-function analyses in microbial ecology. *Appl. Environ. Microbiol.* **65**, 1289–1297.

Lefrancois, P., Euskirchen, G.M., Auerbach, R.K., Rozowsky, J., Gibson, T., Yellman, C.M., Gerstein, M., and Snyder, M. (2009). Efficient yeast ChIP-Seq using multiplex short-read DNA sequencing. *BMC Genomics* **10**, e37.

Levene, M.J., Korlach, J., Turner, S.W., Foquet, M., Craighead, H.G., and Webb, W.W. (2003). Zero-mode waveguides for single-molecule analysis at high concentrations. *Science* **299**, 682–686.

Li, L.L., Victoria, J.G., Wang, C.L., Jones, M., Fellers, G.M., Kunz, T.H., and Delwart, E. (2010). Bat Guano virome: Predominance of dietary viruses from insects and plants plus novel mammalian viruses. *J. Virol.* **84**, 6955–6965.

Li, R.Q., Fan, W., Tian, G., Zhu, H.M., He, L., Cai, J., Huang, Q.F., Cai, Q.L., Li, B., Bai, Y.Q., Zhang, Z.H., Zhang, Y.P., *et al.* (2010). The sequence and de novo assembly of the giant panda genome. *Nature* **463**, 311–317.

Li, R.Q., Zhu, H.M., Ruan, J., Qian, W.B., Fang, X.D., Shi, Z.B., Li, Y.R., Li, S.T., Shan, G., Kristiansen, K., Li, S.G., Yang, H.M., *et al.* (2010). De novo assembly of human genomes with massively parallel short read sequencing. *Genome Res.* **20**, 265–272.

Lindeman, R.L. (1942). The trophic-dynamic aspect of ecology. *Ecology* **23**, 399–418.

Lipson, D., Raz, T., Kieu, A., Jones, D.R., Giladi, E., Thayer, E., Thompson, J.F., Letovsky, S., Milos, P., and Causey, M. (2009). Quantification of the yeast transcriptome by single-molecule sequencing. *Nat. Biotechnol.* **27**, 652–658.

Lister, R., O'Malley, R.C., Tonti-Filippini, J., Gregory, B.D., Berry, C.C., Millar, A.H., and Ecker, J.R. (2008). Highly integrated single-base resolution maps of the epigenome in *Arabidopsis*. *Cell* **133**, 523–536.

Liu, J.M., Livny, J., Lawrence, M.S., Kimball, M.D., Waldor, M.K., and Camilli, A. (2009). Experimental discovery of sRNAs in *Vibrio cholerae* by direct cloning, 5S/tRNA depletion and parallel sequencing. *Nucleic Acids Res.* **37**, e46.

Lo, I., Denef, V.J., VerBerkmoes, N.C., Shah, M.B., Goltsman, D., DiBartolo, G., Tyson, G.W., Allen, E.E., Ram, R.J., Detter, J.C., Richardson, P., Thelen, M.P., *et al.* (2007). Strain-resolved community proteomics reveals recombining genomes of acidophilic bacteria. *Nature* **446**, 537–541.

Madsen, E.L. (2005). Identifying microorganisms responsible for ecologically significant biogeochemical processes. *Nat. Rev. Microbiol.* **3**, 439–446.

Makarieva, A.M., Gorshkov, V.G., Li, B.L., Chown, S.L., Reich, P.B., and Gavrilov, V.M. (2008). Mean mass-specific metabolic rates are strikingly similar across life's major domains: Evidence for life's metabolic optimum. *Proc. Natl. Acad. Sci. USA* **105**, 16994–16999.

Manefield, M., Whiteley, A.S., Griffiths, R.I., and Bailey, M.J. (2002). RNA stable isotope probing, a novel means of linking microbial community function to phylogeny. *Appl. Environ. Microbiol.* **68**, 5367–5373.

Marco, D. (2010). *Metagenomics: Theory, Methods and Applications*. Caister Academic Press, Norfolk, UK.

Margulies, M., Egholm, M., Altman, W.E., Attiya, S., Bader, J.S., Bemben, L.A., Berka, J., Braverman, M.S., Chen, Y.J., Chen, Z.T., Dewell, S.B., Du, L., *et al.* (2005). Genome sequencing in microfabricated high-density picolitre reactors. *Nature* **437**, 376–380.

Martiny, A.C., Kathuria, S., and Berube, P.M. (2009). Widespread metabolic potential for nitrite and nitrate assimilation among *Prochlorococcus* ecotypes. *Proc. Natl. Acad. Sci. USA* **106**, 10787–10792.

McGeogh, M.A., and Gaston, K.J. (2002). Occupancy frequency distributions: Patterns, artefacts and mechanisms. *Biol. Rev.* **77**, 311–331.

McKernan, K.J., Peckham, H.E., Costa, G.L., McLaughlin, S.F., Fu, Y.T., Tsung, E.F., Clouser, C.R., Duncan, C., Ichikawa, J.K., Lee, C.C., Zhang, Z., Ranade, S.S., *et al.* (2009). Sequence and structural variation in a human genome uncovered by short-read, massively parallel ligation sequencing using two-base encoding. *Genome Res.* **19**, 1527–1541.

Medini, D., Donati, C., Tettelin, H., Masignani, V., and Rappuoli, R. (2005). The microbial pan-genome. *Curr. Opin. Genet. Dev.* **15**, 589–594.

Metzker, M.L. (2010). Sequencing technologies – The next generation. *Nat. Rev. Genet.* **11**, 31–46.

Meyer, M., Stenzel, U., Myles, S., Prufer, K., and Hofreiter, M. (2007). Targeted high-throughput sequencing of tagged nucleic acid samples. *Nucleic Acids Res.* **35**, e97.

Meyer, M., Stenzel, U., and Hofreiter, M. (2008). Parallel tagged sequencing on the 454 platform. *Nat. Protoc.* **3**, 267–278.

Miller, W., Drautz, D.I., Ratan, A., Pusey, B., Qi, J., Lesk, A.M., Tomsho, L.P., Packard, M.D., Zhao, F.Q., Sher, A., Tikhonov, A., Raney, B., *et al.* (2008). Sequencing the nuclear genome of the extinct woolly mammoth. *Nature* **456**, 387–390.

Moran, N.A., McLaughlin, H.J., and Sorek, R. (2009). The dynamics and time scale of ongoing genomic erosion in symbiotic bacteria. *Science* **323**, 379–382.

Morin, R.D., Bainbridge, M., Fejes, A., Hirst, M., Krzywinski, M., Pugh, T.J., McDonald, H., Varhol, R., Jones, S.J.M., and Marra, M.A. (2008). Profiling the HeLa S3 transcriptome using randomly primed cDNA and massively parallel short-read sequencing. *Biotechniques* **45**, 81–94.

Morris, R.M., Rappe, M.S., Connon, S.A., Vergin, K.L., Siebold, W.A., Carlson, C. A., and Giovannoni, S.J. (2002). SAR11 clade dominates ocean surface bacterioplankton communities. *Nature* **420**, 806–810.

Mortazavi, A., Williams, B.A., McCue, K., Schaeffer, L., and Wold, B. (2008). Mapping and quantifying mammalian transcriptomes by RNA-Seq. *Nat. Methods* **5**, 621–628.

Mou, X., Sun, S., Edwards, R.A., Hodson, R.E., and Moran, M.A. (2008). Bacterial carbon processing by generalist species in the coastal ocean. *Nature* **451**, 708–711.

Moya-Laraño, J. (2010). Can temperature and water availability contribute to the maintenance of latitudinal diversity by increasing the rate of biotic interactions? *Open Ecol. J.* **3**, 1–13.

Moya-Laraño, J. (in press) Genetic variation, predator-prey interactions and food web structure. *Philos. Trans. R. Soc. B-Biol. Sci.*

Nagalakshmi, U., Wang, Z., Waern, K., Shou, C., Raha, D., Gerstein, M., and Snyder, M. (2008). The transcriptional landscape of the yeast genome defined by RNA sequencing. *Science* **320**, 1344–1349.

Nelson, D.L., and Cox, M.M. (2009). *Lehninger's Principle of Biochemistry*. WH Freeman, New York.

Nolte, V., Pandey, R.V., Jost, S., Medinger, R., Ottenwälder, B., Boenigk, J., and Christian, S. (2010). Contrasting seasonal niche separation between rare and abundant taxa conceals the extent of protist diversity. *Mol. Ecol.* **19**, 2908–2915.

Noonan, J.P., Hofreiter, M., Smith, D., Priest, J.R., Rohland, N., Rabeder, G., Krause, J., Detter, J.C., Paabo, S., and Rubin, E.M. (2005). Genomic sequencing of Pleistocene cave bears. *Science* **309**, 597–600.

Noonan, J.P., Coop, G., Kudaravalli, S., Smith, D., Krause, J., Alessi, J., Platt, D., Paabo, S., Pritchard, J.K., and Rubin, E.M. (2006). Sequencing and analysis of Neanderthal genomic DNA. *Science* **314**, 1113–1118.

Nusbaum, C., Ohsumi, T.K., Gomez, J., Aquadro, J., Victor, T.C., Warren, R.M., Hung, D.T., Birren, B.W., Lander, E.S., and Jaffe, D.B. (2009). Sensitive, specific polymorphism discovery in bacteria using massively parallel sequencing. *Nat. Methods* **6**, 67–69.

Oakley, B.B., Carbonero, F., van der Gast, C.J., Hawkins, R.J., and Purdy, K.J. (2010). Evolutionary divergence and biogeography of sympatric niche-differentiated bacterial populations. *ISME J.* **4**, 488–497.

Oren, A. (2004). Prokaryote diversity and taxonomy: Current status and future challenges. *Philos. Trans. R. Soc. Lond. B* **359**, 623–638.

Ossowski, S., Schneeberger, K., Clark, R.M., Lanz, C., Warthmann, N., and Weigel, D. (2008). Sequencing of natural strains of *Arabidopsis thaliana* with short reads. *Genome Res.* **18**, 2024–2033.

Östman, O., Drakare, S., Kritzberg, E.S., Langenheder, S., Logue, J.B., and Lindstrom, E.S.Regional invariance among microbial communities. *Ecol. Lett.* **13**, 118–127.

Ouverney, C.C., and Fuhrman, J.A. (1999). Combined Microautoradiography-16S rRNA probe technique for determination of radioisotope uptake by specific microbial cell types in situ. *Appl. Environ. Microbiol.* **65**, 1746–1752.

Pace, N.R., Olsen, G.J., and Woese, C.R. (1986). Ribosomal RNA phylogeny and the primary lines of evolutionary descent. *Cell* **45**, 325–326.

Paine, R.T. (1966). Food web complexity and species diversity. *Am. Nat.* **100**, 65–75.
Passalacqua, K.D., Varadarajan, A., Ondov, B.D., Okou, D.T., Zwick, M.E., and Bergman, N.H. (2009). Structure and complexity of a bacterial transcriptome. *J. Bacteriol.* **191**, 3203–3211.
Pelletier, F., Garant, D., and Hendry, A.P. (2009). Eco-evolutionary dynamics. Introduction. *Philos. Trans. R. Soc. B-Biol. Sci.* **364**, 1483–1489.
Perkins, T.T., Kingsley, R.A., Fookes, M.C., Gardner, P.P., James, K.D., Yu, L., Assefa, S.A., He, M., Croucher, N.J., Pickard, D.J., Maskell, D.J., Parkhill, J., *et al.* (2009). A strand-specific RNA-Seq analysis of the transcriptome of the typhoid bacillus *Salmonella typhi*. *PLoS Genet.* **5**, e1000569.
Perkins, D.M., McKie, B.G., Malmqvist, B., Gilmour, S.G., Reiss, J., and Woodward, G. (2010). Environmental warming and biodiversity-ecosystem functioning in freshwater microcosms: Partitioning the effects of species identity, richness and metabolism. *Adv. Ecol. Res.* **43**, 177–209.
Pernthaler, A., Dekas, A.E., Brown, C.T., Goffredi, S.K., Embaye, T., and Orphan, V.J. (2008). Diverse syntrophic partnerships from-deep-sea methane vents revealed by direct cell capture and metagenomics. *Proc. Natl. Acad. Sci. USA* **105**, 7052–7057.
Petchey, O.L., McPhearson, P.T., Casey, T.M., and Morin, P.J. (1999). Environmental warming alters food-web structure and ecosystem function. *Nature* **402**, 69–72.
Petchey, O.L., Beckerman, A.P., Riede, J.O., and Warren, P.H. (2008). Size, foraging, and food web structure. *Proc. Natl. Acad. Sci. USA* **105**, 4191–4196.
Phillips, D.A., Ferris, H., Cook, D.R., and Strong, D.R. (2003). Molecular control points in rhizosphere food webs. *Ecology* **84**, 816–826.
Picardi, E., Horner, D.S., Chiara, M., Schiavon, R., Valle, G., and Pesole, G. (2010). Large-scale detection and analysis of RNA editing in grape mtDNA by RNA deep-sequencing. *Nucleic Acids Res.* **38**, 4755–4767.
Piganeau, G., Desdevises, Y., Derelle, E., and Moreau, H. (2008). Picoeukaryotic sequences in the Sargasso Sea metagenome. *Genome Biol.* **9**, r5.
Proulx, S.R., Promislow, D.E.L., and Phillips, P.C. (2005). Network thinking in ecology and evolution. *Trends Ecol. Evol.* **20**, 345–353.
Ptacnik, R., Moorthi, S.D., and Hillebrand, H. (2010). Hutchinson reversed, or why there need to be so many species. *Adv. Ecol. Res.* **43**, 1–43.
Purdy, K.J., Nedwell, D.B., Embley, T.M., and Takii, S. (1997). Use of 16S rRNA-targeted oligonucleotide probes to investigate the occurrence and selection of sulphate-reducing bacteria in response to nutrient addition to sediment slurry microcosms from a Japanese estuary. *FEMS Microbiol. Ecol.* **24**, 221–234.
Purdy, K.J., Embley, T.M., and Nedwell, D.B. (2002). The distribution and activity of sulphate reducing bacteria in estuarine and coastal marine sediments. *Antonie Van Leeuwenhoek* **81**, 181–187.
Purdy, K.J., Munson, M.A., Embley, T.M., and Nedwell, D.B. (2003). Use of 16S rRNA-targeted oligonucleotide probes to investigate function and phylogeny of sulphate-reducing bacteria and methanogenic archaea in a UK estuary. *FEMS Microbiol. Ecol.* **44**, 361–371.
Purdy, K.J., Nedwell, D.B., and Embley, T.M. (2003). Analysis of the sulphate-reducing bacterial and methanogenic archaeal populations in contrasting Antarctic sediments. *Appl. Environ. Microbiol.* **69**, 3181–3191.
Pushkarev, D., Neff, N.F., and Quake, S.R. (2009). Single-molecule sequencing of an individual human genome. *Nat. Biotechnol.* **27**, 847–850.
Qin, J., Li, R., Raes, J., Arumugam, M., Burgdorf, K.S., Manichanh, C., Nielsen, T., Pons, N., Levenez, F., Yamada, T., Mende, D.R., Li, J., *et al.* (2010). A human gut

microbial gene catalogue established by metagenomic sequencing. *Nature* **464**, 59–65.

Quince, C., Lanzen, A., Curtis, T.P., Davenport, R.J., Hall, N., Head, I.M., Read, L.F., and Sloan, W.T. (2009). Accurate determination of microbial diversity from 454 pyrosequencing data. *Nat. Methods* **6**, 639–641.

Radajewski, S., Ineson, P., Parekh, N.R., and Murrell, J.C. (2000). Stable isotope probing as a tool in microbial ecology. *Nature* **403**, 646–649.

Raffaelli, D. (2007). Food webs, body size and the curse of the latin binomial. In: *From Energetics to Ecosystems: The Dynamics and Structure of Ecological Systems* (Ed. by N. Rooney, K.S. McCann and D.L.G. Noakes), pp. 53–64. Springer, Netherlands.

Raffaelli, D., and Moller, H. (2000). Manipulative field experiments in animal ecology: Do they promise more than they can deliver? *Adv. Ecol. Res.* **30**, 299–338.

Ram, R.J., VerBerkmoes, N.C., Thelen, M.P., Tyson, G.W., Baker, B.J., Blake, R.C., Shah, M., Hettich, R.L., and Banfield, J.F. (2005). Community proteomics of a natural microbial biofilm. *Science* **308**, 1915–1920.

Raskin, L., Rittmann, B.E., and Stahl, D.A. (1996). Competition and coexistence of sulfate-reducing and methanogenic populations in anaerobic biofilms. *Appl. Environ. Microbiol.* **62**, 3847–3857.

Rasmussen, M., Li, Y.R., Lindgreen, S., Pedersen, J.S., Albrechtsen, A., Moltke, I., Metspalu, M., Metspalu, E., Kivisild, T., Gupta, R., Bertalan, M., Nielsen, K., *et al.* (2010). Ancient human genome sequence of an extinct Palaeo-Eskimo. *Nature* **463**, 757–762.

Reiss, J., Bridle, J.R., Montoya, J.M., and Woodward, G. (2009). Emerging horizons in biodiversity and ecosystem functioning research. *Trends Ecol. Evol.* **24**, 505–514.

Reiss, J., Bailey, R.A., Cássio, F., Woodward, G., and Pascoal, C. (2010). Assessing the contribution of micro-organisms and macrofauna to biodiversity-ecosystem functioning relationships in freshwater microcosms. *Adv. Ecol. Res.* **43**, 151–176.

Reiss, J., Forster, J., Hirst, A.G., Cássio, F., Pascoal, C., Stewart, R., and Hirst, A.G. (2010). When microscopic organisms inform general ecological theory. *Adv. Ecol. Res.* **43**, 45–85.

Renaut, S., Nolte, A.W., and Bernatchez, L. (2010). Mining transcriptome sequences towards identifying adaptive single nucleotide polymorphisms in lake whitefish species pairs (*Coregonus* spp. *Salmonidae*). *Mol. Ecol.* **19**, 115–131.

Rocap, G., Larimer, F.W., Lamerdin, J., Malfatti, S., Chain, P., Ahlgren, N.A., Arellano, A., Coleman, M., Hauser, L., Hess, W.R., Johnson, Z.I., Land, M., *et al.* (2003). Genome divergence in two *Prochlorococcus* ecotypes reflects oceanic niche differentiation. *Nature* **424**, 1042–1047.

Rodriguez-Brito, B., Li, L.L., Wegley, L., Furlan, M., Angly, F., Breitbart, M., Buchanan, J., Desnues, C., Dinsdale, E., Edwards, R., Felts, B., Haynes, M., *et al.* (2010). Viral and microbial community dynamics in four aquatic environments. *ISME J.* **4**, 739–751.

Rodriguez-Valera, F., Martin-Cuadrado, A.B., Rodriguez-Brito, B., Pasic, L., Thingstad, T.F., Rohwer, F., and Mira, A. (2009). Explaining microbial population genomics through phage predation. *Nat. Rev. Microbiol.* **7**, 828–836.

Ronaghi, M., Uhlen, M., and Nyren, P. (1998). A sequencing method based on real-time pyrophosphate. *Science* **281**, 363–365.

Rosset, R., Julien, J., and Monier, R. (1966). Ribonucleic acid composition of bacteria as a function of growth rate. *J. Mol. Biol.* **18**, 308–320.

Rusch, D.B., Halpern, A.L., Sutton, G., Heidelberg, K.B., Williamson, S., Yooseph, S., Wu, D.Y., Eisen, J.A., Hoffman, J.M., Remington, K., Beeson, K.,

Tran, B., *et al.* (2007). The Sorcerer II Global Ocean Sampling expedition: Northwest Atlantic through eastern tropical pacific. *PLOS Biol.* **5**, 398–431.
Saiki, R.K., Bugawan, T.L., Horn, G.T., Mullis, K.B., and Erlich, H.A. (1986). Analysis of enzymatically amplified beta-globin and HLA-DQalpha DNA with allele-specific oligonucleotide probes. *Nature* **324**, 163–166.
Saiki, R.K., Gelfand, D.H., Stoffel, S., Scharf, S.J., Higuchi, R., Horn, G.T., Mullis, K.B., and Erlich, H.A. (1988). Primer-directed enzymatic amplification of DNA with a thermostable DNA polymerase. *Science* **239**, 487–491.
Sanger, F., and Coulson, A.R. (1975). Rapid method for determining sequences in DNA by primed synthesis with DNA polymerase. *J. Mol. Biol.* **94**, 441–448.
Sanger, F., Air, G.M., Barrell, B.G., Brown, N.L., Coulson, A.R., Fiddes, J.C., Hutchison, C.A., Slocombe, P.M., and Smith, M. (1977). Nucleotide sequence of bacteriophage $\psi\chi$174 DNA. *Nature* **265**, 687–695.
Sanger, F., Nicklen, S., and Coulson, A.R. (1977). DNA sequencing with chain-terminating inhibitors. *Proc. Natl. Acad. Sci. USA* **74**, 5463–5467.
Sarin, S., Prabhu, S., O'Meara, M.M., Pe'er, I., and Hobert, O. (2008). *Caenorhabditis elegan*s mutant allele identification by whole-genome sequencing. *Nat. Methods* **5**, 865–867.
Schütte, U.M.E., Abdo, Z., Foster, J., Ravel, J., Bunge, J., Solheim, B., and Forney, L.J. (2010). Bacterial diversity in a glacier foreland of the high Arctic. *Mol. Ecol.* **19**, 54–66.
Sharma, V.K., Kumar, N., Prakash, T., and Taylor, T.D. (2010). MetaBioME: A database to explore commercially useful enzymes in metagenomic datasets. *Nucleic Acids Res.* **38**, D468–D472.
Shen, Y.F., Sarin, S., Liu, Y., Hobert, O., and Pe'er, I. (2008). Comparing platforms for *C. elegans* mutant identification using high-throughput whole-genome sequencing. *PLoS ONE* **3**, e4012.
Shi, Y.M., Tyson, G.W., and DeLong, E.F. (2009). Metatranscriptomics reveals unique microbial small RNAs in the ocean's water column. *Nature* **459**, 266–269.
Shuster, S.M., Lonsdorf, E.V., Wimp, G.M., Bailey, J.K., and Whitham, T.G. (2006). Community heritability measures the evolutionary consequences of indirect genetic effects on community structure. *Evolution* **60**, 991–1003.
Sinigalliano, C.D., Kuhn, D.N., and Jones, R.D. (1995). Amplification of the amoA gene from diverse species of ammonium-oxidizing bacteria and from an indigenous bacterial population from seawater. *Appl. Environ. Microbiol.* **61**, 2702–2706.
Smith, L.M., Sanders, J.Z., Kaiser, R.J., Hughes, P., Dodd, C., Connell, C.R., Heiner, C., Kent, S.B.H., and Hood, L.E. (1986). Fluorescence detection in automated DNA sequence analysis. *Nature* **321**, 674–679.
Smith, C.J., Nedwell, D.B., Dong, L.F., and Osborn, A.M. (2007). Diversity and abundance of nitrate reductase genes (*narG* and *napA*), nitrite reductase genes (*nirS* and *nrfA*), and their transcripts in estuarine sediments. *Appl. Environ. Microbiol.* **73**, 3612–3622.
Sogin, M.L., Morrison, H.G., Huber, J.A., Mark Welch, D., Huse, S.M., Neal, P.R., Arrieta, J.M., and Herndl, G.J. (2006). Microbial diversity in the deep sea and the underexplored "rare biosphere". *Proc. Natl. Acad. Sci. USA* **103**, 12115–12120.
Srivatsan, A., Han, Y., Peng, J.L., Tehranchi, A.K., Gibbs, R., Wang, J.D., and Chen, R. (2008). High-precision, whole-genome sequencing of laboratory strains facilitates genetic studies. *PLoS Genet.* **4.**
Stanier, R.Y., and Smith, J.H.C. (1960). The chlorophylls of green bacteria. *Biochim. Biophys. Acta* **41**, 478–484.

Stanier, R.Y., and van Niel, C.B. (1962). The concept of a bacterium. *Arch. Microbiol.* **42**, 17–35.

Stevens, C.E., and Hume, I.D. (1998). Contributions of microbes in vertebrate gastrointestinal tract to production and conservation of nutrients. *Physiol. Rev.* **78**, 393–427.

Stewart, F.J., Ottesen, E.A., and DeLong, E.F. (2010). Development and quantitative analyses of a universal rRNA-subtraction protocol for microbial metatranscriptomics. *ISME J.* **4**, 896–907.

Stoeck, T., Bass, D., Nebel, M., Christen, R., Jones, M.D.M., Breiner, H.W., and Richards, T.A. (2010). Multiple marker parallel tag environmental DNA sequencing reveals a highly complex eukaryotic community in marine anoxic water. *Mol. Ecol.* **19**, 21–31.

Strous, M., Pelletier, E., Mangenot, S., Rattei, T., Lehner, A., Taylor, M.W., Horn, M., Daims, H., Bartol-Mavel, D., Wincker, P., Barbe, V., Fonknechten, N., *et al.* (2006). Deciphering the evolution and metabolism of an anammox bacterium from a community genome. *Nature* **440**, 790–794.

Tettelin, H., Riley, D., Cattuto, C., and Medini, D. (2008). Comparative genomics: The bacterial pan-genome. *Curr. Opin. Microbiol.* **11**, 472–477.

Throbäck, I.N., Enwall, K., Jarvis, A., and Hallin, S. (2004). Reassessing PCR primers targeting *nirS*, *nirK* and *nosZ* genes for community surveys of denitrifying bacteria with DGGE. *FEMS Microbiol. Ecol.* **49**, 401–417.

Treusch, A.H., Leininger, S., Kletzin, A., Schuster, S.C., Klenk, H.P., and Schleper, C. (2005). Novel genes for nitrite reductase and *amo*-related proteins indicate a role of uncultivated mesophilic crenarchaeota in nitrogen cycling. *Environ. Microbiol.* **7**, 1985–1995.

Tringe, S.G., and Rubin, E.M. (2005). Metagenomics: DNA sequencing of environmental samples. *Nat. Rev. Genet.* **6**, 805–814.

Tringe, S.G., von Mering, C., Kobayashi, A., Salamov, A.A., Chen, K., Chang, H.W., Podar, M., Short, J.M., Mathur, E.J., Detter, J.C., Bork, P., Hugenholtz, P., *et al.* (2005). Comparative metagenomics of microbial communities. *Science* **308**, 554–557.

Turnbaugh, P.J., Ley, R.E., Mahowald, M.A., Magrini, V., Mardis, E.R., and Gordon, J.I. (2006). An obesity-associated gut microbiome with increased capacity for energy harvest. *Nature* **444**, 1027–1031.

Tyler, H.L., Roesch, L.F.W., Gowda, S., Dawson, W.O., and Triplett, E.W. (2009). Confirmation of the sequence of '*Candidatus* Liberibacter asiaticus' and assessment of microbial diversity in huanglongbing-infected citrus phloem using a metagenomic approach. *Mol. Plant Microbe Interact.* **22**, 1624–1634.

Tyson, G.W., and Banfield, J.F. (2005). Cultivating the uncultivated: A community genomics perspective. *Trends Microbiol.* **13**, 411–415.

Tyson, G.W., Chapman, J., Hugenholtz, P., Allen, E.E., Ram, R.J., Richardson, P.M., Solovyev, V.V., Rubin, E.M., Rokhsar, D.S., and Banfield, J. F. (2004). Community structure and metabolism through reconstruction of microbial genomes from the environment. *Nature* **428**, 37–43.

Tyson, G.W., Lo, I., Baker, B.J., Allen, E.E., Hugenholtz, P., and Banfield, J.F. (2005). Genome-directed isolation of the key nitrogen fixer *Leptospirillum ferrodiazotrophum* sp. nov. from an acidophilic microbial community. *Appl. Environ. Microbiol.* **71**, 6319–6324.

Urban, M.C., Leibold, M.A., Amarasekare, P., De Meester, L., Gomulkiewicz, R., Hochberg, M.E., Klausmeier, C.A., Loeuille, N., de Mazancourt, C., Norberg, J.,

Pantel, J.H., Strauss, S.Y., *et al.* (2008). The evolutionary ecology of metacommunities. *Trends Ecol. Evol.* **23**, 311–317.
Valouev, A., Ichikawa, J., Tonthat, T., Stuart, J., Ranade, S., Peckham, H., Zeng, K., Malek, J.A., Costa, G., McKernan, K., Sidow, A., Fire, A., *et al.* (2008). A high-resolution, nucleosome position map of *C. elegans* reveals a lack of universal sequence-dictated positioning. *Genome Res.* **18**, 1051–1063.
Van Bers, N.E.M., Van Oers, K., Kerstens, H.H.D., Dibbits, B.W., Crooijmans, R., Visser, M.E., and Groenen, M.A.M. (2010). Genome-wide SNP detection in the great tit *Parus major* using high throughput sequencing. *Mol. Ecol.* **19**, 89–99.
Vandenkoornhuyse, P., Dufresne, A., Quaiser, A., Gouesbet, G., Binet, F., Francez, A.J., Mahe, S., Bormans, M., Lagadeuc, Y., and Couee, I. (2010). Integration of molecular functions at the ecosystemic level: Breakthroughs and future goals of environmental genomics and post-genomics. *Ecol. Lett.* **13**, 776–791.
van der Gast, C.J., Lilley, A.K., Ager, D., and Thompson, I.P. (2005). Island size and bacterial diversity in an archipelago of engineering machines. *Environ. Microbiol.* **7**, 1220–1226.
van Vliet, A.H.M. (2010). Next generation sequencing of microbial transcriptomes: Challenges and opportunities. *FEMS Microbiol. Lett.* **302**, 1–7.
VanGuilder, H.D., Vrana, K.E., and Freeman, W.M. (2008). Twenty-five years of quantitative PCR for gene expression analysis. *Biotechniques* **44**, 619–626.
Venter, J.C., Remington, K., Heidelberg, J.F., Halpern, A.L., Rusch, D., Eisen, J.A., Wu, D., Paulsen, I., Nelson, K.E., Nelson, W., Fouts, D.E., Levy, S., *et al.* (2004). Environmental genome shotgun sequencing of the Sargasso Sea. *Science* **304**, 66–74.
Wagner, M., Roger, A.J., Flax, J.L., Brusseau, G.A., and Stahl, D.A. (1998). Phylogeny of dissimilatory sulfite reductases supports an early origin of sulfate respiration. *J. Bacteriol.* **180**, 2975–2982.
Wang, J., Wang, W., Li, R.Q., Li, Y.R., Tian, G., Goodman, L., Fan, W., Zhang, J.Q., Li, J., Zhang, J.B., Guo, Y.R., Feng, B.X., *et al.* (2008). The diploid genome sequence of an Asian individual. *Nature* **456**, 60–65.
Ward, B.B., and O'Mullan, G.D. (2005). Community level analysis: Genetic and biogeochemical approaches to investigate community composition and function in aerobic ammonia oxidation. *Environmental Microbiology*. Elsevier Academic Press, San Diegopp. 395–413.
Whitaker, R.J., Grogan, D.W., and Taylor, J.W. (2003). Geographic barriers isolate endemic populations of hyperthermophilic archaea. *Science* **301**, 976–978.
Whittaker, R.H. (1969). New concepts of kingdoms of organisms. *Science* **163**, 150–159.
Wilhelm, B.T., Marguerat, S., Watt, S., Schubert, F., Wood, V., Goodhead, I., Penkett, C.J., Rogers, J., and Bahler, J. (2008). Dynamic repertoire of a eukaryotic transcriptome surveyed at single-nucleotide resolution. *Nature* **453**, 1239–1243.
Woese, C.R. (1983). The primary lines of descent and the universal ancestor. In: *Evolution from Molecules to Men* (Ed. by D.S. Bendall), pp. 209–233. Cambridge University Press, Cambridge.
Woese, C.R., and Fox, G.E. (1977). Phylogenetic structure of the prokaryotic domain: The primary kingdoms. *Proc. Natl. Acad. Sci. USA* **74**, 5088–5090.
Wolf, J.B.W., Bayer, T., Haubold, B., Schilhabel, M., Rosenstiel, P., and Tautz, D. (2010). Nucleotide divergence vs. gene expression differentiation: Comparative transcriptome sequencing in natural isolates from the carrion crow and its hybrid zone with the hooded crow. *Mol. Ecol.* **19**, 162–175.
Woodward, G. (2009). Biodiversity, ecosystem functioning and food webs in fresh waters: Assembling the jigsaw puzzle. *Freshw. Biol.* **54**, 2171–2187.

Woodward, G., Benstead, J.P., Beveridge, O.S., Blanchard, J., Brey, T., Brown, L., Cross, W.F., Friberg, N., Ings, T.C., Jacob, U., Jennings, S., Ledger, M.E., *et al.* (2010). Ecological networks in a changing climate. *Adv. Ecol. Res.* **42**, 71–138.

Woodward, G., Perkins, D.M., and Brown, L.E. (2010). Climate change and freshwater ecosystems: Impacts across multiple levels of organization. *Philos. Trans. R. Soc. B-Biol. Sci.* **365**, 2093–2106.

Woodward, G., Blanchard, J., Lauridsen, R.B., Edwards, F.K., Jones, J.I., Figueroa, D., Warren, P.H., and Petchey, O.L. (2010). Individual-based food webs: Species identity, body size and sampling effects. *Adv. Ecol. Res.* **43**, 211–266.

Woyke, T., Teeling, H., Ivanova, N.N., Huntemann, M., Richter, M., Gloeckner, F.O., Boffelli, D., Anderson, I.J., Barry, K.W., Shapiro, H.J., Szeto, E., Kyrpides, N.C., *et al.* (2006). Symbiosis insights through metagenomic analysis of a microbial consortium. *Nature* **443**, 950–955.

Yoder-Himes, D.R., Chain, P.S.G., Zhu, Y., Wurtzel, O., Rubin, E.M., Tiedje, J.M., and Sorek, R. (2009). Mapping the *Burkholderia cenocepacia* niche response via high-throughput sequencing. *Proc. Natl. Acad. Sci. USA* **106**, 3976–3981.

Yoshida, T., Jones, L.E., Ellner, S.P., Fussmann, G.F., and Hairston, N.G. (2003). Rapid evolution drives ecological dynamics in a predator-prey system. *Nature* **424**, 303–306.

Yvon-Durocher, G., Montoya, J.M., Trimmer, M., and Woodward, G. (2010). Warming alters the size spectrum and shifts the distribution of biomass in freshwater ecosystems. *Glob. Chang. Biol.* (in press).

Yvon-Durocher, G., Reiss, J., Blanchard, J., Ebenman, B., Perkins, D.M., Reuman, D.C., Thierry, A., Woodward, G., and Petchey, O.L. (2010). Across ecosystem comparisons of size structure: Methods, approaches, and prospects. *Oikos* (in press).

Yvon-Durocher, G., Allen, A.P., Montoya, J.M., Trimmer, M., and Woodward, G. (2010). The temperature dependence of the carbon cycle in aquatic systems. *Adv. Ecol. Res.* **43**, 267–313.

Zak, D.R., Blackwood, C.B., and Waldrop, M.P. (2006). A molecular dawn for biogeochemistry. *Trends Ecol. Evol.* **21**, 288–295.

Zaura, E., Keijser, B.J.F., Huse, S.M., and Crielaard, W. (2009). Defining the healthy "core microbiome" of oral microbial communities. *BMC Microbiol.* **9**, e259.

Zuckerkandl, E., and Pauling, L. (1965). Molecules as documents of evolutionary history. *J. Theor. Biol.* **8**, 357–366.

Assessing the Contribution of Micro-Organisms and Macrofauna to Biodiversity–Ecosystem Functioning Relationships in Freshwater Microcosms

JULIA REISS, R.A. BAILEY, FERNANDA CÁSSIO, GUY WOODWARD AND CLÁUDIA PASCOAL

SUMMARY

A large body of research has revealed (often) positive biodiversity–ecosystem functioning (B–EF) relationships in manipulative experiments. The vast majority of such studies have focused on either micro- or macro-organisms, and none we are aware of have manipulated the diversity of both simultaneously under controlled laboratory conditions. We performed a microcosm experiment in which we manipulated species richness of aquatic fungi and

ADVANCES IN ECOLOGICAL RESEARCH VOL. 43
© 2010 Elsevier Ltd. All rights reserved
0065-2504/10 $35.00
DOI: 10.1016/S0065-2504(10)43004-4

invertebrates, two taxonomically distant sets of consumers that contribute to the same key ecosystem process in freshwaters, the decomposition of terrestrial leaf litter. We used a novel statistical design to maximize parsimony and analytical power in an experiment with three levels of species richness (seven mono-culture, 21 di-culture, and seven tri-culture treatments). Litter decomposition was measured as both leaf mass loss and the production of fine particulate organic matter (FPOM). We tested whether species richness affected these two processes or whether polycultures performed as predicted from their component mono-cultures. Further, we calculated assemblage metabolism in each microcosm to test whether the processes were driven by the metabolic demands of fungi and invertebrates.

In general, across the 35 treatments, most species combinations performed in an additive fashion and we found no effect of species richness on either process. There was evidence of assemblage identity effects (i.e. certain species combinations not performing as expected), with instances of significant differences for species combinations that contained both caddis larvae and fungi. These assemblages performed worse than expected, which might have been due to dual vertical and horizontal interactions, with the possibility that although both consumed litter directly the former may also have grazed on the latter. Apart from these particular species combinations, overall performance of a species in polyculture was effectively the same as in mono-culture and reflected its metabolic demands. This suggests that even taxonomically distant consumers might exhibit a degree of functional redundancy for certain processes provided the remaining species can attain sufficient population biomass (and hence metabolic capacity) to compensate for the loss of other species, although whether such compensatory mechanisms operate in the field remains unknown. Further species contribute to a multitude of ecosystem processes and progressively more species are needed to sustain the sum of them. Our experiment highlights how, by taking metabolic demands into account, future B–EF studies could help to disentangle how species contribute to ecosystem processes both separately and in combination, and to help partition the effects of taxonomic and functional diversity.

I. INTRODUCTION

Ecosystem processes, such as resource consumption, are driven to a large extent by the biomass and metabolic needs of those organisms that sustain them (Reiss *et al.*, 2009), and because small organisms have higher mass-specific metabolic demands than larger organism they should support faster process rates per unit biomass (Brown *et al.*, 2004). In addition, species identity and diversity (Cardinale *et al.*, 2002) also play a role, as revealed by numerous biodiversity–ecosystem functioning (B–EF) experiments. The latter point has received substantial attention in recent years, with the underlying rationale being that species-rich assemblages perform better than species poor ones at driving ecosystem processes because of, for instance, complementarity or facilitation (Cardinale *et al.*, 2002). However, some experimental studies have shown that species richness or identity does not always affect ecosystem processes (e.g. Dang *et al.*, 2005, see Thompson and Starzomski, 2007), suggesting a degree of redundancy, with different species performing similar roles. Such redundancy could help to compensate for species loss, if the remaining species respond by increasing their abundance or biomass. However, species contribute to more than a few ecosystem processes at a time and biodiversity becomes more important the more processes are assessed (Gamfeldt *et al.*, 2008; Reiss *et al.*, 2009; Zavaleta *et al.*, 2010). Many researchers have also argued that the functional diversity of an assemblage is a more suitable predictor of ecosystem functioning than is taxonomic richness *per se* (Hooper *et al.*, 2002; Reiss *et al.*, 2009; Woodward, 2009).

Functional diversity of consumer assemblages is often closely coupled with taxonomic similarity, and taxa that are not closely related might have very different mechanisms of resource consumption (Woodward, 2009). Decomposition of leaf litter, for example, is driven by both micro- and macro-organism assemblages, suggesting that these organisms might not be functionally identical and could potentially complement each other and consume the resource faster and more efficiently in combination than on their own: certainly, their modes of resource acquisition are very different from one another (Hieber and Gessner, 2002).

B–EF laboratory experiments have often involved detritivore assemblages, but they have been conducted involving assemblages of either microbial decomposers (the main decomposers are fungi; Graça *et al.*, 2005; Pascoal and Cássio, 2004) or invertebrate detritivores but, to the best of our knowledge, none have explicitly manipulated the biodiversity of both and measured their impacts on more than one process simultaneously. Leaf-eating fauna (shredders) have been shown to benefit from the presence of aquatic fungi (aquatic hyphomycetes) which condition leaf litter, thereby improving its palatability (e.g. Graça, 2001; Graça *et al.*, 2005), and previous studies have demonstrated that shredders often prefer conditioned over unconditioned leaves (Graça *et al.*, 2001). Whether shredders have a reciprocal effect by stimulating fungal growth remains a moot point. Some studies have shown that leaf-eating invertebrates do not influence fungal activity in field trials (Chung and Suberkropp, 2008; Ferreira and Graça, 2006). There is also evidence of detritivores either feeding on unconditioned litter (if less effectively; Graça *et al.*, 2001) or feeding primarily on decomposers rather than the litter itself with some caddis larvae instars deriving up to 100% of their biomass from fungal decomposers and not the leaves themselves

(Chung and Suberkropp, 2009). Together, these studies suggest that a wide range of B–EF relationships (e.g. negative or positive) might be manifested when decomposers and detritivores co-exist (Gessner *et al.*, 2010).

Although functionally distinct species might exhibit different mechanisms for resource consumption, the rate of resource depletion will be driven, to a large extent, by the biomass and metabolic demands of the consumer assemblage as a whole (Brown *et al.*, 2004), but this has rarely been addressed within a B–EF context (but see Perkins *et al.*, this volume). By integrating information on biomass and metabolism into novel experimental designs the respective importance of species identity, richness and functional traits (e.g. body size and metabolic capacity (MC)) can be explored and potentially partitioned to provide new insights into B–EF relationships.

A few B–EF studies have included the MC of invertebrates, based on their assemblage biomass, and found it to be a significant predictor of processes (e.g. McKie *et al.*, 2008; Perkins *et al.*, this volume; Ruesink and Srivastava, 2001). However, it remains unclear if these findings for macrofauna also apply to fungi, which are the main micro-consumers of leaf litter (Graça *et al.*, 2005; Hieber and Gessner, 2002; Pascoal and Cássio, 2004). It is likely that, as with macro-organisms, litter decomposition by fungi is also mostly determined by their metabolic demands, but this is potentially difficult to quantify because their mycelia do not have a clearly defined individual body mass.

When manipulating the biodiversity of macrofauna and fungi simultaneously, we might expect to see a range of potential B–EF relationships being manifested. Firstly, leaf decomposition could be driven by the interaction of these functionally different groups, with richness effects arising from complementarity or facilitation and/or assemblage identity effects. These interactions could occur within or between groups (e.g. vertical effects because shredders can feed on fungi). Secondly, the MC of consumers might be the ultimate driver of leaf decomposition irrespective of taxonomy *per se*, with species identity effects being manifested primarily via differences in body mass.

In the current study, we tested for B–EF relationships by measuring litter decomposition in microcosms with predetermined assemblages of fungal (four species) and macrofaunal consumers (three species). Our principal aims were to test for potential species richness effects, whether fungi and macrofauna perform in an additive fashion, and whether certain species combinations are facilitative (i.e. whether certain polycultures perform better than their component mono-cultures). We tested for these potential effects by employing a combination of linear models that accounted for all these effects. We chose to focus on these three biodiversity effects and did not explicitly address the full range of potential vertical versus horizontal diversity effects with our experimental design. In addition to investigating these

B–EF relationships, we also sought to test for the potential underlying effects of biomass and MC of the consumer assemblages. Decomposers contribute to many aspects of leaf decomposition and therefore we evaluated not only leaf mass loss but also the production of fine particulate organic matter (FPOM), which gave us two response variables.

II. MATERIAL AND METHODS

A. Organisms and Set-Up

Three invertebrate detritivore species (the amphipod *Gammarus pulex* (L.), the isopod *Asellus aquaticus* (L.) and larvae of the caddis-fly *Potamophylax rotundipennis* (Brauer) and four aquatic hyphomycete species (*Tricladium splendens* Ingold, *Alatospora acuminata* Ingold, *Anguillospora filiformis* Greath. and *Heliscus lugdunensis* Sacc. and Thérry) were used in the experiment. All seven species occur within the Iberian Peninsula, where the current study was carried out (Cortes *et al.*, 1998; Duarte *et al.*, 2006; Graça *et al.*, 1994; Medeiros *et al.*, 2009; Pascoal *et al.*, 2005c). Fungi were isolated from single spores collected in rivers in NW Portugal, and the invertebrates were collected from the field 2 weeks prior to the experiment. Fungal species were cultured on agar plates at room temperature until the mycelia had covered the entire surface of the plate (after Duarte *et al.*, 2006). In the experiment, alder litter (*Alnus glutinosa* L.) was used as the food resource for both macrofauna and fungi. Air-dried alder leaves were weighed (3 g for each microcosm) and leached in 105 labeled containers filled with 500 ml distilled water for 2 days. The leaves were then wrapped in labeled aluminum foil and sterilized in an autoclave for 40 min. We prepared 105 sterile microcosms (500 ml Erlenmeyer flasks) containing a mixture of 1:3 parts stream to distilled water by autoclaving the flasks (containing the water and sealed with cotton wool at the neck). To maintain aseptic conditions and to prevent potential contamination, we autoclaved all airstones (a Pasteur pipette filled with cotton wool to filter air) in aluminum foil. These were then inserted individually into each microcosm and held in place with sterile cotton wool plugs, which sealed the neck of each Erlenmeyer flask. The experiment was run in a controlled-temperature room (18 °C) with a 12 h light:12 h dark cycle.

To create different levels of species richness, all seven species were assembled in mono-, di-, and tri-cultures (Figure 1A, Table 1). For logistical reasons, we reduced the total number of tri-culture treatments with an experimental design that provided more statistical power than using random combinations: in tri-culture, each species was in combination with one of the other six species once (Figure 1B). Any collection of tri-cultures with

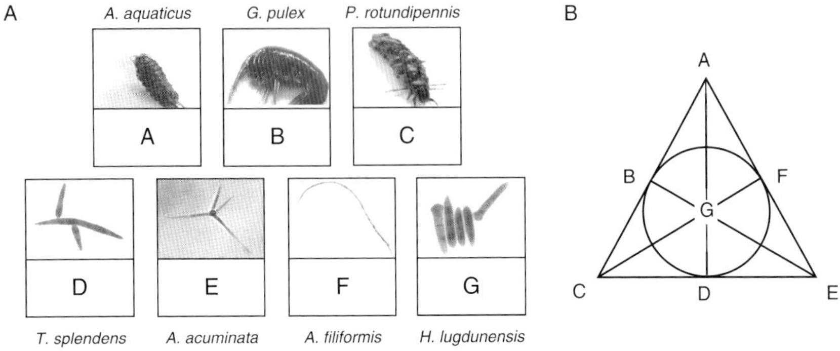

Figure 1 (A) Labels used for all seven species; (B) in tri-culture, the seven species (A–G) were not assembled in all possible combinations but instead we used a reduced design where every species was in combination with the other six species once. The diagram shows how this can be visualized: every straight line (and the circle) in the triangle will give the necessary species combination (e.g. AFE, BGE or BDF).

Table 1 The experimental design using three shredder species (A–C) and four fungal species (D–G)

	Mono-culture	Di-culture	Tri-culture
No. individuals and/or plug slice	12	6+6	4+4+4
Assemblage identity	A–G	AB, AC, AD, . . ., FG	ABC, AFE, AGD, BGE, CDE, CGF, BDF
No. of treatments	7	21	7
Rationale	All possible mono-cultures	All possible combinations	Each species in combination with another of the six species once

In the case of shredders, 12 individuals were added in mono-culture and for fungi 12 agar plug slices containing mycelia were added. In di-culture, we added half of the numbers in mono-culture and in tri-culture we added 1/3.

this property is called a ‘balanced incomplete-block design’ (Bailey, 2008). We used the combinations shown in Figure 1B (see also Bailey, 2008). Since each species is present in some tri-cultures, information on their performance is available for all three richness levels. Moreover, since each pair of species is present in some tri-cultures, we can obtain information on their pairwise interactions. Statistical power in a balanced incomplete-block design is higher than for random combinations (Bailey, 2008) because (i) each species is present equally often and (ii) each pair of species is present equally often. This design gave us seven mono-cultures, 21 ‘di-cultures’ and seven (instead of 35) 'tri-culture' treatments (Table 1). Thus, there were 35 treatments (assemblage identities) in total, which were randomly assigned to microcosms in one block. All treatments were replicated three times (three blocks and 105 microcosms in total).

Fungal species were added to the microcosms 7 days prior to the introduction of the macrofauna enabling them to establish under sterile conditions. Adding the fungi prior to the fauna was necessary, because macrofauna can carry bacteria and fungal spores in their guts that cannot be excluded entirely by starving the fauna. We rinsed and left all invertebrates in sterile water (autoclaved 1:3 parts stream and distilled water) 2 hours before they were added to the microcosms: a longer time period was not used, in order to avoid stressing the animals unduly. We used small cores (diameter of 5.5 mm) of the agar the fungi were cultured on, which were cut into four pieces ('plug slice'; see Duarte *et al.*, 2006) (Table 1). We also froze whole plugs from each fungal species for initial ergosterol determination, as this provides a proxy measure of fungal biomass (e.g. Hladyz *et al.*, 2009). Because we wanted to test for additive effects, we added either 12, 6 or 4 individuals for fauna or 12, 6 or 4 plug slices for fungi to the mono-, di-, and tri-cultures they had been assigned to (Table 1). All fungal species established visible biomass on the leaves within 1 week. Macrofaunal species were allowed to feed for 21 days and then the experiment was terminated.

B. Biomass Determination and Response Variables

To determine faunal biomass in the microcosms, high-resolution digital photographs were taken at $\times 100$ magnification of the individuals added to each microcosm at the start of the experiment. The body length of each individual was then measured with the image analysis software Image-Pro® Plus (Media Cybernetics, Inc.) and converted into dry body mass using log–log length–body mass relationships, which we derived for each species (*A. aquaticus*: $y = 2.652x - 1.841$, $r^2 = 0.94$, $P < 0.0001$; *G. pulex*: $y = 3.015x - 2.242$, $r^2 = 0.94$, $P < 0.0001$; *P. rotundipennis*: $y = 0.190x + 1.508$, $r^2 = 0.74$, $P < 0.0001$) by measuring the length of 25 randomly chosen individuals which were then dried at 40 °C to constant mass and weighed. The length of *P. rotundipennis* was defined as a straight line linking front to back on the concave (i.e. dorsal) surface of its protective case and body mass was measured as mg dry mass of the insect with the case removed. Total faunal biomass per microcosm was derived by summing the body mass of all the individuals within each assemblage.

At the end of the experiment, we cut 10 small leaf disks (diameter 8 mm) from each microcosm for ergosterol determination, which were then frozen at -80 °C. The remaining leaf material was dried at 80 °C to constant mass and weighed. Leaf mass loss was then calculated after correcting for losses caused by leaching (after Hladyz *et al.*, 2009). In addition, the FPOM in the microcosms, consisting of faeces and finely shredded leaf material (<1 mm diameter), was dried and weighed separately.

To estimate fungal biomass, we determined the amount of mycelium (i.e. ergosterol) in the plugs we added at the beginning of the experiment and also on the leaves at the end of the experiment. We freeze-dried the mycelium plugs and frozen leaf disks and used the ergosterol extraction method described in Pascoal *et al.* (2010). Ergosterol was converted to fungal biomass using a conversion factor of 5.5 μg ergosterol mg^{-1} fungal mycelium (Gessner and Chauvet, 1993). We used these initial and final estimates of fungal biomass to establish whether biomass of fungal species would differ between the four mono-culture treatments, to potentially correct our estimate for assemblage metabolism (see below).

In summary, our response variables were leaf mass loss and FPOM produced over 28 days. Both response variables were not expressed as rates, because the duration of leaf mass decomposition differed for fungi and invertebrates.

C. Predictors

Our predictors for the response were species richness and its associated attributes (e.g. 'assemblage identity', see statistical models below), macrofaunal biomass in the microcosms (summed body masses in each microcosm) and 'assemblage metabolism' (Perkins *et al.*, this volume). We included the latter predictor because we hypothesized that the metabolic demands of the consumers would be a key driver of resource consumption (McKie *et al.*, 2008; Perkins *et al.*, this volume). We calculated the metabolic demand for one unit of mass for each species and then extrapolated this to the MC of each microcosm (see Perkins *et al.*, this volume).

To do this for fauna, we assumed that metabolic demand would scale with body mass to the power of 3/4, following a general allometric scaling relationship that applies to a wide range of organisms, from unicells to large metazoans (Brown *et al.*, 2004; Peters, 1983). We used this to derive the MC of each assemblage, by summing the MC of all individuals in a microcosm (McKie *et al.*, 2008; Perkins *et al.*, this volume).

$$\text{MC(Metabolic capacity)} = \sum (M^{3/4}) \tag{1}$$

where M is individual body mass (mg).

To derive an estimate for MC of fungi, we assumed that spore size would determine how quickly the species would grow and consume resources, also following the same metabolic power law. Such an attempt to calculate metabolic demand for fungi has not been tested before, but it is supported by the fact that spore mass (or spore size) is negatively related to sporulation rates (Chauvet and Suberkropp, 1998; Duarte *et al.*, 2006), that is species with small spores produce more of them per unit biomass than species with larger spores. Spore mass had been calculated previously (Pascoal and Cássio, 2004) for all four species we used in our experiment (1,715, 219, 952 and 647 pg spore^{-1} for *T. splendens*, *A. acuminata*, *A. filiformis* and *H. lugdunensis*, respectively). Initial (plug) and final biomass in all fungal mono-cultures showed that total fungal biomass did not differ significantly between the four species (analysis of variance, ANOVA, with Tukey test: $F_{3,8}=0.02$, $P=0.99$), that is, all species were added with the same biomass and accumulated comparable biomass throughout the experiment. We therefore did not have to correct MC for the difference in biomass or needed spore counts. The MC of each species in mono-culture was (MC_i) and was calculated from information on average spore mass as:

$$MC_i(\text{Metabolic capacity of individual species}) = M_i^{3/4}, \quad (2)$$

where M_i is the average spore mass of the species. The four values from mono-cultures then gave MC for all species mixtures.

We adjusted MC to resource consumption by applying a correction factor (i_0 [the antilog of the regression between log MC and log leaf mass loss]; Perkins *et al.*, this volume) to derive our estimates of assemblage metabolism. We then summed MC in each microcosm, which gave us an estimate for the total metabolic demand of the assemblage (fungi only, fauna only, or fungi and fauna together, depending on the treatment). We employ the term 'assemblage metabolism' *sensu* Perkins *et al.* (this volume).

$$\text{Assemblage metabolism} = i_0 \sum(\text{per capita } M^{3/4}), \quad (3)$$

where i_0 is an empirically derived constant (after Brown *et al.*, 2004; Perkins *et al.*, this volume). This equation yields predicted values for the expected value for resource consumption by each assemblage, based upon the sum of the metabolic capacities of all consumers.

D. Statistical Analysis

Average leaf mass loss and production of FPOM in the mono-cultures was compared between all seven species in a one-way ANOVA with a Tukey HSD test as a *post hoc* comparison.

To assess B–EF relations in the two experiments, we performed a classical nested ANOVA. We tested a collection of linear models on the data from the

35 assemblage identities (see Table 1). Because the models are related and can be ordered in a hierarchy, ANOVA compares the goodness of fit of the linked models and tests for whether the difference between a model and its related smaller ones can explain the data significantly better or not. The experiment was run simultaneously in three blocks (one replicate per treatment in each block), so block effects were added when ANOVA was performed.

Our combination of all models simultaneously tested for a range of possible biodiversity effects: richness, sampling and identity effects (i.e. where certain species combinations perform differently from predictions derived from their respective mono-cultures), additive mechanisms and interactions between species (i.e. if performance is altered when species are in polyculture).

The models were designed to explain the relationship between richness and the response variables, starting from a small and simple model ('Richness', Table 2), over intermediate models to the largest one ('Assemblage Identity', Table 2). For example, the model 'Richness' would be significant if the average in mono-, di- and tri-cultures was different and the model 'Assemblage Identity' would be significant if certain species combinations would not perform as expected from mono-culture (Table 2). The model 'Type' would be significant if species perform the same in di-culture (or tri-culture) relative to mono-culture. We called this model 'Type' and not 'Species' or 'Additive Effects of Species', to avoid confusion with models used in other B–EF studies. The collection of models accounts for density effects as it can identify whether species change their performance in di- or tri-culture and generate a predicted value for this changed performance, for example effects that would arise because species are released from interspecific competition.

All statistical tests were carried out using the statistical software R 2.5.1 (R Development Core Team 2007).

III. RESULTS

A. Mono-Cultures

In mono-culture, leaf mass loss was higher for the invertebrates than for the fungal species, although *H. lugdunensis* showed leaf mass loss comparable to that of *G. pulex* (Figure 2). Larvae of the caddis *P. rotundipennis* consumed significantly more leaf material than any other species and also produced most FPOM (ANOVA with Tukey test: $F_{6,14}=58.93$, $P<0.0001$ for leaf mass loss and $F_{6,14}=16.89$, $P<0.0001$ for FPOM; Figure 2). The other two invertebrates (*G. pulex* and *A. aquaticus*) produced similar quantities of FPOM to those produced by the fungi (Figure 2). The high leaf mass loss in *H. lugdunensis* cultures was reflected in relatively high amounts of FPOM (Figure 2).

Table 2 Statistical models for the response leaf mass loss

Question	Statistical model	Response depends on:
(1) Is there an effect of species richness?	(1) 'Richness'	Species richness (mono-, di- and tri-culture = 3 levels)
(2) Do species perform in an additive fashion or do they interact?	(2) 'Type'	$y = a_1x_1 + a_2x_2 + a_3x_3 + a_4x_4 + a_5x_5 + a_6x_6 + a_7x_7$
	(3) 'Richness + Type'	level 1: $y = \mathbf{f} + a_1x_1 + a_2x_2 + a_3x_3 + a_4x_4 + a_5x_5 + a_6x_6 + a_7x_7$
		level 2: $y = \mathbf{g} + a_1x_1 + a_2x_2 + a_3x_3 + a_4x_4 + a_5x_5 + a_6x_6 + a_7x_7$
		level 3: $y = \mathbf{h} + a_1x_1 + a_2x_2 + a_3x_3 + a_4x_4 + a_5x_5 + a_6x_6 + a_7x_7$
	(4) 'Richness × Type'	level 1: $y = b_1x_1 + b_2x_2 + b_3x_3 + b_4x_4 + b_5x_5 + b_6x_6 + b_7x_7$
		level 2: $y = c_1x_1 + c_2x_2 + c_3x_3 + c_4x_4 + c_5x_5 + c_6x_6 + c_7x_7$
		level 3: $y = d_1x_1 + d_2x_2 + d_3x_3 + d_4x_4 + d_5x_5 + d_6x_6 + d_7x_7$
(3) Do certain species combinations (assemblage identities) perform better than expected from mono-culture?	(5) 'Assemblage identity'	Assemblage identities: A, B, C, D, E, F, G, AB, AC, etc.

a is the fitted parameter for each species (species A-G) calculated from all the data on the assumption that all species always behave the same as they do in mono-culture; x is the number of individuals (or plugs) in a culture.

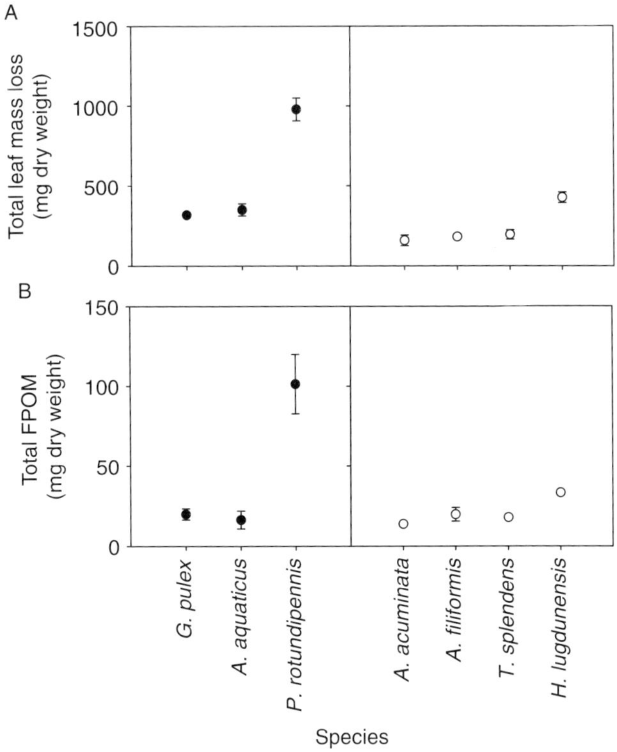

Figure 2 The average (±SE) for the response leaf mass loss (A) and FPOM (B) for all mono-cultures of the seven species (black and white symbols for the invertebrates and fungi, respectively) used in the experiment.

B. Richness Effect, Additive Performance and Assemblage Identity Effects

Leaf mass loss was not affected by species richness; that is mono-, di- and tri-cultures did not vary significantly in their response and average leaf mass loss was very similar between the three levels of richness (Model 'Richness' not significant; Table 3). Average total leaf mass loss was 372 ± 60 SE, 415 ± 32 SE and 421 ± 32 SE mg for mono-, di- and tri-cultures, respectively, at the end of the experiment. In general, species performed in an additive fashion (model 'Type' highly significant; Table 3), meaning leaf mass loss in poly-culture could be predicted from mono-culture. However, effects were not always additive, as a few species combinations showed leaf mass loss that was not predicted from mono-cultures, as revealed by the statistical significance of our 'Assemblage Identity' model (Table 3). Regressing the predicted

Table 3 ANOVA table for response leaf mass loss and FPOM

		Leaf decomposition				FPOM production			
Source	d.f.	SS	MS	F	P	SS	MS	F	P
Richness	2	33,812	16,906	1.27	n.s.	113	56	0.57	n.s.
Type	6	4,220,578	703,430	53.04	< 0.0005	24,949	4158	41.89	< 0.0005
Richness × Type	12	107,612	8968	0.68	n.s.	2448	204	2.05	n.s.
Assemblage identity	14	730,206	52,158	3.93	< 0.005	3807	272	2.74	< 0.05
Block	2	122,075	61,038	–	–	122	61	–	–
Error	68	901,754	13,261	–	–	6750	99	–	–
Total	104	6,116,037	–	–	–	38,189	4851	–	–

values of our 'Type' model against observed leaf mass loss values showed that these assemblage identity effects were due to five species combinations that all contained *P. rotundipennis*. When this species was placed in combination with (1) *H. lugdunensis*, (2) *A. aquaticus*, (3) *A. filiformis* and *H. lugdunensis*, (4) *T. splendens* and *A. acuminata*, assemblage performances were worse than expected, whereas its di-cultures with *A. acuminata* performed better than expected (see Appendix A).

Testing the effects of species richness on FPOM production yielded very similar results to those observed for leaf mass loss; assemblage identity effects were less pronounced but still highly significant (Table 3). These, again, were due to three species combinations containing the caddis fly larvae: *P. rotundipennis* with *H. lugdunensis*; *A. filiformis* and *H. lugdunensis* and with *T. splendens* and *A. acuminata* (see Appendix B) showed unexpected FPOM values. Overall, however, species produced the equivalent amounts of FPOM in mono-, di- and tri-culture (model 'Type' highly significant, Table 3).

C. Shredder Biomass and Whole Assemblage Metabolism as Explanatory Variables

Additive performances were explained largely by the biomass of consumers and the MC of the assemblage: there was, for instance, a positive relationship between log shredder biomass and log leaf mass loss ($y = 0.57x - 0.78$, $r^2 = 0.55$, $F = 88.52$, $P < 0.0001$; Figure 3). The calculation of assemblage

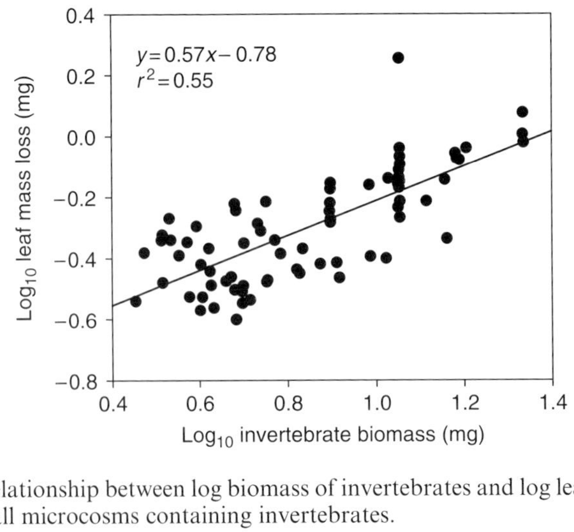

Figure 3 Relationship between log biomass of invertebrates and log leaf mass loss as observed in all microcosms containing invertebrates.

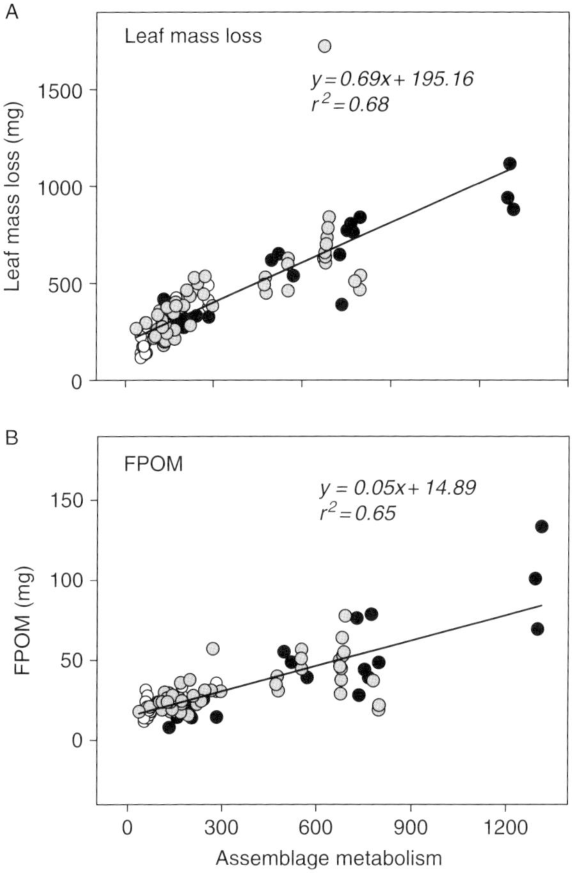

Figure 4 Relationship between assemblage metabolism and leaf mass loss in all microcosms. Microcosms containing fungi only, shredders only and a combination of shredders and fungi are represented with white, black and grey symbols, respectively.

metabolism offered a way to test this for fungi and shredders together (i.e. assemblage metabolism in each microcosm) this accounted for more than two-thirds of the variation in the data ($y = 0.69x + 195.16$, $r^2 = 0.68$, $F = 221.06$, $P < 0.0001$; Figure 4 A). Production of FPOM was also similarly well-explained by assemblage metabolism ($y = 0.05x + 14.89$, $r^2 = 0.65$, $F = 187.90$, $P < 0.0001$; Figure 4B).

However, the occurrence of assemblage identity effects showed that both FPOM production and leaf mass loss were not solely determined by metabolic demands, but also to some extent by species traits that led to non-additive effects in certain combinations (e.g. some assemblages containing *P. rotundipennis*).

IV. DISCUSSION

A. Species Richness, Identity and Metabolism as Drivers of B–EF Relationships

In our experiment, species richness was not a significant predictor of either litter mass loss or FPOM production, which was explained by two effects: additive performance and assemblage identity effects. For example, for leaf mass loss five combinations of species behaved differently from predictions derived from their respective mono-cultures, but these were effectively a minority within an overall additive performance. Additive performance here means that species all had a characteristic way of consuming leaves (and of producing FPOM) and that this did not change when they were in combination with each other.

This additive performance, in terms of both leaf mass loss and FPOM production, reflected the metabolic demands of both the microbial and macrofaunal consumers, which were overlain with other species-specific traits. Our results therefore suggest redundancy between fungi and invertebrates, such that, once population biomass and MC are taken into account, consumers may be largely interchangeable in their effects on the two ecosystem processes we measured, even though they are taxonomically distant from one another. However, in reality, it might not always be possible for such compensation to be expressed, and because they differ in size (and MC) species will have different (but largely predictable) *per capita* effects. Whether these species are truly redundant remains to be seen, as each will contribute to many more ecosystem processes than the two measured here. This caveat also applies to most other B–EF studies, which have typically measured only one or two response variables. As the number of processes measured increases so the probability of redundancy decreases when 'multi-dimensional functioning' is considered (Gamfeldt *et al.*, 2008; Reiss *et al.*, 2009; Zavaleta *et al.*, 2010), and future studies need to consider these additional dimensions of functional diversity more explicitly.

In our study, assemblage identity effects were evident, which can generally be due to effects caused by characteristic interactions between species. In our case, the caddis *P. rotundipennis* caused a significant assemblage identity effect because it behaved differently when it was in combination with certain fungi (combinations). Species identity effects have been reported in other B–EF studies of aquatic fungi (Bärlocher and Corkum, 2003; Duarte *et al.*, 2006; Pascoal *et al.*, 2010; Raviraja *et al.*, 2006), suggesting that species traits could have greater impacts on ecosystem processes than diversity *per se*. Identification of potential keystone species may therefore be important for predicting the consequences of species loss, but this is not always a simple task. For example, in the study of Duarte *et al.* (2006), *Articulospora tetracladia* was the species responsible for the greatest mass loss of alder leaves, while in that of Bärlocher and Corkum (2003), the same species on oak leaves was only dominant under high nutrient concentrations. Thus a species that drives decomposition in one set of circumstances might be redundant in another, lending support to the idea that diversity effects may be highly context- and scale-dependent, for example depend on the environmental conditions, the study period or the number of processes measured (Cardinale *et al.*, 2000; Reiss *et al.*, 2010). Nevertheless, despite this potential for contingency, our experiment identified important and predictable metabolic mechanisms behind B–EF relationships (cf. Perkins *et al.*, this volume). These need to be addressed in future experiments to test their generality and, ultimately, measured under field conditions. Despite the fact that it was a significant predictor of leaf decomposition, the assumption that spore mass determines metabolic demands of fungi (and that a three-quarter law applies to this relationship) needs to be tested more explicitly in laboratory trials.

B. The Roles of Invertebrates and Fungi in Decomposition Processes

In general, leaf-eating invertebrates are the dominant consumers of leaf litter in natural systems (Hieber and Gessner, 2002). Typically, large species are often assumed to be lost first when ecosystems are placed under environmental stress (Raffaelli, 2004); hence, it is important to understand whether smaller species could mitigate their loss. Large species have lower PB (production divided by biomass) ratios than smaller species and they are less abundant. Secondary production is hypothesized to be independent of body size (Brown *et al.*, 2004), which suggests that small organisms might be able to perform the role of larger organisms if the former are able to compensate for the loss of the latter by increasing in abundance. However, if entire body size classes (across species) vanish from a system, the smaller organisms might not be able to compensate fully for this loss in terms of secondary production and other process rates.

Field studies assessing the relative contribution of microbes and invertebrates to leaf decomposition have repeatedly highlighted that it is faster when shredders have access to the leaves (e.g. Hladyz *et al.*, 2009). Leaf processing by shredders can be extremely effective: for instance, Azevedo-Pereira *et al.* (2006) estimated that the shredder *Lepidostoma hirtum* alone could consume the average annual input of litter within a stream. Similarly, others (Feio and Graça, 2000; González and Graça, 2003) have also concluded that another caddis, *Sericostoma vittatum*, can consume most of the allochthonous leaf input in Portuguese streams. However, shredders might be less effective in systems exposed to environmental stress: for instance, under acidic (Larrañaga *et al.*, 2010) or eutrophic (Pascoal *et al.*, 2005a) conditions the importance of micro-organisms may be far more pronounced.

Despite the ubiquity of micro-organisms and their crucial role in carbon and nutrient cycling, we still have a limited understanding of the relationships between microbial diversity and ecological processes. Fungal diversity is believed to be important for decomposition and related processes in terrestrial and aquatic systems (Bell *et al.*, 2009), and novel molecular techniques are starting to provide important new insights into much of the previously hidden biodiversity of the microbial world (Hurd *et al.*, this volume, 2010). Natural leaf-associated fungal assemblages in streams typically have higher species richness (e.g. 8–28 species, Duarte *et al.*, 2009) than the maximum tested in microcosm experiments (up to eight species, Dang *et al.*, 2005). However, they are usually dominated by two to three species (Duarte *et al.*, 2009; Fernandes *et al.*, 2009; Pascoal *et al.*, 2005b), as in our microcosms, so the level of diversity used in our study is realistic, at least in terms of the number of the main drivers involved.

Positive relationships between aquatic fungal species richness and leaf-litter decomposition have been reported from previous microcosm studies (Bärlocher and Corkum, 2003; Duarte *et al.*, 2006; Pascoal *et al.*, 2010; Raviraja *et al.*, 2006; Treton *et al.*, 2004), but in other studies there is evidence for redundancy (Dang *et al.*, 2005), with responses saturating at low levels of diversity (Woodward, 2009). In most studies, however, individual fungal performances were weighted by initial inocula of the constituent species, which ignores the potentially disproportionate activity and consumption of some species and therefore their ultimate contribution to leaf decomposition. In our study, however, fungal biomass did not vary significantly between species and we were also able to account for their MC.

Most of the researches have demonstrated the significance of species identity or traits, highlighting the importance of focusing research on broader aspects of biodiversity (e.g. Gamfeldt *et al.*, 2008). Ascribing the relative contribution of each fungal species in mixtures has been a major hindrance where attempts have been made to separate complementarity and selection effects unequivocally. Application of molecular techniques (denaturing gradient gel electrophoresis – DGGE) to fungi on decomposing leaves has recently made it possible to overcome this problem (Pascoal *et al.*, 2010), although like other PCR-based techniques, DGGE is subject to biases during DNA extraction and amplification, and newer sequencing techniques coupled with Quantitative PCR could potentially quantify the role of each fungal species more accurately (Hurd *et al.*, 2010).

Complementarity effects among aquatic hyphomycete species have been clearly demonstrated for biomass production (Duarte *et al.*, 2006; Pascoal *et al.*, 2010) and litter decomposition (Treton *et al.*, 2004), although we found no evidence of them in our study. As pointed out by Gessner *et al.* (2010), there appears to be great potential for complementarity among microbial decomposers, and niche differentiation may result from the heterogeneity of leaf material, allowing fungi to colonize different areas in decomposing leaves, from fungal succession during leaf decomposition (Bärlocher, 1992; Gessner *et al.*, 1993) or from the production by different species of different arrays of extracellular enzymes against a range of plant polymers (Suberkropp *et al.*, 1983). Conversely, competitive interaction between aquatic hyphomycete species growing on litter has been detected, via reduced conidium production (Pascoal *et al.*, 2010; Treton *et al.*, 2004).

Despite the wealth of theoretical developments of B–EF laboratory studies that have been conducted in recent years, the impact of biodiversity loss in nature is difficult to predict due to the presence of potentially confounding variables and spatial and temporal variability (Loreau *et al.*, 2001). The impoverishment of fungal assemblages due to anthropogenic stress has had variable effects on leaf decomposition in streams, but there is evidence that within assemblages, increases in the biomass or abundance of tolerant species can compensate for the loss of sensitive species (Pascoal *et al.*, 2005b). According to Niyogi *et al.* (2002), changes in functioning are predicted to follow changes in biomass and suppression of functioning is related to suppression of biomass, rather than to loss of biodiversity, and our experiment supports this view, as processes were largely driven by the total amount of metabolic activity, irrespective of species richness.

So far, relationships between biodiversity and plant litter decomposition rates have only been examined at short-term and small scales, and further research is needed to understand diversity–function relationships when scaling experiments up from small homogeneous patches to large heterogeneous spatial scales (Giller *et al.*, 2004). Because microbes have short generation times relative to macrofauna, they are particularly well-suited for addressing the long-term consequences of biodiversity loss at the intergenerational scales where the mechanisms that underpin B–EF relations are most likely to be manifested (Reiss *et al.*, 2010).

C. Interactions Between Decomposers

The assemblage identity effects in this study deserve some further attention, because they show that leaf-eating invertebrates and fungi did interact, albeit only for a limited number of species combinations and not as an overall trend. Interestingly, these species combination all contained the caddis species,

P. rotundipennis. These assemblage identity effects were mainly negative, meaning we did not observe the 'classic' facilitation effects often reported in other B–EF studies. One possible reason for this is that both vertical and horizontal interactions might have been in operation, that is the caddis fed mostly on fungal biomass instead of leaf material and therefore leaf mass loss was reduced in microcosms containing these species combinations. The 'grazing' of fungi is known to be employed by other caddis (e.g. Chung and Suberkropp, 2009), so it is conceivable that *P. rotundipennis* might be inhibiting microbial activity in this way in our study. Lecerf *et al.* (2007) have highlighted the importance of fungal species identity to leaf consumption by shredders and found that *Gammarus fossarum* consumed oak leaves inoculated with *Goniopila monticola* more than twice as fast as those inoculated with other fungal species. Indeed, the influence of fungal identity on consumption by shredders has been demonstrated in other studies, although not in a B–EF context (Arsuffi and Suberkropp, 1985; Bärlocher and Kendrick, 1973), and Krauss *et al.* (2005) suggest that shredder preference for fungi depends primarily on their activity (amount of spores produced) rather than their biomass. Although fungi can enhance leaf-litter decomposition through positive effects on resource quality for shredders (Graça *et al.*, 1993), few data are available on how fungal diversity might affect the richness and composition of invertebrate detritivores (but see Ferreira and Graça, 2006).

We did not assess bacteria in our experiment and although they can contribute to leaf decomposition their contribution seems to be minor relative to fungi (Gulis and Suberkropp, 2003; Pascoal and Cássio, 2004; Pascoal *et al.*, 2005a). Although both synergistic and antagonistic relationships between fungi and bacteria have been reported, the former typically provide the latter with resources that they cannot acquire on their own (Gulis and Suberkropp, 2003): it would be instructive, therefore, to carry out more sophisticated B–EF experiments in the future that also characterize and manipulate bacterial diversity, and the new generation of sequencing technologies that have emerged recently could prove invaluable in this regard (Hurd *et al.*, this volume).

Interactions between decomposers and detrivores may be horizontal and/or vertical and our results suggested that at least one shredder may have fed primarily on fungi, rather than litter *per se*. In future B–EF studies, it would be desirable to develop more complex designs that can test more explicitly if species richness manipulations across trophic levels produce effects that are fundamentally different from those within trophic levels. Although fungi and shredders both contribute to the same process, they can also be regarded as separate trophic levels with bottom-up effects flowing from fungi to shredders, and top-down effects of shredders on fungi (Woodward, 2009). The statistical models would have to account for this by testing whether vertical diversity effects are significantly different from horizontal diversity effects. In our relatively simple experiment, regressing our predicted values of the model

'Type' against observed values showed that most species combinations that did interact were indeed 'vertical' combination (i.e. shredder–fungi combinations), but more sophisticated designs are needed to test this particular hypothesis more rigorously.

D. Conclusions

Our experiment showed that an underlying driver of ecosystem processes was the total metabolic activity of the consumer assemblages, largely irrespective of species identity or richness: this predictor variable could be incorporated into future B–EF studies to aid identification of the mechanisms that underlie biodiversity effects (Reiss *et al.*, 2009). Our reduced statistical design could be applied in such studies to help simplify these logistically challenging experiments, which otherwise may contain dozens of treatments: we were able to reduce the set-up by 28 treatments while maintaining statistical power that would have been lost when using a random design. In addition, the incorporation of metabolism can be used to assess the impacts of not only biomass but also environmental temperature in B–EF studies, which could be key to predicting the potential effects of global warming (e.g. Perkins *et al.*, this volume). Although our data suggested the potential for redundancy, this could be overestimated because, as in most other B–EF studies, we examined only two processes, whereas in reality ecosystem functioning is composed of many more dimensions. Consequently, future B–EF research needs to incorporate a wider variety of both response and predictor variables to gauge the true potential of redundancy within species assemblages.

ACKNOWLEDGEMENTS

This work was funded by a grant from the Natural Environment Research Council, UK (grant reference: NE/D013305/1), awarded to G. W. We thank Aitor Larrañaga and Brendan McKie for their comments on an earlier version of this manuscript.

APPENDIX

Predicted leaf mass loss (A) and FPOM (B) as generated by the statistical model 'Type' regressed against observed values. Assemblage identities not performing as predicted by the model are highlighted with letters (see Figure 1 for key).

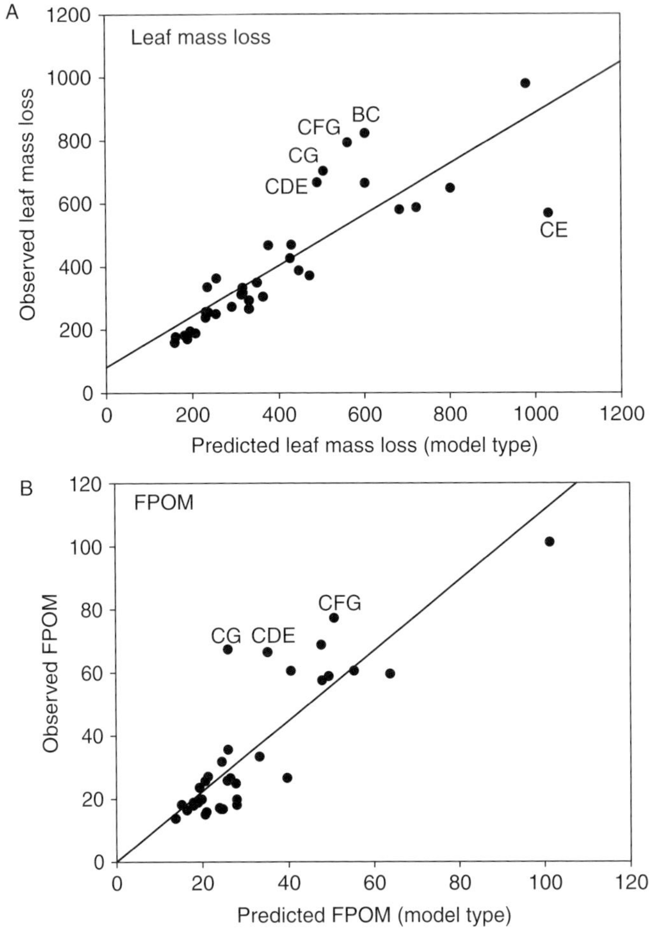

REFERENCES

Arsuffi, T.L., and Suberkropp, K. (1985). Selective feeding by stream caddisfly (trichoptera) detritivores on leaves with fungal-colonized patches. *OIKOS* **45**, 50–58.

Azevedo-Pereira, H.V.S., Graça, M.A.S., and Gonzalez, J.M. (2006). Life history of *Lepidostoma hirtum* in an Iberian stream and its role in organic matter processing. *Hydrobiologia* **559**, 183–192.

Bailey, R.A. (2008). *Design of Comparative Experiments*. Cambridge University Press, Cambridge.

Bärlocher, F. (1992). Effects of drying and freezing autumn leaves on leaching and colonization by aquatic hyphomycetes. *Freshw. Biol.* **28**, 1–7.
Bärlocher, F., and Corkum, M. (2003). Nutrient enrichment overwhelms diversity effects in leaf decomposition by stream fungi. *OIKOS* **101**, 247–252.
Bärlocher, F., and Kendrick, B. (1973). Fungi and food preferences of *Gammarus pseudolimnaeus*. *Arch. Hydrobiol.* **72**, 501–516.
Bell, T., Gessner, M.O., Grifths, R.I., McLaren, J.R., Morin, P.J., van der Heijden, M., and van der Putten, W.H. (2009). Microbial biodiversity and ecosystem functioning under controlled conditions and in the wild. In: *Biodiversity, Ecosystem Functioning, and Human Wellbeing: An Ecological and Economic Perspective* (Ed. by S. Naeem, D. Bunker, M. Loreau, A. Hector and C. Perring). Oxford University Press.
Brown, J.H., Gillooly, J.F., Allen, A.P., Savage, V.M., and West, G.B. (2004). Toward a metabolic theory of ecology. *Ecology* **85**, 1771–1789.
Cardinale, B.J., Nelson, K., and Palmer, M.A. (2000). Linking species diversity to the functioning ecosystems: On the importance of environmental context. *OIKOS* **91**, 175–183.
Cardinale, B.J., Palmer, M.A., and Collins, S.L. (2002). Species diversity enhances ecosystem functioning through interspecific facilitation. *Nature* **415**, 426–429.
Chauvet, E., and Suberkropp, K. (1998). Temperature and sporulation of aquatic hyphomycetes. *Appl. Environ. Microbiol.* **64**, 1522–1525.
Chung, N., and Suberkropp, K. (2008). Influence of shredder feeding and nutrients on fungal activity and community structure in headwater streams. *Fundam. Appl. Limnol.* **173**, 35–46.
Chung, N., and Suberkropp, K. (2009). Contribution of fungal biomass to the growth of the shredder, *Pycnopsyche gentilis* (Trichoptera: Limnephilidae). *Freshw. Biol.* **54**, 2212–2224.
Cortes, R.M.V., Ferreira, M.T., Oliveira, S.V., and Godinho, F. (1998). Contrasting impact of small dams on the macroinvertebrates of two Iberian mountain rivers. *Hydrobiologia* **389**, 51–61.
Dang, C.K., Chauvet, E., and Gessner, M.O. (2005). Magnitude and variability of process rates in fungal diveristy-litter decomposition relationships. *Ecol. Lett.* **8**, 1129–1137.
Duarte, S., Pascoal, C., Cássio, F., and Bärlocher, F. (2006). Aquatic hyphomycete diversity and identity affect leaf litter decomposition in microcosms. *Oecologia* **147**, 658–666.
Duarte, S., Pascoal, C., Garabetian, F., Cássio, F., and Charcosset, J.Y. (2009). Microbial decomposer communities are mainly structured by trophic status in circumneutral and alkaline streams. *Appl. Environ. Microbiol.* **75**, 6211–6221.
Feio, M.J., and Graça, M.A.S. (2000). Food consumption by the larvae of *Sericostoma vittatum* (Trichoptera), an endemic species from the Iberian Peninsula. *Hydrobiologia* **439**, 7–11.
Fernandes, I., Duarte, S., Cássio, F., and Pascoal, C. (2009). Mixtures of zinc and phosphate affect leaf litter decomposition by aquatic fungi in streams. *Sci. Total Environ.* **407**, 4283–4288.
Ferreira, V., and Graça, M.A.S. (2006). Do invertebrate activity and current velocity affect fungal assemblage structure in leaves? *Int. Rev. Hydrobiol.* **91**, 1–14.
Gamfeldt, L., Hillebrand, H., and Jonsson, P.R. (2008). Multiple functions increase the importance of biodiversity for overall ecosystem functioning. *Ecology* **89**, 1223–1231.

Gessner, M.O., and Chauvet, E. (1993). Ergosterol-to-biomass conversion factors for aquatic hyphomycetes. *Appl. Environ. Microbiol.* **59**, 502–507.

Gessner, M.O., Thomas, D.N., Jeanlouis, A.M., and Chauvet, E. (1993). Stable successional patterns of aquatic hyphomycetes on leaves decaying in a summer cool stream. *Mycol. Res.* **97**, 163–172.

Gessner, M.O., Swan, C.M., Dang, C.K., McKie, B.G., Bardgett, R.D., Wall, D.H., and Hättenschwiler, S. (2010). Diversity meets decomposition. *Trends Ecol. Evol.* **25**, 372–380.

Giller, P.S., Hillebrand, H., Berninger, U.G., Gesser, M.O., Hawkins, S., Inchausti, P., Inglis, C., Leslie, H., Malmqvist, B., Monaghan, P.J., Morin, P.J., and O'Mullan, G. (2004). Biodiversity effects on ecosystem functioning: Emerging issues and their experimental test in aquatic environments. *OIKOS* **104**, 423–436.

González, J.M., and Graça, M.A.S. (2003). Conversion of leaf litter to secondary production by a shredding caddis-fly. *Freshw. Biol.* **48**, 1578–1592.

Graça, M.A.S. (2001). The role of invertebrates on leaf litter decomposition in streams – A review. *Int. Rev. Hydrobiol.* **86**, 383–393.

Graça, M.A.S., Maltby, L., and Calow, P. (1993). Importance of fungi in the diet of *Gammarus pulex* and *Asellus aquaticus*. II. Effects on growth, reproduction and physiology. *Oecologia* **96**, 304–309.

Graça, M.A.S., Maltby, L., and Calow, P. (1994). Comparative ecology of *Gammarus pulex* (L.) and *Asellus aquaticus* (L.) I: Population dynamics and microdistribution. *Hydrobiologia* **281**, 155–162.

Graça, M.A.S., Cressa, C., Gessner, M.O., Feio, M.J., Callies, K.A., and Barrios, C. (2001). Food quality, feeding preferences, survival and growth of shredders from temperate and tropical streams. *Freshw. Biol.* **46**, 947–957.

Graça, M.A.S., Bärlocher, F., and Gessner, M.O. (2005). *Methods to Study Litter Decomposition: A Practical Guide*. Springer, Dodrecht, The Netherlands.

Gulis, V., and Suberkropp, K. (2003). Leaf litter decomposition and microbial activity in nutrient-enriched and unaltered reaches of a headwater stream. *Freshwater Biol.* **48**, 123–134.

Hieber, M., and Gessner, M.O. (2002). Contribution of stream detrivores, fungi, and bacteria to leaf breakdown based on biomass estimates. *Ecology* **83**, 1026–1038.

Hladyz, S., Gessner, M.O., Giller, P.S., Pozo, J., and Woodward, G. (2009). Resource quality and stoichiometric constraints on stream ecosystem functioning. *Freshw. Biol.* **54**, 957–970.

Hooper, D.U., Solan, M., Symstad, A.J., Diaz, S., Gesser, M.O., Buchmann, N., Degrange, V., Grime, P., Hulot, F., Mermillod-Blondin, F., Roy, J., Spehn, E., *et al.* (2002). Species diversity, functional diversity, and ecosystem functioning. In: *Biodiversity and Ecosystem Functioning. Synthesis and Perspectives* (Ed. by M. Loreau, S. Naeem and P. Inchausti), pp. 195–208. Oxford University Press, Oxford.

Hurd et al., this volume

Krauss, G., Sridhar, K.R., and Bärlocher, F. (2005). Aquatic hyphomycetes and leaf decomposition in contaminated groundwater wells in Central Germany. *Arch. Hydrobiol.* **162**, 416–428.

Larrañaga, A., Layer, K., Basaguren, A., Pozo, J., and Woodward, G. (2010). Consumer body composition and community structure in a stream is altered by pH. *Freshwater Biol.* **55**, 670–680.

Lecerf, A., Risnoveanu, G., Popescu, C., Gessner, M.O., and Chauvet, E. (2007). Decomposition of diverse litter mixtures in streams. *Ecology* **88**, 219–227.

Loreau, M., Naeem, S., Inchausti, P., Bengtsson, J., Grime, J.P., Hector, A., Hooper, D.U., Huston, M.A., Raffaelli, D., Schmid, B., Tilman, D., and Wardle, D.A. (2001). Ecology – Biodiversity and ecosystem functioning: Current knowledge and future challenges. *Science* **294**, 804–808.

McKie, B.G., Woodward, G., Hladyz, S., Nistorescu, M., Preda, E., Popescu, C., Giller, P.S., and Malmqvist, B. (2008). Ecosystem functioning in stream assemblages from different regions: Contrasting responses to variation in detritivore richness, evenness and density. *J. Anim. Ecol.* **77**, 495–504.

Medeiros, A.O., Pascoal, C., and Graça, M.A.S. (2009). Diversity and activity of aquatic fungi under low oxygen conditions. *Freshw. Biol.* **54**, 142–149.

Niyogi, D.K., McKnight, D.M., and Lewis, W.M. (2002). Fungal communities and biomass in mountain streams affected by mine drainage. *Arch. Hydrobiol.* **155**, 255–271.

Pascoal, C., and Cássio, F. (2004). Contribution of fungi and bacteria to leaf litter decomposition in a polluted river. *Appl. Environ. Microbiol.* **70**, 5266–5273.

Pascoal, C., Cássio, F., Marcotegui, A., Sanz, B., and Gomes, P. (2005a). Role of fungi, bacteria, and invertebrates in leaf litter breakdown in a polluted river. *J. North Am. Benthol. Soc.* **24**, 784–797.

Pascoal, C., Cássio, F., and Marvanova, L. (2005b). Anthropogenic stress may affect aquatic hyphomycete diversity more than leaf decomposition in a low-order stream. *Arch. Hydrobiol.* **162**, 481–496.

Pascoal, C., Marvanová, L., and Cássio, F. (2005c). Aquatic hyphomycetes diversity in streams of Northwest Portugal. *Fungal Divers.* **19**, 109–128.

Pascoal, C., Cássio, F., Nikolcheva, L., and Bärlocher, F. (2010). Realized fungal diversity increases functional stability of leaf litter decomposition under zinc stress. *Microb. Ecol.* **59**, 84–93.

Perkins, D.M., McKie, B.G., Malmqvist, B. Gilmour, S.G., Reiss, J., and Woodward, G. (2010). Environmental warming and biodiversity-ecosystem functioning in freshwater microcosms: partitioning the effects of species identity, richness and metabolism *Adv. Ecol. Res.* **43,** 177–209.

Peters, R.H. (1983). *The Ecological Implications of Body Size*. Cambridge University Press, Cambridge.

Raffaelli, D. (2004). How extinction patterns affect ecosystems. *Science* **306**, 1141–1142.

Raviraja, N.S., Nikolcheva, L.G., and Bärlocher, F. (2006). Fungal growth and leaf decomposition are affected by amount and type of inoculum and by external nutrients. *Sydowia* **58**, 98–104.

Reiss, J., Bridle, J.R., Montoya, J.M., and Woodward, G. (2009). Emerging horizons in biodiversity and ecosystem functioning research. *Trends Ecol. Evol.* **24**, 505–514.

Reiss, J., Forster, A., Cássio, J., Pascoal, C., Stewart, R., and Hirst, A.G. (2010). When microscopic organisms inform general ecological theory. *Adv. Ecol. Res.* **43,** 45–85.

Ruesink, J.L., and Srivastava, D.S. (2001). Numerical and per capita responses to species loss: Mechanisms maintaining ecosystem function in a community of stream insect detritivores. *Oikos* **93**, 221–234.

Suberkropp, K., Arsuffi, T.L., and Anderson, J.P. (1983). Comparison of degradative ability, enzymatic activity, and palatability of aquatic hyphomycetes grown in leaf litter. *Environ. Microbiol.* **46**, 237–244.

Thompson, R., and Starzomski, B.M. (2007). What does biodiversity actually do? A review for managers and policy makers. *Biodivers. Conserv.* **16**, 1359–1378.

Treton, C., Chauvet, E., and Charcosset, J.-Y. (2004). Competitive interaction between two aquatic hyphomycete species and increase in leaf litter breakdown. *Microb. Ecol.* **48**, 439–446.

Woodward, G. (2009). Biodiversity, ecosystem functioning and food webs in freshwaters: Assembling the jigsaw puzzle. *Freshwater Biol.* **54**, 2171–2187.

Zavaleta, E.S., Pasari, J.R., Hulvey, K.B., and Tilman, G.D. (2010). Sustaining multiple ecosystem functions in grassland communities requires higher biodiversity. *Proc. Natl. Acad. Sci. USA* **107**, 1443–1446.

Environmental Warming and Biodiversity–Ecosystem Functioning in Freshwater Microcosms: Partitioning the Effects of Species Identity, Richness and Metabolism

DANIEL M. PERKINS, BRENDAN G. MCKIE,
BJÖRN MALMQVIST, STEVEN G. GILMOUR, JULIA REISS
AND GUY WOODWARD

ADVANCES IN ECOLOGICAL RESEARCH VOL. 43
© 2010 Elsevier Ltd. All rights reserved
0065-2504/10 $35.00
DOI: 10.1016/S0065-2504(10)43005-6

SUMMARY

Predicting the effects of global warming on biodiversity–ecosystem functioning (B–EF) relationships is complicated by potential interactions among abiotic and biotic variables at multiple levels of organisation, including adaptation within regional species populations and changes in community composition and species richness. We investigated the capacity for assemblages of three freshwater invertebrate consumer species (*Asellus aquaticus*, *Nemoura cinerea* and *Sericostoma personatum*) from temperate (southern England) and boreal (northern Sweden) regions to respond to expected shifts in temperature and basal resources, and quantified rates of a key ecosystem process (leaf-litter decomposition). Predictions of assemblage metabolism, derived from allometric-body size and temperature scaling relationships, accounted for approximately 40% of the variance in decomposition rates. Assemblage species composition accounted for further variance, but species richness *per se* had no discernible effect. Regional differences were evident in rates of leaf decomposition across temperature and resource manipulations, and in terms of the processing efficiency of temperate and boreal consumers of the same species (i.e. after correcting for body size and metabolic capacity), suggesting that intraspecific variation among local populations could modulate B–EF effects. These differences have implications for extrapolating how environmental warming and other aspects of climate change (e.g. species range shifts) might affect important drivers of ecosystem functioning over large biogeographical scales.

I. INTRODUCTION

A. Climate Change and Biodiversity–Ecosystem Functioning Relationships

The effects of global warming on community composition and species richness have now been described for a wide range of systems (Burgmer *et al.*, 2007; Parmesan and Yohe, 2003; Thomas *et al.*, 2004; Walther *et al.*, 2002). Far less is known, however, about how climate change and global biodiversity loss, both of which are occurring at rates that are unprecedented in recent history, are likely to interact to affect ecosystem functioning (Chapin *et al.*, 1997; Perkins *et al.*, 2010). This is especially true for fresh waters (Giller *et al.*,

2004; Woodward, 2009), where extinction rates can be as high as those observed in tropical rainforests (Ricciardi and Rasmussen, 1999).

Positive relationships between biodiversity measures (typically species richness) and key ecosystem processes, such as decomposition, have emerged from numerous "biodiversity–ecosystem functioning" (B–EF) experiments (see reviews by Chapin *et al.*, 1997; Gessner *et al.*, 2010; Giller *et al.*, 2004; Ptacnik *et al.*, 2010). Very few studies have investigated how B–EF relationships might be altered by climate change. The few exceptions from fresh waters have focused primarily on microbial (e.g. Dang *et al.*, 2009; Petchey *et al.*, 1999) rather than macrofaunal systems (Woodward, 2009). Since climate change and biodiversity loss could have synergistic effects across many levels of biological organisation (IPCC, 2007), we need to be able to gauge a range of biotic responses to warming, from changes in the physiology of individuals to higher-level B–EF relationships (Ings *et al.*, 2009; Purdy *et al.*, 2010; Reiss *et al.*, 2009, 2010b; Woodward *et al.*, 2010a,b).

Species richness is often positively associated with faster ecosystem process rates, due to a range of mechanisms including: (1) complementary resource use, via niche differentiation; (2) facilitation, whereby the activities of some species enhance those of others; (3) selection effects, in which diverse assemblages are more likely to include species that have disproportionately strong effects (e.g. Cardinale *et al.*, 2002; Emmerson *et al.*, 2001; Naeem *et al.*, 1994). Further, biodiversity can act as an insurance mechanism for process maintenance in changing environments, because diverse communities are more likely to include tolerant species that can compensate for those negatively affected by a given stressor (Yachi and Loreau, 1999).

The importance of species richness in determining ecosystem functioning appears, however, to be neither unequivocal nor ubiquitous. Numerous studies have reported idiosyncratic (i.e. context dependent; Emmerson *et al.*, 2001; McKie *et al.*, 2008), neutral (e.g. species redundancy; Huston and McBride, 2002), or negative B–EF relationships (McKie *et al.*, 2009) and biological drivers may be swamped by abiotic drivers (Huston and McBride, 2002). In the latter case, environmental warming is likely to have particularly pervasive effects on B–EF relationships in fresh waters because these systems are dominated by ectotherms, whose metabolic rates are determined largely by body size and temperature (Perkins *et al.*, 2010; Woodward, 2009). If warming is sufficiently rapid, some species may be forced beyond their normal thermal envelope, thus impairing their physiological performance and ability to contribute to ecosystem processes, and ultimately compromising their long-term persistence (Burgmer *et al.*, 2007; Deutsch *et al.*, 2008; Huey and Stevenson, 1979; Thomas *et al.*, 2004; Vannote and Sweeney, 1980). Even if species richness remains constant, changes in relative abundance and community composition may alter ecosystem processes profoundly (Emmerson *et al.* 2001; McKie *et al.*, 2008).

B. Temperature and Body Size as Drivers of B–EF Relationships

Recent studies have suggested that environmental warming often favours smaller organisms, both within and among species, due to metabolic and physiological constraints that scale allometrically with individual body size (see reviews Perkins *et al.*, 2010; Woodward *et al.*, 2010a,b and references therein). Warming could therefore alter both intra- and interspecific interactions, which together ultimately determine B–EF relationships, but this potentially critical aspect of climate change has been largely overlooked to date (Woodward *et al.*, 2010a).

Potentially important new insights could be gained if the effects of body size and the "metabolic capacity" (MC) of consumers (*sensu* Ruesink and Srivastava, 2001) can be integrated into B–EF experiments, for instance, via the application of the "Metabolic Theory of Ecology" (MTE) (*sensu* Brown *et al.*, 2004). Building upon well-established body size allometries (e.g. Peters, 1983) and temperature scaling (e.g. Gillooly *et al.*, 2001), the MTE can be used to make quantitative predictions about a wide range of biological processes across different levels of organisation, from individuals to ecosystems (Brown *et al.*, 2004; Woodward *et al.*, 2010b; Yvon-Durocher *et al.*, 2010a-d).

Our study focused on rates of detrital processing by invertebrate consumers, a key ecosystem process in stream food webs, many of which are fuelled largely by terrestrial leaf-litter inputs (e.g. Cummins *et al.*, 1973; Gessner *et al.*, 2010; Wallace *et al.*, 1997). Consequently, the rate of leaf-litter decomposition has been the principal response variable measured in most freshwater B–EF research to date (e.g. Jonsson and Malmqvist, 2000, 2003; McKie *et al.*, 2008), although very few studies have explicitly addressed the effects of consumer body mass and/or temperature on this process (but see McKie *et al.*, 2008; Ruesink and Srivastava, 2001). Given that the allometric scaling coefficient between individual body mass and basal metabolic rate is less than 1 (Brown *et al.*, 2004), the size distribution of individuals within and across species within a given assemblage should have a strong influence on process rates: that is, an assemblage composed of small individuals will have higher mass-specific process rates than one composed of larger individuals (e.g. Peters, 1983). Most B–EF research to date, however, has either applied an average body size per species (e.g. Jonsson and Malmqvist, 2000, 2003) or controlled for biomass *a priori* (e.g. Jonsson *et al.*, 2002; Ruesink and Srivastava, 2001). In the few cases where allometric scaling has been incorporated explicitly within B–EF studies (e.g. Mckie *et al.*, 2008), a traditional species averaging approach has been applied, thus losing much of the individual-based information required to scale up accurately to the assemblage level (Woodward *et al.*, 2010a,c).

C. Local and Regional Effects of Climate Change on B–EF Relationships

Local acclimation or adaptation within species populations may also play a role in determining the impacts of climate change (Lucassen *et al.*, 2006; Portner *et al.*, 2006) especially at large biogeographical scales, yet this too has been largely ignored within a B–EF context (Schmid *et al.*, 2002; Woodward *et al.*, 2010a,b; but see Gamfeldt *et al.*, 2005). For example, boreal and temperate populations of the same species could exhibit different responses to warming if they are adapted to their respective local thermal regimes. Consequently, gauging the responses of species populations from different regions should improve our ability to predict the effects of warming at large spatial scales.

We used model experimental systems (i.e. laboratory microcosms) to measure B–EF responses of freshwater assemblages to warming in temperate and boreal latitudes, whilst incorporating information on consumer body size and metabolism. As with all microcosm experiments, there is an unavoidable trade-off between replication and realism (Lawton, 1995), but such an approach still offers one of the few ways in which ecological theory can be tested and B–EF relationships characterised in a controlled and replicated manner (Benton *et al.*, 2007; Giller *et al.*, 2004; Woodward *et al.*, 2010b) as a step towards understanding the potential impacts of climate change in multispecies systems. The microcosms we used each contained leaf-shredding macroinvertebrates and terrestrial leaf litter (cf. Jonsson and Malmqvist, 2000, 2003; McKie *et al.*, 2008; Reiss *et al.*, 2010a) from either a temperate (southern England) and a boreal (northern Sweden) region.

The interpretation of B–EF experiments often requires the application of complex statistical models to disentangle the different mechanisms that may be at play (Reiss *et al.*, 2009; Schmid *et al.*, 2002). We used a factorial treatment design and statistical models that incorporated information on body mass- (Peters, 1983) and temperature-scaling relationships (Gillooly *et al.*, 2001) to test for the effects of richness, assemblage composition, species identity and individual-based traits (i.e. MC). In addition, we conducted the experiments at two different latitudes (southern England and northern Sweden) to also test for potential biogeographical effects resulting from differences in resource type and possible local thermal adaptation among consumer populations. We hypothesised that, after standardising for the effects of consumer size and resource type, temperate consumers (which typically experience greater mean annual temperatures than their boreal counterparts) should be better able to exploit resources at elevated temperatures. We predicted that individual body mass and temperature would determine process rates, due to the constraints they exert on assemblage metabolism, and that these would have primacy over the potential effects of species identity, richness and regional variation in the performance of species populations.

II. METHODS

A. Study Organisms

We used three widespread leaf-shredding species as model organisms in our experimental study: the caddisfly larva *Sericostoma personatum* (Kirby and Spence), the stonefly nymph *Nemoura cinerea* (Retzius) and the crustacean *Asellus aquaticus* (L.), representing three commonly occurring orders (Trichoptera, Plecoptera and Isopoda, respectively) in fresh waters. These species differ in body size and feeding mode: stoneflies and isopods tend to scrape and graze leaf surfaces, whilst caddisflies possess robust biting mouthparts (Graca *et al.*, 1993; Jonsson *et al.*, 2002). All three species are distributed over a broad latitudinal range within Europe (http://www.faunaeur.org/; http://www.freshwaterecology.info/), although stoneflies are generally more strongly associated with the cooler headwaters within a given region than are the other taxa (http://www.freshwaterecology.info/). Individual consumers were collected from field sites in southern England (51°32′N 0°5′W) and near Umeå in northern Sweden (63°8′N 20°2′E) (referred to as "temperate" or "boreal" consumers from here on, respectively), constituting populations with distinct biogeographical histories. Laboratory trials were conducted in the UK (March 2007) and Sweden (May 2007), to coincide with mean spring stream temperatures of approximately 10 and 5 °C in collection sites, respectively. In the laboratory, photoperiod was set to 8:16 h light:dark, and consumers were fed *ad libitum* with *Populus nigra* (L.) litter for 4 weeks prior to the start of the experiments.

B. Laboratory Experiments

Glass microcosms (400 ml), sealed with a mesh net cover (1 mm), were used as experimental units. They were filled with 3 g of air-dried alder leaves and submerged in aerated water baths ($1.0 \times 0.5 \times 0.5$ m) containing diluted stream water in the ratio 1:3 parts stream:degassed and dechlorinated tap water that had comparable pH and nutrient concentrations between regions (see Appendix I). Water baths were maintained at 5, 10 and 15 °C (±1 °C range over the duration of the experiment; Table 1) to simulate the mean spring stream temperatures of approximately 5 and 10 °C in collection sites

Table 1 Experimental design showing the predictors and response used in the trials

	Predictors					
	Region	Temperature	Resource	Richness	Assemblage identity	Response
(a) Description of predictors and response	Temperate,	5, 10, 15	*A. glutinosa,*	1, 2, 3	Monocultures (12 ind.)	Leaf decomposition (mg mass loss day^{-1})
	Boreal		*A. incana*		*A.s; N.c; S. p*	Leaf Processing Efficiency (mg mass loss day^{-1} MC^{-1})
					Dicultures (6+6 ind.) *A. s+N. c; A. s+S. p; N. c+S. p* Triculture (4+4+4 ind.): *A. a+N. c+S. p*	
(b) Number of parameters for each predictor and number of responses estimated	2	3	2	3	7	2

ind., individuals of each species in each microcosm. *A. s*, *Asellus aquaticus*; *N. c*, *Nemoura cinerea*; *S. p*, *Sericostoma personatum*.

in Sweden and the UK, respectively, plus 5 °C above each, in line with predicted 2100 warming scenarios (IPCC, 2007).

To simulate regional differences in riparian vegetation and to account for potential adaptations to local food resources, we ran the above set-up with leaves from two alder species in both the Swedish and English trials. *Alnus glutinosa* [Gaertn] is the dominant species within this genus in southern England, whereas *Alnus incana* [Moench] is dominant in northern Sweden (Tallantire, 1974) (Table 1). The latter typically replaces the former species at colder, higher latitudes and *A. glutinosa* is predicted to continue expanding its range polewards rapidly in response to global warming, to become increasingly dominant in northern Scandinavia by the end of the century (Kullman, 2008). Nutrient quality of the litter was characterised for both leaf types using a C:N elemental analyser (Flask EA Thermo-Finnigan, Bremen, Germany) to determine initial molar carbon-to-nitrogen ratios, which can be an important determinant of decomposition rates (Hladyz *et al.*, 2009). Our boreal leaves (*A. incana*) had significantly lower C:N (mean 22.7 ± 0.60 SE), indicating a better quality resource, than did our temperate leaves (*A. glutinosa* [(24.7 ± 0.36 SE]) ($t = -3.92$, d.f. = 4, $P = 0.016$), although the magnitude of the difference was relatively small (e.g. cf. Hladyz *et al.*, 2009). Leaf-litter was conditioned in the microcosms for 1 week prior to the addition of consumers, giving sufficient time for microbial biofilms to become well established on the leaf surfaces (Cummins *et al.*, 1973; McKie *et al.*, 2008).

We employed a factorial treatment design to partition the possible effects of assemblage identity (i.e. monocultures of individual species and their specific combinations in polycultures) and species richness *per se*. Consumer treatments were designated as follows: three single-species monocultures (12 individuals of each species); three two-species mixtures (6 individuals of each species, equaling 12 in total); one three-species polyculture treatment (four individuals of each species) and a control treatment (lacking macroinvertebrate consumers) to account for physical leaching and microbially mediated leaf decomposition (Table 1). These densities of consumers are comparable to those commonly recorded in the field (e.g. Dangles and Malmqvist, 2004; Hladyz *et al.*, 2009; McKie *et al.*, 2008). Consumer treatments at each laboratory were replicated twice for each resource type, and this was repeated for all three temperatures, to give a total of 168 microcosms (Table 1).

The biomass of consumer assemblages in the microcosms was determined by taking high resolution digital photographs at 100× magnification of the 12 individuals in each microcosm. The body length of each individual was measured using image analysis software (Image Pro Plus 6.3. Media Cybernetics, Inc®) and then converted into dry body mass using power law length–mass regressions derived from independent measurements of 50 randomly selected individuals from a pool of organisms not used in the experiment, for each species. Such length–mass relationships are widely used to provide reliable estimates of invertebrate body mass (cf. Benke *et al.*, 1999) and the equations we derived for each species are as follows: *A. aquaticus*, $y=2.65x-1.84$, $r^2=0.94$; *S. personatum*, $y=1.82x-1.03$, $r^2=0.77$; *N. cinerea*, $y=1.18x-0.76$, $r^2=0.69$. The length of *S. personatum* was defined as a straight line linking front to back along the concave surface of its protective case, and body mass was measured as mg dry mass of the insect with the case removed. In addition, initial elemental C:N content of consumer tissues were measured using three replicates comprising a pooled subset of twelve randomly selected individuals from laboratory stock cultures: that is, 36 individuals of each regional species population, to assess potential stoichiometric imbalances between consumers and their resources (after Hladyz *et al.*, 2009).

Consumers were added to microcosms and checked halfway through the experiment for any emerged or dead individuals, which were removed, photographed and replaced with equivalent-sized organisms from the laboratory stock cultures. Experiments ran for 28 days in total for both Swedish and English trials, at which point leaf mass loss in the fastest monocultures was near 50%. All consumers were then separated from the remaining leaf litter, photographed and counted. The leaf litter remaining at the end of the trial was removed, oven-dried to constant mass at 60 °C and weighed.

C. Predictor Variables

To relate differences in consumer body mass within our biodiversity manipulations (i.e. within and between species populations) and their estimated metabolic demands, we calculated the MC of each consumer assemblage based upon general allometric-body size scaling relationships:

$$Metabolic\ capacity = \sum \left(per\ capita\ M^{3/4}\right) \tag{1}$$

where M is individual body mass (dry weight, mg). The 3/4 exponent used here describes a general allometric scaling relationship between body mass and basal metabolic rate (BMR) that can be applied to a wide range of organisms (see Brown *et al.*, 2004; Peters, 1983). Within species the scaling exponent may vary somewhat from this average value (Glazier, 2005), in part because the range of body masses is far narrower than in the more general relationship that spans many species. Consequently, the use of this general exponent was corroborated by empirical data from the literature detailing intraspecific size–metabolism relationships for the caddisfly and the isopod used in our trials, and in the case of the stonefly from a confamilial species (Appendix II).

We also used MTE to investigate temperature–metabolism relationships in our experiment because it offers a general framework that allows quantitative predictions to be made about the influence of temperature on metabolic rates (e.g. Brown *et al.*, 2004; Gillooly *et al.*, 2001). These scaling relationships are based upon three variables: individual body mass, ambient temperature and an exponential temperature function, the strength of which is determined by an activation energy (AE) of metabolism (typically between 0.6 and 0.7 eV) which describes the temperature dependence of metabolic processes. We first verified that our data met certain criteria relating to allometric-body size scaling and temperature dependence of leaf decomposition as a proxy for community metabolism: that is, (1) MC should scale with decomposition rate, with a slope of unity and (2) the temperature dependence of rates of decomposition should approximate to theoretically expected values of AE (Appendices 3–4). We then calculated an estimate of the capacity of each consumer assemblage to exploit resources based upon the metabolic demands of its constituent members at a given temperature [using the MC calculation (Eq. 1)] and the temperature dependence of metabolism (Brown *et al.*, 2004; Gillooly *et al.*, 2001). The joint effects of body mass and temperature on individual BMR, I, are given by:

$$I = i_0 M^{3/4} \mathrm{e}^{-Ei/kT}, \tag{2}$$

where i_0 is a normalisation constant that converts mass to energy, M is body mass, E_i is the activation energy of respiration (0.63 eV; after Gillooly *et al.*, 2001), k is the Boltzmann constant and T is temperature in Kelvin. This equation provides a prediction of "assemblage metabolism", which is calculated thus:

$$\text{Assemblage metabolism} = i_0 \sum (\textit{per capita } M^{3/4}) \mathrm{e}^{-E/kT} \tag{3}$$

where i_0 is an empirically derived constant (after Brown *et al.*, 2004 and see Appendix III). This equation yields predicted values for the expected rate at which resources should be consumed by each assemblage, based upon the sum of the metabolic capacities of all individual consumers in addition to the environmental temperature, which we used as covariates in the statistical analysis.

D. Response Variables

Two response variables were used in our analyses to investigate different aspects of ecosystem functioning based on traditionally measured bulk rates of leaf decomposition, and rates that were corrected for the MC of consumer assemblages.

(1) Bulk rates of decomposition of leaf litter (i.e. combined microbial- and invertebrate-mediated leaf decomposition) were calculated as the difference in oven dry mass (mg) before and after the experiments, after adjustment for initial leaching for both leaf types (after McKie *et al.*, 2008).
(2) A measure of leaf processing efficiency (LPE) was attained by expressing decomposition rates relative to the MC of each invertebrate consumer assemblage, after adjustments for microbial and leaching contributions (after McKie *et al.*, 2008). This allows the processing efficiency of the assemblages to be estimated relative to their size-related MC (McKie *et al.*, 2008, 2009; Ruesink and Srivastava, 2001).

E. Statistical Methods

The effects of biodiversity and other experimental variables on the response variables were tested using a collection of general linear models. Since variances were homogenous in the Swedish and English trials, the two data sets were analysed together, with "region" fitted as a main effect. The collection of models included mainly factorial effects terms and tested for all experimental predictors on responses (Table 1), but excluded control treatments (which were analysed separately: see Appendix V). In the analysis, the initial models included all factors and all possible higher order interactions. To maximise parsimony, backwards selection was performed and factors were subsequently dropped until the model was minimally adequate (Draper and Smith, 1998): that is, high explanatory power was maintained as measured via coefficient of determination values (r^2), with no cost to inflated deviance measures [here, we used the Akaike Information Criterion (AIC)]. Data from all 168 experimental microcosms were used in the statistical analyses, which were performed on untransformed data using R version 2.9.0 (R Development Core Team, 2009).

We employed a traditional type of factorial effects analysis, which has been used, for example, by Jonsson and Malmqvist (2000) in previous B-EF experiments. The main effects and interactions of experimental variables on rates of leaf decomposition were tested via analysis of variance (ANOVA), based on Type I (sequential) sums of squares. Factors relating to diversity treatments were fitted first: specifically, richness fitted first followed by assemblage identity, followed by assemblage identity nested within richness (e.g. Jonsson and Malmqvist, 2000). We chose our species to represent specific functional types (chewing caddisflies vs. abrading stoneflies and isopods), and so fitted assemblage identity as a fixed rather than a random factor in order to compare their responses explicitly, rather than regarding them as a subset of all possible species. This enabled us to test for the effects of individual species and their specific combinations across regions. Next, region was fitted as a fixed factor, to test for potential divergence between the temperate and boreal assemblages. Subsequently, temperature and resource type were fitted as categorical fixed factors. Assemblage metabolism (i.e. values derived from Eq. 3) was fitted as a covariate term, last of the main factors, to specifically test for additional variation explained by this factor (Draper and Smith, 1998). Following model simplification, richness was eventually dropped from the final model and assemblage metabolism was fitted first (Draper and Smith, 1998). Since Type I ANOVA can be sensitive to the order that factors are fitted, analyses were run with different ordering of factors to ensure a robust interpretation of results.

In addition, we repeated this analysis outlined above, using LPE as the response variable. This enabled us to compare, for instance, the relative efficiency of regional consumer populations, after accounting for differences in MC (Table 2) and microbial contributions (Appendix V) between regions.

III. RESULTS

A. Microbial-Only Mediated Leaf Decomposition

Analysis of the microbial-only treatments (Appendix V) revealed that leaf decomposition increased with temperature ($F_{1,26}=4.82$, $P=0.017$) and was faster for *A. incana* (the better quality resource) than for *A. glutinosa* ($F_{1,26}=42.3$, $P<0.001$). A significant temperature × resource interaction revealed that *A. glutinosa* decomposition rates increased more rapidly with temperature than did *A. incana* ($F_{2,26}=4.76$ $P=0.017$). In addition a region × resource interaction ($F_{1,26}=12.63$, $P<0.001$), indicated that *A. incana* decomposition was faster than for *A. glutinosa* in temperate than boreal locations. There were no significant differences in microbial-only decomposition rates across regions, or interactions between region and temperature (Appendix V).

B. Differences Between Consumers

Consumer survival was high in both the UK (mean ± SE: 84.2 ± 1.6%) and Swedish (79.1 ± 2.5%) experiments (cf. 75–85% survival in Jonsson and Malmqvist, 2003). Mean initial body masses of *A. aquaticus* and *S. personatum* (but not *N. cinerea*) differed between the two regions, with individuals in the temperate populations being on average 31% and 15% larger than those from boreal populations, respectively (Table 2). The initial C:N ratios of consumers were 11% lower for *N. cinerea* in the temperate populations than

Table 2 Taxonomic identity and characteristics of consumer species populations from temperate and boreal regions: mean pre-experiment body mass and C:N ratio, used for *t*-test comparisons between regional populations

Species	Population	Body mass: mean ± SE (dry mass mg)	*t*	d.f.	*P*	Molar C:N: mean ± SE	*t*	d.f.	*P*
Asellus aquaticus (L.)	Temperate	4.90 ± 0.30	−3.84	198	< 0.001	4.76 ± 0.08	1.26	4	n.s.
	Boreal	3.40 ± 0.25				5.33 ± 0.44			
Sericostoma personatum (Kirby and Spence)	Temperate	8.01 ± 0.25	−2.95	198	0.0017	5.72 ± 0.32	1.02	4	n.s.
	Boreal	6.79 ± 0.33				6.05 ± 0.09			
Nemoura cinerea (*Retzius*)	Temperate	0.90 ± 0.03	−0.33	198	n.s.	5.73 ± 0.01	−13.43	4	< 0.001
	Boreal	0.91 ± 0.02				5.08 ± 0.05			

in their boreal counterparts, but there were no significant regional differences for *S. personatum* or *A. aquaticus* populations (Table 2).

C. Leaf Decomposition: Within Regions

Species richness had no effect on leaf decomposition ($F_{2,145}=0.87$, $P=0.42$; Appendix VI): this was a robust result and not due to a Type II error, as even a very small effect size ≥ 0.067 (i.e. a 7% change in the mean) would have been detected with satisfactory statistical power (≥ 0.8). Subsequently, to maximise parsimony, species richness was removed from the final factorial effects models (Table 3a).

The covariate term relating to assemblage metabolism (derived from body size and temperature scaling) was by far the strongest predictor of decomposition rates ($F_{1,145}=332.95$, $P<0.001$; Table 3a). Estimated assemblage metabolism was positively correlated with observed decomposition rates ($r=0.64$, $n=168$, $P<0.0001$), and these data did not differ significantly from an observed:predicted 1:1 line ($y=0.91$, 95% CI$=0.74$–1.08; Figure 1): i.e., decomposition rates matched predicted values based upon information on consumer body sizes and temperature.

Assemblage identity had significant effects on decomposition rates ($F_{6,145}=8.88$, $P<0.001$), even with the assemblage metabolism covariate fitted, indicating the presence of identity effects above those related to body mass *per se* (Table 3a). Essentially, decomposition was fastest for monocultures of *S. personatum* and all assemblages that contained this species

Table 3 Factorial effects ANOVA testing the effects of assemblage metabolism, assemblage identity, temperature, resource identity and region for (a) leaf decomposition ($r^2=0.82$) and (b) leaf processing efficiency ($r^2=0.45$)

		(a) Leaf decomposition (mg mass loss day^{-1})			(b) LPE (mg mass loss day^{-1} MC^{-1})		
	Factor	d.f.	*F*	*P*	d.f.	*F*	*P*
Within region	Assemblage metabolism	1	322.78	<0.001	1	0.12	0.726
	Assemblage identity	6	8.88	<0.001	6	9.18	<0.001
	Temperature	2	7.95	<0.001	2	10.75	<0.001
	Resource	1	139.53	<0.001	1	2.45	0.120
	Temperature × resource	2	7.63	<0.001	2	6.23	0.003
Across regions	Region	1	60.07	<0.001	1	27.83	<0.001
	Region × temperature	2	3.52	0.032	2	2.71	0.069
	Region × resource	1	28.06	<0.001	1	0.04	0.850
	Residual	151	–	–	151	–	–

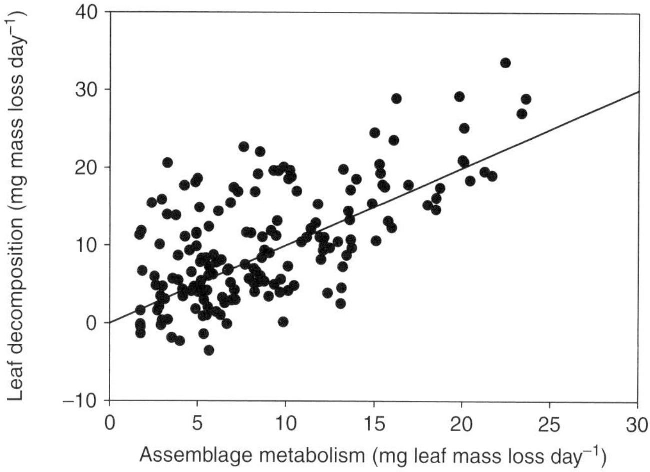

Figure 1 Relationship between estimated rates of assemblage metabolism (mg leaf mass loss day^{-1}) derived from metabolic scaling theory (see Section II) and observed rates of leaf decomposition (mg mass loss day^{-1}) in 168 aquatic microcosms, with 1:1 line fitted.

(Figure 2a), whereas rates in *A. aquaticus* and *N. cinerea* monocultures were statistically identical (Figure 2a).

Resource identity was a strong predictor of decomposition rates ($F_{1,145}=139.6$, $P<0.001$; Table 3a), which were higher for *A. incana* (the better quality resource) than for *A. glutinosa*. Decomposition rates also differed among the three temperatures ($F_{2,145}=7.95$, $P<0.001$), highlighting an additional effect of thermal variation on the performance of the species over and above that incorporated in the estimates of invertebrate assemblage metabolism (Table 3a). In addition to these main effects, a two-way interaction ($F_{2,145}=7.53$ $P<0.001$) revealed that decomposition increased more rapidly for *A. glutinosa* than for *A. incana* with increasing temperature, as was also found for the microbial-only treatments.

D. Leaf Decomposition: Across Regions

Leaf decomposition rates were faster for temperate than boreal consumer assemblages ($F_{1,145}=60.10$, $P<0.001$; Table 3a and Figure 3). A region × resource type interaction ($F_{1,145}=28.01$, $P<0.001$) highlighted that temperate consumers processed the better quality resource (boreal alder) faster than their boreal counterparts. A region × temperature interaction ($F_{2,145}=3.52$, $P=0.032$) revealed further regional differences, with decomposition rates for temperate consumers increasing more rapidly with temperature than was the case for the boreal consumers (Table 3a and Figure 3a).

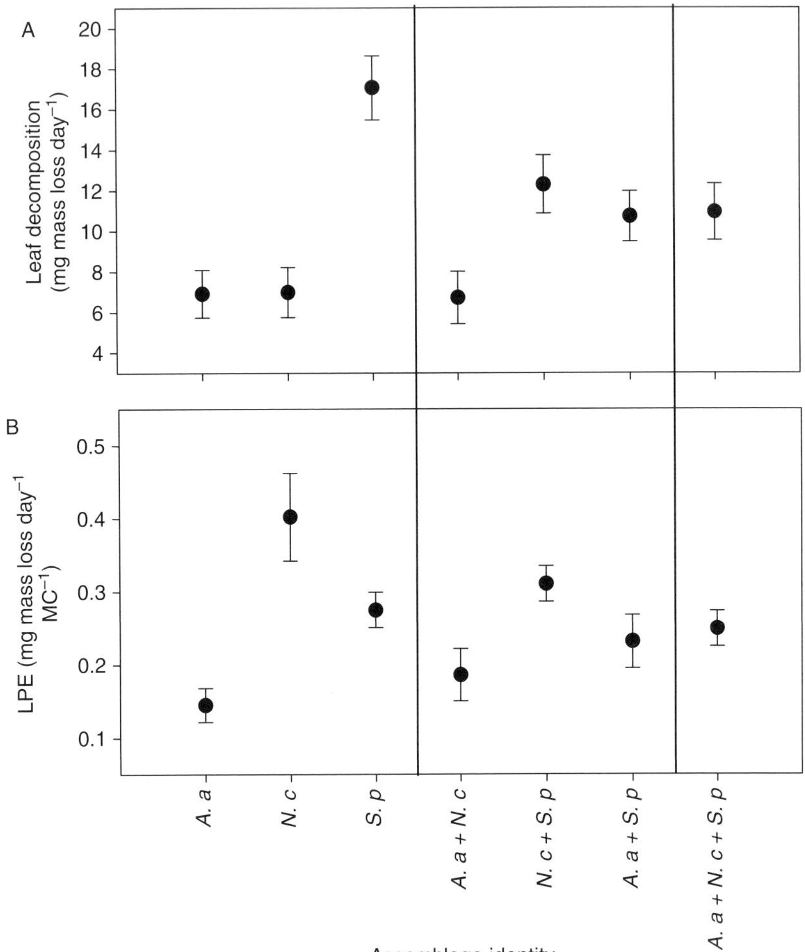

Figure 2 The effects of assemblage identity (pooled over region, temperature and resource treatments, $n=24$ microcosms in total) on (A) leaf decomposition (mg mass loss day^{-1}) and (B) leaf processing efficiency (mg mass loss day^{-1} MC^{-1}). Error bars represent ±SE. Consumer abbreviations *A. a*, *Asellus aquaticus*; *N. c*, *Nemoura cinerea*; *S. p*, *Sericostoma personatum*.

E. LPE: Within Regions

Assemblage identity ($F_{6,145}=9.18$, $P=<0.001$), but not assemblage metabolism, was a significant predictor of LPE (Table 3b), indicating that differences among the assemblages depended on more than body mass alone (since

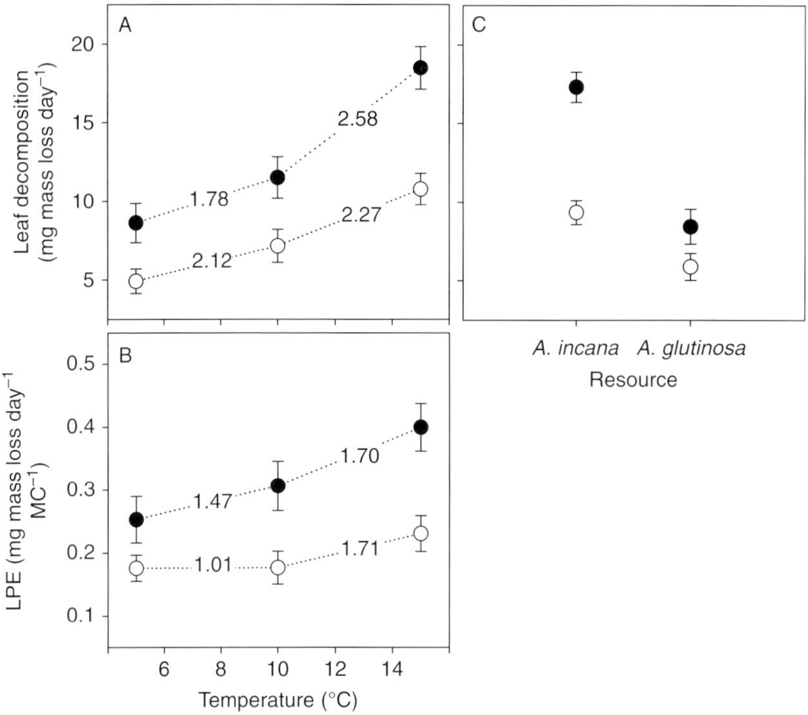

Figure 3 Across region comparisons: (A) Leaf decomposition (mg mass loss day^{-1}) and (B) leaf processing efficiency (mg mass loss day^{-1} MC^{-1}), across thermal regimes. Numbers correspond to Q_{10} values based upon mean values between adjacent temperatures, with values in bold representing statistical differences between regions (region × temperature) and (C) leaf decomposition (mg mass loss day^{-1}) of *A. glutinosa* (native to temperate) and *A. incana* (native to boreal, regions) leaf litter. Black circles represent mean values for temperate consumer populations and clear circles, boreal populations. Error bars represent ±SE.

MC was accounted for in the response). *N. cinerea* displayed highest LPE, followed by *S. personatum* and *A. aquaticus* (Figure 2b).

Temperature had a strong positive effect on LPE ($F_{2,145} = 10.75$, $P < 0.001$) (Figure 3b). In contrast to bulk decomposition rates (Table 3a), LPE did not differ with resource type (Table 3b), indicating the potential influence of microbial contributions to bulk rates. A temperature × resource identity interaction ($F_{2,145} = 6.23$, $P = 0.003$) was evident, with LPE increasing with temperature at a faster rate for *A. glutinosa* than for *A. incana*.

F. LPE: Across Regions

Region was a strong predictor in the collection of models ($F_{1,145}=27.83$, $P<0.001$; Table 3b), with LPE being consistently higher for temperate consumers than for their boreal counterparts. These effects were consistent across resource types and temperature, as revealed by the non-significant the region × resource type and region × temperature interactions (Table 3b).

IV. DISCUSSION

Overall, species richness *per se* had no effect in our study, regardless of whether we corrected for body mass and MC (cf. Reiss *et al.*, 2010a). Ecosystem functioning was, however, determined primarily by several parameters associated with climate change and biogeographical effects: resource identity and temperature had particularly marked effects, with consumer body mass, and to lesser extent consumer identity, also exerting significant influences. Regional differences were evident between the boreal and temperate consumers of the same species, in terms of their processing efficiency (i.e. even after correcting for MC), suggesting that intraspecific variation among local populations could modulate B–EF effects, although these effects were relatively weak when compared with those of temperature *per se*. Together, these results suggest that climate change could influence B–EF relationships in fresh waters in both profound and yet also subtle ways across multiple levels of organisation and at different trophic levels (e.g. resource vs. consumer identity effects). Although this study was out of necessity limited to only a small number of species and carried out in just two locations, it highlights the potential for climate change to influence B–EF relationships in ways that have not been explored previously.

A. Drivers of Ecosystem Functioning

The changes in thermal regimes simulated in our study, in line with predicted global warming scenarios for the next century (IPCC, 2007), had marked influences on leaf decomposition rates, with Q_{10} values ranging from 1.8 to 2.6 for a 5 °C increase. These changes were well predicted by three variables: the MC of assemblages, the environmental temperature and the temperature dependence of metabolism (Brown *et al.*, 2004; Gillooly *et al.*, 2001). The combination of these variables explained approximately 40% of the variation in the dataset as a whole, demonstrating that a considerable, but not exhaustive, portion of the biological complexity of our model systems could be reduced to just a few parameters.

In addition, the effects of resource identity were overlain on these responses to temperature. Overall, decomposition of the boreal *A. incana* litter was about 41% faster than for temperate *A. glutinosa*, matching observations from the microbe-only treatments (Appendix V). These differences may reflect the small, yet statistically significant, differences in the molar C:N content of these resources: *A. incana* had lower C:N relative to *A. glutinosa* and this ratio is a major determinant of leaf-litter decomposition rates (Hladyz *et al.*, 2009).

Temperature × resource identity interactions were also evident, with the poorer quality temperate litter being processed faster at higher temperatures, relative to the better quality boreal litter. This effect remained even after applying microbial and MC corrections and might therefore indicate compensatory feeding among the consumers in order to maximise acquisition of nutrients from a poor-quality resource (Sterner and Elser, 2002), when metabolic demands are high. If this is the case, warming-induced shifts in the taxonomic composition (Kullman, 2008; Tallantire, 1974) and C:nutrient content (Reich and Oleksyn, 2004) of riparian vegetation could have important consequences for freshwater ecosystems that have been overlooked in previous research, especially if the increases in atmospheric CO_2 that are predicted to accompany global warming (IPCC, 2007) are translated into changes in leaf-litter quality at the base of the food web (Ferreira *et al.*, 2010; Woodward *et al.*, 2010b).

Process rates also differed across regions for the same combinations of consumer and resource species, and our experimental design helped us to isolate the main factors behind these biogeographical effects. Since we were able to hold species identity constant, the observed divergences likely reflected intraspecific variation in the performances of regional invertebrate populations within a given assemblage. Intriguingly, regional differences explained more variation in the data than was accounted for by the different species combinations overall (Table 3), suggesting that intraspecific effects might be at least as important as interspecific effects. Our approach also allowed us to correct for differences in consumer body size (and thus MC) across regions, which revealed that temperate consumers were more efficient at leaf-litter processing than were their boreal counterparts. These differences in the capacity to exploit resources might reflect seemingly subtle shifts in consumer-resource stoichiometric imbalances observed across regional populations (Table 2), which could influence decomposition rates (cf. Hladyz *et al.*, 2009). Further studies are needed, however, to assess whether this is a general effect across a wider pool of species.

Finally, there were also marked differences between the two regions: decomposition rates in the temperate systems displayed a stronger and more rapidly accelerating increase with temperature than did those in boreal systems. This was partly, but not wholly, attributable to regional differences in body sizes of consumers (Table 3), suggesting possible intrinsic differences in the potential processing rates of consumers from different areas. Our temperate populations were likely to experience greater annual thermal variability than those in the far north of Sweden, which might favour the evolution of greater flexibility in their responses to temperature (McKie *et al.*, 2004) and hence the maintenance of more "efficient" functioning across a wider thermal gradient. Whilst we found some evidence to lend support to this suggestion, it needs to be tested more rigorously in future experiments, and with a wider range of both consumer and resource species.

B. Species Richness and Identity Effects

We found no effects of increasing consumer richness on ecosystem functioning, despite the high statistical power of our analyses. The low taxonomic relatedness of our study species should have increased the potential scope for niche differentiation to have occurred over evolutionary time relative to more closely related taxa (Petchey and Gaston, 2006). This lack of a significant richness effect suggests functional redundancy in our assemblages, despite their phylogenetic divergence, though the scope for niche differentiation and facilitation may have been limited by the lack of marked resource heterogeneity in our study (cf. Gessner *et al.*, 2010). In contrast, some previous studies using an experimental set-up and duration similar to ours have reported 23–94% increases in process rates in three-species polycultures versus monocultures (Jonsson and Malmqvist, 2000, 2003; Jonsson *et al.*, 2002; McKie *et al.*, 2008). Taken together, these contrasting results highlight the apparent context dependency of B–EF relationships (McKie *et al.*, 2008, 2009). There were also no richness × temperature or assemblage identity × temperature interactions, which should have been detected if the way species interacted with one another (e.g. via competition or facilitation) was temperature dependent. Our experiment confirmed that all our species, which coexist in many European fresh waters, from the temperate to the boreal zone, exhibited essentially similar and predictable responses to temperature, irrespective of whether they were in monoculture or polyculture.

It is possible that richness effects might have become apparent if we had used more diverse assemblages or if we had measured a wider range of ecosystem processes (Gamfeldt *et al.*, 2008; Hector and Bagchi, 2007; Reiss *et al.*, 2009). However, one-to-three leaf-shredding species is typical of the local diversity of the dominant members of this guild in many temperate and boreal streams (e.g. Dangles and Malmqvist, 2004; Woodward *et al.*, 2008). In addition, leaf litter is the predominant energy input to many stream food webs (e.g. Wallace *et al.*, 1997), so its decomposition provides a useful proxy of overall "ecosystem functioning". Running the experiment for longer might have increased the chances of detecting a significant richness effect (Jonsson, 2006), but this also runs the risk of potentially exaggerating competitive interactions as food resources become limiting, which might not be representative of conditions in the field (e.g. Dobson and Hildrew, 1992).

Ecosystem processes are determined by both biotic and abiotic (e.g. temperature) factors and their interactions, but the relative contribution of these drivers has rarely been considered simultaneously in B–EF experiments (Hillebrand and Matthiessen, 2009; Woodward *et al.*, 2010a,b), especially in fresh waters (but see Dang *et al.*, 2009; Petchey *et al.*, 1999). In general, although we found far stronger responses to temperature than to biotic drivers, there were still statistically significant effects for many of the main effects and their interactions. For instance, decomposition by temperate consumers on native leaves increased by 180% from 10 to 15 °C, and for boreal consumers on non-native resources, by 128% from 5 to 10 °C. The effects of climate change might therefore be mediated primarily through temperature change, but there are clearly a range of additional, more subtle, effects that are also in operation.

C. Incorporating Metabolic Constraints in B–EF Experiments

Our study is consistent with others that have emphasised the role of assemblage composition and species identity (rather than diversity *per se*) in driving process rates (Jonsson *et al.*, 2002; Straub and Snyder, 2006; see Cardinale *et al.*, 2006 and references therein). We were able to gain additional insight into the mechanisms behind these effects by using allometric scaling to examine the underlying functional traits (i.e. body mass and MC), which accounted for most of the species-specific differences in performance (see also Reiss *et al.*, 2010a). As with any small-scale study, however, a degree of caution needs to be taken when attempting to extrapolate to natural systems, as a host of other parameters not manipulated in our study could also be important in the "real world" (McKie *et al.*, 2008; Rall *et al.*, 2010).

Some further caveats are associated with the application of the MTE, which, although it has gained extensive support in some recent studies (e.g. Acuna *et al.*, 2008; DeLong *et al.*, 2010; Yvon-Durocher *et al.*, 2010), has also been challenged on various grounds, including its mechanistic basis, potential discrepancies between BMR and field metabolic rate (FMR), the exact values of the scaling exponents and its universality (Clarke and Fraser, 2004; DeLong *et al.*, 2010; Glazier, 2005). Whilst a discussion of the mechanistic explanation and the possible reasons for quarter-power scaling is beyond the scope of this chapter, the potential for discrepancies to occur between field and basal rates merits further consideration. For instance, it has been debated whether differences between FMR and BMR will be reflected primarily in elevation (intercepts), that is, FMR > BMR (Brown *et al.*, 2004; Savage *et al.*, 2004), or in the scaling exponents (Clarke and Fraser, 2004; Glazier, 2005) of body mass–metabolism relationships. FMR might be expected to differ from BMR when animals are subjected to some form of environmental stress absent under controlled laboratory conditions (i.e. where BMR is typically measured). For example, resource limitation, predator avoidance and thermal regimes beyond those normally encountered could all be potential influences in the real world: but these conditions were absent in our laboratory microcosm study. Our results were indeed consistent with the scaling exponents associated with BMR, since MC and assemblage metabolism had a close fit to the observed values of leaf mass loss for all species combinations (see Appendix III), and BMR was a good predictor of decomposition rates in general (Figure 1).

Whilst the MTE is still the subject of ongoing debate, the general framework it offers provides a simple, mechanistic model for testing the predicted effects of temperature effects on biological processes (Woodward *et al.*, 2010a,b; Yvon-Durocher *et al.*, 2010a-d). Our study highlights how, in combination with relatively simple experiments, this approach might be developed to help disentangle some of the seemingly complex drivers that operate in natural systems and to explore the possible underlying mechanisms in B–EF relationships. Notwithstanding the potential caveats associated with the MTE and microcosm experiments in general, our results represent some of the first steps towards gauging the relative importance of a range of abiotic and biotic drivers in multispecies systems under controlled and replicated simulations of projected climate change scenarios.

V. CONCLUSION

Our results suggest that as the climate changes, environmental warming and shifts in the identity and quality of basal resources could be key determinants of detrital processing in fresh waters. In addition to these overriding influences, we also detected more subtle effects associated with non-size-related compositional change of consumers at both the population and assemblage level, with the latter being dependent on the additive performance of species rather than richness effects. Consequently, the order in which consumers (individuals or species) and their associated traits (especially their body size) are lost from local communities in the coming decades could have profound effects on ecosystem functioning (Petchey *et al.*, 2004; Raffaelli, 2004).

Whilst our results suggest that in the medium term (i.e. within this century, as simulated in our experiments) environmental warming may lead to general increases in decomposition rates, significant differences in the performance of consumer populations indicate that species identity and regional effects also need to be considered. As such, this study highlights several new and potentially important synergies that could emerge as our climate changes over the coming century.

ACKNOWLEDGEMENTS

We would like to thank Paul Fletcher (UK), Kai Lehmann and Johan Baudau (Sweden) for assistance in the field and the Natural Environment Research Council for financial support awarded to G. W. (Grant reference: NE/D013305/1), with additional funding from the BioCycle project (Swedish Research Council Grant 2006-375 to B. M.), part of the European Science Foundation's Eurodiversity Program. We would also like to thank Ulrich Brose, Sally Hladyz and two anonymous referees for their comments, which greatly improved this chapter.

APPENDIX I. WATER NUTRIENT COMPARISON ACROSS REGIONS

Five 1 l water samples were taken from stream water used in both experimental trails to determine dissolved inorganic nutrients. From these samples, aliquots (7 ml) were extracted and filtered through a pre-flushed (20 ml deionised water) polyethersulphone membrane filter (0.2 mm Minisart Plustm Sartorius UK, Ltd.). Samples were analysed using a segmented flow auto analyser (Skalar, Netherlands) and standard techniques (Grasshoff *et al.*, 1999). Concentrations of ammonium (NH_4^+), nitrate (NO_3^-), nitrite (NO_2^-) and phosphate (PO_4^{3-}) were quantified (μg/l).

Table S1 Nutrient values for stream water used in trials across regions

	Nutrient concentration (μg/l) ±SE			
Region	Ammonium (NH_4^+)	Nitrate (NO_3^-)	Nitrite (NO_2^-)	Phosphate (PO_4^{3-})
UK	15.1 ± 1.8	23.4 ± 3.5	2.85 ± 3.7	7.2 ± 2.0
Sweden	8.63 ± 1.7	27.6 ± 3.5	1.75 ± 3.7	5.29 ± 2.1

APPENDIX II. INTRASPECIFIC BODY MASS–METABOLISM RELATIONSHIPS FOR FRESHWATER INVERTEBRATES

Table S2 Summary of previous studies of body mass–metabolism relationships for two of the three species of freshwater invertebrates used in the current study (highlighted in bold)

Taxa	Order, family	Source	Metabolism measurement	Body mass range	Temperature range (°C)	Exponent range	Exponent mean
***Asellus aquaticus* (L.)**	Isopoda, Asellidae	Adcock (1982)	Oxygen consumption	0.1–10 mg DW	2–18	0.62–0.84 (9)	0.74
***Sericostoma personatum* (Kirby and Spence)**	Trichoptera, Sericostomatidae	Iversen (1979)	Oxygen consumption	0.5–15 mg DW	3–18	0.68–0.75 (4)	0.70
			Ingestion of leaf material			0.65–0.75 (4)	0.69
Pteronarcys scotti (Ricker)	Plecoptera, Nemouridae	Mcdiffet (1970)	Oxygen consumption	0.5–1.5 g WW	5–15	0.71–0.84 (3)	0.79

DW, dry weight; WW, wet weight.

No data were available for the stonefly *N. cinerea* used in the current study, so data are shown for a related taxon *P. scotti*, from the same family. Numbers in parentheses correspond to the number of body mass–metabolism exponents reported by the study derived from measurements made on greater than 12 individuals. The theoretically predicted exponent of 0.75 (Brown *et al.*, 2004) falls within the ranges of exponents reported for all studies.

APPENDIX III. TESTING THE VALIDITY OF MC CALCULATIONS AND DERIVING AN EMPIRICAL NORMALISATION CONSTANT FOR EQ. 3

To test the validity of applying the approach relating to MC calculations (Eq. 1), we regressed the mean MC of all consumer assemblages against temperature normalised leaf decomposition on a log–log plot (Figure S1). A slope of 1 would reveal isometric scaling between assemblage MC and decomposition (a measure of community metabolism), that is the capacity for and actual metabolic rates are realised. The regression equation $y = 1.02x + 24.5$, $r^2 = 0.84$ reveals that indeed the slope of 1.02 is statistically indistinguishable from that of the predicted value of unity ($t = -0.19$, d.f. $= 24$, $P = 0.85$), so the assumption of a general 3/4 scaling exponent (e.g. after Brown *et al.*, 2004) relating to BMRs was upheld for the conditions of our study.

An empirical normalisation constant for Eq. 3 is derived through the antilog of the intercept (24.5) for the relationship between metabolic capacity and temperature corrected decomposition on a log–log plot (Figure S2). This empirical constant converts mass to energy, which allowed us to calculate predicted values for assemblage metabolism in Eq. 3 (in main text) in units: mg mass loss day^{-1}.

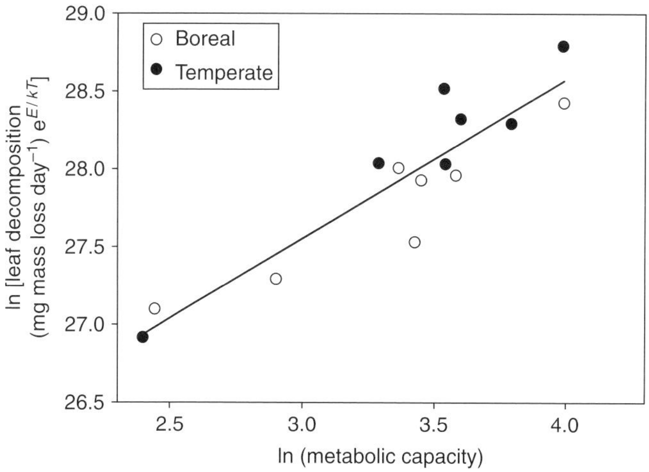

Figure S1 Log–log relationship between metabolic capacity (MC) and temperature normalised leaf decomposition (mg mass loss day^{-1}) for temperate and boreal consumer assemblages (filled and open circles, respectively), where E is the mean activation energy for heterotrophic metabolism (–0.63 after Brown *et al.*, 2004), k is the Boltzmann constant and T is treatment temperature in Kelvin.

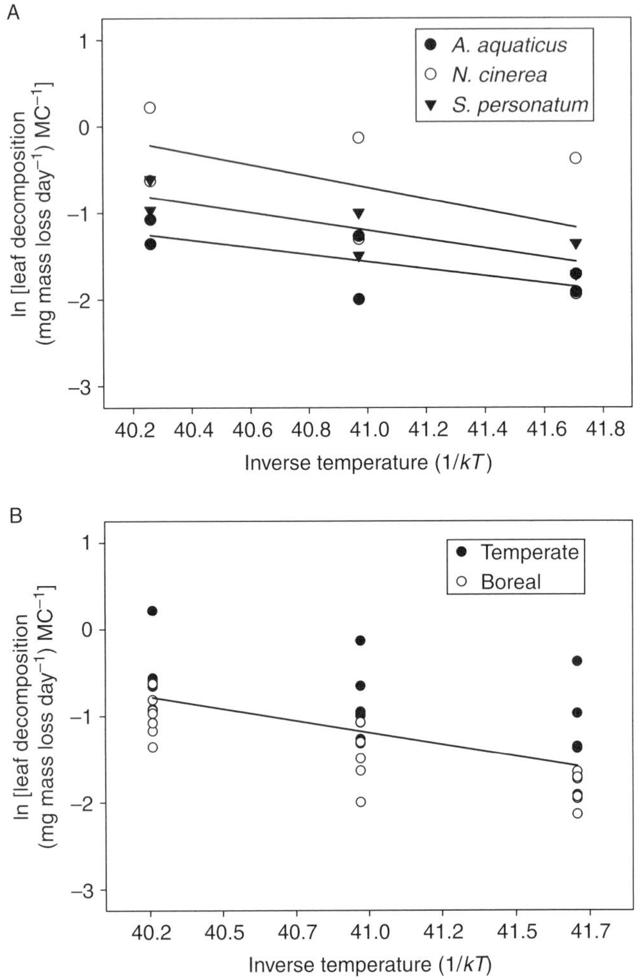

Figure S2 (A) Arrhenius plot of the natural logarithm of mass corrected leaf decomposition (mg mass loss day^{-1}) for monocultures of the three species populations used in the experimental trials as a function of inverse temperature ($1/kT$), where k is the Boltzmann constant and T is the treatment temperature in Kelvin. Analysis of covariance (ANCOVA) revealed significant differences between intercepts (normalisation constants) ANCOVA $F_{2,12}=4.69$, $P=0.031$, but not slopes $F_{2,12}=0.13$, $P=0.87$. Holding slopes constant yields a mean slope of -0.53 (CI $=-0.16$ to -0.88). The mean expected AE for heterotrophic metabolism of -0.63 (Gillooly *et al.*, 2001) falls within these 95% confidence intervals. (B) Arrhenius relationship between the inverse of absolute temperature ($1/kT$) and the natural logarithm of mass corrected leaf decomposition (mg mass loss day^{-1}) across species assemblages from both boreal and temperate populations (open and shaded circles, respectively) $r^2=0.38$, $n=42$, $P<0.001$ regression equation $y=-0.55x+21.3y$. The slope of the regression—0.55 (CI $=-0.33$ to -0.77) corresponds to the activation energy (eV). The mean expected AE for heterotrophic metabolism of -0.63 (Gillooly *et al.*, 2001) falls within these 95% confidence intervals.

APPENDIX IV. TEMPERATURE DEPENDENCE OF LEAF DECOMPOSITION: COMPARING INTRA- AND INTERSPECIFIC RELATIONSHIPS AND EXPECTED VALUES, AS PREDICTED FROM METABOLIC SCALING RELATIONSHIPS

The temperature dependence for rates of leaf decomposition was tested via Arrhenius relationships between the inverse of absolute temperature and mass-corrected decomposition to see if the resulting AE values (slope of this relationship) follows an expected mean value −0.63 (Gillooly *et al.*, 2001 range 0.6–0.7 eV) typically observed for heterotrophic metabolism (Brown *et al.*, 2004; Gillooly *et al.*, 2001).

APPENDIX V. RESULTS FOR MICROBIAL-ONLY TREATMENTS

To gain information on physical leaching and microbial-mediated leaf decomposition, we ran additional treatments lacking macroinvertebrate consumers. Unlike consumer treatments which had two replicates per treatment combination, there were three replicates for microbial-only treatments. The main effects and interactions of experimental variables were tested via analysis of variance (ANOVA) for the response variable leaf decomposition (mg mass loss day^{-1}). Region, temperature and resource identity were fitted as fixed categorical factors. Initial models included all possible high order interactions and model simplification subsequently performed (as described in Section II).

Table S3 ANOVA table for microbe-only treatments, $R^2 = 0.76$

	Factor	Leaf decomposition (mg mass loss day^{-1})		
		d.f.	*F*	*P*
Within region	Temperature	2	4.817	0.017
	Resource	1	42.30	<0.001
	Temperature × resource	2	4.76	0.017
Across regions	Region	1	3.84	0.061
	Region × temperature	2	1.74	0.196
	Region × resource	1	12.63	0.001
	Residual	26	–	–

APPENDIX VI. ANOVA WITH FACTOR "RICHNESS" RETAINED

Table S4 Factorial effects ANOVA testing the effects of richness, assemblage identity, temperature, resource, region and assemblage metabolism for (a) leaf decomposition (mg mass loss day^{-1}) and (b) LPE (mg mass loss day^{-1} MC^{-1})

		(a) Leaf decomposition (mg mass loss day^{-1})			(b) LPE (mg mass loss day^{-1} MC^{-1})		
	Factor	d.f.	*F*	*P*	d.f.	*F*	*P*
Within region	Richness	2	0.88	0.418	2	0.89	0.411
	Assemblage identity	4	44.56	<0.001	4	11.83	<0.001
	Temperature	2	78.86	<0.001	2	7.33	<0.001
	Resource	1	139.72	<0.001	1	2.47	0.118
	Assemblage metabolism	1	12.42	<0.001	1	9.11	0.003
	Temperature × resource	2	7.63	<0.001	2	6.23	0.003
Across regions	Region	1	101.66	<0.001	1	31.66	<0.001
	Region × temperature	2	3.52	0.032	2	2.71	0.069
	Region × resource	1	28.06	<0.001	1	0.04	0.850
	Residual	151	–	–	151	–	–

Richness which was subsequently dropped in main analysis through model simplification and the covariate term assemblage metabolism (which was fitted last of the main factors to specifically test for additional variation explained by this factor) fitted first.

REFERENCES

Acuna, V., Wolf, A., Uehlinger, U., and Tockner, K. (2008). Temperature dependence of stream benthic respiration in an Alpine river network under global warming. *Freshwater Biol.* **53**, 2076–2088.

Adcock, J.A. (1982). Energetic of a population of *Asellus-aquaticus* (Crustacea, Isopoda)—Respiration and energy budgets. *Freshwater Biol.* **12**, 257–269.

Benke, A.C., Huryn, A.D., Schmock, L.A., and Wallace, J.B. (1999). Length-mass relationships for freshwater macroinvertebrates in North America with particular reference to the southeastern United States. *J. N. Am. Benthol. Soc.* **18**, 308–343.

Benton, T.G., Solan, M., Travis, J.M.J., and Sait, S.M. (2007). Microcosm experiments can inform global ecological problems. *Trends Ecol. Evol.* **22**, 516–521.

Brown, J.H., Gillooly, J.F., Allen, A.P., Savage, V.M., and West, G.B. (2004). Toward a metabolic theory of ecology. *Ecology* **85**, 1771–1789.

Burgmer, T., Hillebrand, H., and Pfenninger, M. (2007). Effects of climate-driven temperature changes on the diversity of freshwater macroinvertebrates. *Oecologia* **151**, 93–103.

Cardinale, B.J., Palmer, M.A., and Collins, S.L. (2002). Species diversity enhances ecosystem functioning through interspecific facilitation. *Nature* **415**, 426–429.

Cardinale, B.J., Srivastava, D.S., Duffy, J.E., Wright, J.P., Downing, A.L., Sankaran, M., and Jouseau, C. (2006). Effects of biodiversity on the functioning of trophic groups and ecosystems. *Nature* **443**, 989–992.

Chapin, F.S., Walker, B.H., Hobbs, R.J., Hooper, D.U., Lawton, J.H., Sala, O.E., and Tilman, D. (1997). Biotic control over the functioning of ecosystems. *Science* **277**, 500–504.

Clarke, A., and Fraser, K.P.P. (2004). Why does metabolism scale with temperature? *Funct. Ecol.* **18**, 243–251.

Cummins, K.W., Petersen, R.C., Howard, F.O., Wuycheck, J.C., and Holt, V.I. (1973). Utilization of leaf litter by stream detritivores. *Ecology* **54**, 336–345.

Dang, C.K., Schindler, M., Chauvet, E., and Gessner, M.O. (2009). Temperature oscillation coupled with fungal community shifts can modulate warming effects on litter decomposition. *Ecology* **90**, 122–131.

Dangles, O., and Malmqvist, B. (2004). Species richness-decomposition relationships depend on species dominance. *Ecol. Lett.* **7**, 395–402.

DeLong, J.P., Okie, J.G., Moses, M.E., Sibly, R.M., and Brown, J.H. (2010). Shifts in metabolic scaling, production, and efficiency across major evolutionary transitions of life. *Proceedings of the National Academy of Sciences of the United States of America* **107**, 12941–12945.

Deutsch, C.A., Tewksbury, J.J., Huey, R.B., Sheldon, K.S., Ghalambor, C.K., Haak, D.C., and Martin, P.R. (2008). Impacts of climate warming on terrestrial ectotherms across latitude. *Proc. Natl. Acad. Sci. USA* **105**, 6668–6672.

Dobson, M., and Hildrew, A.G. (1992). A test of resource limitation among shredding detritivores in low order streams in southern England. *J. Anim. Ecol.* **61**, 69–77.

Draper, N.R., and Smith, H. (1998). Applied Regression Analysis. Wiley, New York.

Emmerson, M.C., Solan, M., Emes, C., Paterson, D.M., and Raffaelli, D. (2001). Consistent patterns and the idiosyncratic effects of biodiversity in marine ecosystems. *Nature* **411**, 73–77.

Ferreira, V., GonÇalves, A.L., Godbold, D.L., and Canhoto, C. (2010). Effect of increased atmospheric CO_2 on the performance of an aquatic detritivore through changes in water temperature and litter quality. *Glob. Change Biol.* 10.1111/j.1365-2486.2009.02153.x.

Gamfeldt, L., Wallen, J., Jonsson, P.R., Berntsson, K.M., and Havenhand, J.N. (2005). Increasing intraspecific diversity enhances settling success in a marine invertebrate. *Ecology* **86**, 3219–3224.

Gamfeldt, L., Hillebrand, H., and Jonsson, P.R. (2008). Multiple functions increase the importance of biodiversity for overall ecosystem functioning. *Ecology* **89**, 1223–1231.

Gessner, M.O., Swan, C.M., Dang, C.K., Mckie, B.G., Bardgett, R.D., Wall, D.H., and Hättenschwiler, S. (2010). Diversity meets decomposition. *Trends Ecol. Evol.* **25**, 372–380.

Giller, P.S., Hillebrand, H., Berninger, U.G., Gessner, M.O., Hawkins, S., Inchausti, P., Inglis, C., Leslie, H., Malmqvist, B., Monaghan, P.J., Morin, P.J., and O'Mullan, G. (2004). Biodiversity effects on ecosystem functioning: Emerging issues and their experimental test in aquatic environments. *Oikos* **104**, 423–436.

Gillooly, J.F., Brown, J.H., West, G.B., Savage, V.M., and Charnov, E.L. (2001). Effects of size and temperature on metabolic rate. *Science* **293**, 2248–2251.

Glazier, D.S. (2005). Beyond the '3/4-power law': Variation in the intra- and interspecific scaling of metabolic rate in animals. *Biol. Rev.* **80**, 611–662.

Graca, M.A.S., Maltby, L., and Calow, P. (1993). Importance of fungi in the diet of gammarus pulex and asellus aquaticus. 1. Feeding strategies. *Oecologia* **93**, 139–144.

Grasshoff, K., Kremling, K. and Ehrhardt, M. (Eds.) (1999). Methods of Seawater Analysis Wiley-Vch Verlag GmbH, Weinheim.

Hector, A., and Bagchi, R. (2007). Biodiversity and ecosystem multifunctionality. *Nature* **448**, 188–191.

Hillebrand, H., and Matthiessen, B. (2009). Biodiversity in a complex world: consolidation and progress in functional biodiversity research. *Ecol. Lett.* **12**, 1405–1419.

Hladyz, S., Gessner, M.O., Giller, P.S., Pozo, J., and Woodward, G. (2009). Resource quality and stoichiometric constraints on stream ecosystem functioning. *Freshwater Biol.* **54**, 957–970.

Huey, R.B., and Stevenson, R.D. (1979). Integrating thermal physiology and ecology of ectotherms—Discussion of approaches. *Am. Zool.* **19**, 357–366.

Huston, M., and McBride, A. (2002). Evaluating the relative strength of biotic versus abiotic controls on ecosystem processes. In: *Biodiversity and Ecosystem Functioning: Synthesis and Perspectives* (Ed. by M. Loreau, S. Naeem and I. Inchausti), pp. 47–60. Oxford University Press, Oxford.

Ings, T.C., Montoya, J.M., Bascompte, J., Bluthgen, N., Brown, L., Dormann, C.F., Edwards, F., Figueroa, D., Jacob, U., Jones, J.I., Lauridsen, R.B., Ledger, M.E., *et al.* (2009). Ecological networks—Beyond food webs. *J. Anim. Ecol.* **78**, 253–269.

IPCC (2007). *Contribution of Working Group I to the Fourth Assessment Report of the Intergovernmental Panel on Climate Change.* Cambridge University Press, Cambridge, UK.

Iversen, T.M. (1979). Laboratory energetics of larvae of *Sericostoma personatum* (Trichoptera). *Holarctic Ecol.* **2**, 1–5.

Jonsson, M. (2006). Species richness effects on ecosystem functioning increase with time in an ephemeral resource system. *Acta Oecol.* **29**, 72–77.

Jonsson, M., and Malmqvist, B. (2000). Ecosystem process rate increases with animal species richness: Evidence from leaf-eating, aquatic insects. *Oikos* **89**, 519–523.

Jonsson, M., and Malmqvist, B. (2003). Importance of species identity and number for process rates within different stream invertebrate functional feeding groups. *J. Anim. Ecol.* **72**, 453–459.

Jonsson, M., Dangles, O., Malmqvist, B., and Guerold, F. (2002). Simulating species loss following perturbation: Assessing the effects on process rates. *Proc. Roy. Soc. Lond. B. Biol.* **269**, 1047–1052.

Kullman, L. (2008). Thermophilic tree species reinvade subalpine Sweden—Early responses to anomalous late Holocene climate warming. *Arct. Antarct. Alp. Res.* **40**, 104–110.

Lawton, J.H. (1995). Ecological experiments with model systems. *Science* **269**, 328–331.

Lucassen, M., Koschnick, N., Eckerle, L.G., and Portner, H.O. (2006). Mitochondrial mechanisms of cold adaptation in cod (*Gadus morhua* L.) populations from different climatic zones. *J. Exp. Biol.* **209**, 2462–2471.

McDiffet, W.F. (1970). The transformation of energy by a stream detritivore *Pteronarcys-scotti* (Plecoptera). *Ecology* **51**, 975–988.

McKie, B.G., Cranston, P.S., and Pearson, R.G. (2004). Gondwanan mesotherms and cosmopolitan eurytherms: Effects of temperature on the development and survival of Australian Chironomidae (Diptera) from tropical and temperate populations. *Mar. Freshwater Res.* **55**, 759–768.

McKie, B.G., Woodward, G., Hladyz, S., Nistorescu, M., Preda, E., Popescu, C., Giller, P.S., and Malmqvist, B. (2008). Ecosystem functioning in stream assemblages from different regions: Contrasting responses to variation in detritivore richness, evenness and density. *J. Anim. Ecol.* **77**, 495–504.

McKie, B.G., Schindler, M., Gessner, M.O., and Malmqvist, B. (2009). Placing biodiversity and ecosystem functioning in context: Environmental perturbations and the effects of species richness in a stream field experiment. *Oecologia* **160**, 757–770.

Naeem, S., Thompson, L.J., Lawler, S.P., Lawton, J.H., and Woodfin, R.M. (1994). Declining biodiversity can alter the performance of ecosystems. *Nature* **368**, 734–737.

Parmesan, C., and Yohe, G. (2003). A globally coherent fingerprint of climate change impacts across natural systems. *Nature* **421**, 37–42.

Perkins, D.M., Reiss, J., Yvon-Durocher, G., and Woodward, G. (2010). Global change and food webs in running waters. *Hydrobiologia* **657**, 181–198.

Petchey, O.L., and Gaston, K.J. (2006). Functional diversity: Back to basics and looking forward. *Ecol. Lett.* **9**, 741–758.

Petchey, O.L., McPhearson, P.T., Casey, T.M., and Morin, P.J. (1999). Environmental warming alters food-web structure and ecosystem function. *Nature* **402**, 69–72.

Petchey, O.L., Downing, A.L., Mittelbach, G.G., Persson, L., Steiner, C.F., Warren, P.H., and Woodward, G. (2004). Species loss and the structure and functioning of multitrophic aquatic systems. *Oikos* **104**, 467–478.

Peters, R.H. (1983). The Ecological Implications of Body Size. Cambridge University Press, Cambridge.

Portner, H.O., Bennett, A.F., Bozinovic, F., Clarke, A., Lardies, M.A., Lucassen, M., Pelster, B., Schiemer, F., and Stillman, J.H. (2006). Trade-offs in thermal adaptation: The need for a molecular to ecological integration. *Physiol. Biochem. Zool.* **79**, 295–313.

Ptacnik, R., Moorthi, S.D., and Hillebrand, H. (2010). Hutchinson reversed, or why there need to be so many species. *Adv. Ecol. Res.* **43**, 1–43.

Purdy, K.J., Hurd, P.J., Moya-Laraño, J., Trimmer, M., and Woodward, G. (2010). Systems biology for ecology. *Adv. Ecol. Res.* **43**, 87–149.

R Core Development Team (2009). *R: A Language and Environment for Statistical Computing*. R Foundation for Statistical Computing, Vienna.

Raffaelli, D. (2004). How extinction patterns affect ecosystems. *Science* **306**, 1141–1142.

Rall, B.C., Vucic-Pestic, O., Ehnes, R.B., Emmerson, M., and Brose, U. (2010). Temperature, predator-prey interaction strength and population stability. *Glob. Change Biol.* **16**, 2145–2157.

Reich, P.B., and Oleksyn, J. (2004). Global patterns of plant leaf N and P in relation to temperature and latitude. *Proc. Natl. Acad. Sci. USA* **101**, 11001–11006.

Reiss, J., Bridle, J.R., Montoya, J.M., and Woodward, G. (2009). Emerging horizons in biodiveristy-ecosystem functioning research. *Trends Ecol. Evol.*

Reiss, J., Bailey, R.A., Cássio, F., Woodward, G., and Pascoal, C. (2010a). Assessing the contribution of micro-organisms and macrofauna to biodiversity-ecosystem functioning relationships in freshwater microcosms. *Adv. Ecol. Res.* **43**, 151–176.

Reiss, J., Forster, J., Hirst, A.G., Cássio, F., Pascoal, C., Stewart, R., and Hirst, A.G. (2010b). When microscopic organisms inform general ecological theory. *Adv. Ecol. Res.* **43**, 45–85.

Ricciardi, A., and Rasmussen, J.B. (1999). Extinction rates of North American freshwater fauna. *Conserv. Biol.* **13**, 1220–1222.

Ruesink, J.L., and Srivastava, D.S. (2001). Numerical and per capita responses to species loss: Mechanisms maintaining ecosystem function in a community of stream insect detritivores. *Oikos* **93**, 221–234.

Savage, V.M., Gillooly, J.F., Woodruff, W.H., West, G.B., Allen, A.P., Enquist, B.J., and Brown, J.H. (2004). The predominance of quarter-power scaling in biology. *Func. Ecol.* **18**, 257–282.

Schmid, B., Hector, A., Huston, M.A., Inchausti, P., Nijs, I., Leadley, P., and Tilman, D. (2002). The design and analysis of biodiversity experiments. In: *Biodiversity and Ecosystem Functioning: Synthesis and Perspectives* (Ed. by M. Loreau, S. Naeem and P. Inchausti), pp. 61–75. Oxford University Press, Oxford.

Sterner, R.W., and Elser, J.J. (2002). Ecological Stoichiometry: The Biology of Elements from Molecules to the Biosphere. Princeton University Press, Princeton, NJ.

Straub, C.S., and Snyder, W.E. (2006). Species identity dominates the relationship between predator biodiversity and herbivore suppression. *Ecology* **87**, 277–282.

Tallantire, P. (1974). Paleohistory of grey alder (*Alnus incana* (L.) *moench*) and black alder (*A. glutinosa* (L.) *gaertn*) in Fennoscandia. *New Phytol.* **73**, 529–546.

Thomas, C.D., Cameron, A., Green, R.E., Bakkenes, M., Beaumont, L.J., Collingham, Y.C., Erasmus, B.F.N., De Siqueira, M.F., Grainger, A., Hannah, L., Hughes, L., Huntley, B., *et al.* (2004). Extinction risk from climate change. *Nature* **427**, 145–148.

Vannote, R.L., and Sweeney, B.W. (1980). geographic analysis of thermal equilibria—A conceptual model for evaluating the effect of natural and modified thermal regimes on aquatic insect communities. *Am. Nat.* **115**, 667–695.

Wallace, J.B., Eggert, S.L., Meyer, J.L., and Webster, J.R. (1997). Multiple trophic levels of a forest stream linked to terrestrial litter inputs. *Science* **277**, 102–104.

Walther, G.R., Post, E., Convey, P., Menzel, A., Parmesan, C., Beebee, T.J.C., Fromentin, J.M., Hoegh-Guldberg, O., and Bairlein, F. (2002). Ecological responses to recent climate change. *Nature* **416**, 389–395.

Woodward, G. (2009). Biodiversity, ecosystem functioning and food webs in fresh waters: Assembling the jigsaw puzzle. *Freshwater Biol.*

Woodward, G., Papantoniou, G., Edwards, F.E., and Lauridsen, R.B. (2008). Trophic trickles and cascades in a complex food web: Impacts of a keystone predator on stream community structure and ecosystem processes. *Oikos* **117**, 683–692.

Woodward, G., Perkins, D.M., and Brown, L.C. (2010a). Climate change and freshwater ecosystems: Impacts across multiple levels of organization. *Philos. Trans. R. Soc. B* **365**, 2093–2106.

Woodward, G., Benstead, J.P., Beveridge, O.S., Blanchard, J., Brey, T., Brown, L., *et al.* (2010b). Ecological networks in a changing climate. *Adv. Ecol. Res.* **42**, 71–138.

Woodward, G., Blanchard, J., Lauridsen, R.B., Edwards, F.K., Jones, J.I., Figueroa, D., Warren, P.H., and Petchey, O.L. (2010c). Individual-based food webs: Species identity, body size and sampling effects. *Adv. Ecol. Res.* **43**, 211–266.

Yachi, S., and Loreau, M. (1999). Biodiversity and ecosystem productivity in a fluctuating environment: The insurance hypothesis. *Proc. Natl. Acad. Sci. USA* **96**, 1463–1468.

Yvon-Durocher, G., Jones, J.I., Trimmer, M., Woodward, G., and Montoya, J.M. (2010a). Warming alters the metabolic balance of ecosystems. *Philos. Trans. R. Soc. B* **365**, 2117–2126.

Yvon-Durocher, G., Montoya, J.M., Trimmer, M., and Woodward, G. (2010b). Warming alters the size spectrum and shifts the distribution of biomass in freshwater ecosystems. Global Change Biology, in press. DOI: 10.1111/j.1365-2486.2010.02321.x

Yvon-Durocher, G., Montoya, J.M., Jones, J.I., Woodward, G., and Trimmer, M. (2010c). Warming alters the fraction of primary production released as methane in freshwater mesocosms. Global Change Biology, in press. DOI: 10.1111/j.1365-2486.2010.02289.x

Yvon-Durocher, G., Allen, A.P., Montoya, J.M., Trimmer, M., and Woodward, G. (2010d). The temperature dependence of the carbon cycle in aquatic systems. *Adv. Ecol. Res.* **43,** 267–313.

Individual-Based Food Webs: Species Identity, Body Size and Sampling Effects

GUY WOODWARD, JULIA BLANCHARD, RASMUS B. LAURIDSEN, FRANCOIS K. EDWARDS, J. IWAN JONES, DAVID FIGUEROA, PHILIP H. WARREN AND OWEN L. PETCHEY

ADVANCES IN ECOLOGICAL RESEARCH VOL. 43 0065-2504/10 $35.00
© 2010 Elsevier Ltd. All rights reserved
DOI: 10.1016/S0065-2504(10)43006-8

SUMMARY

The study of food webs has been a central theme within ecology for decades, and their structure and dynamics have been used to assess a range of key properties of communities (e.g. complexity–stability relationships) and ecosystems (e.g. fluxes of energy and nutrients). However, many food web parameters are sensitive to sampling effort, which is rarely considered, and further, most studies have used either species- or size-averaged data for both nodes and links, rather than individual-based data, which is the level of organisation at which trophic interactions occur. This practice of aggregating data hides a considerable amount of biologically meaningful variation and could, together with potential sampling effects, create methodological artefacts. New individual-based approaches could improve our understanding of, and ability to predict, food web structure and dynamics, particularly if they are derived from simple metabolic and foraging constraints. We explored the effect of species-averaging in four highly-resolved individual-based aquatic food webs (Broadstone Stream, the Afon Hirnant, Tadnoll Brook and the Celtic Sea) and found that it obscured structural regularities resulting from intraspecific size variation. The individual-based approach provided clearer insights into seasonal and ontogenetic shifts, highlighting the importance of the temporal component of size-structuring in ecological networks. An extension of the Allometric Diet Breadth Model predicted the structure of the empirical food webs almost twice as accurately as the equivalent species-based webs, with the best-fitting model predicting 83% of the links correctly in the Broadstone Stream size-based web, and the few mismatches between the model and data were explained largely by sampling effects. Our results highlight the need for theoretical explanations to correspond closely with methods of data collection and aggregation, which is the exception rather than the rule at present. We suggest how this situation can be improved by including individual-level data and more explicit information on sampling effort when constructing food webs in future studies.

I. INTRODUCTION

A. Recent Advances in Food Web Data and Theory

Trophic interactions are key determinants of population abundance and dynamics, the structure and persistence of communities, and the rate and sustainability of ecosystem processes (Ings *et al.*, 2009; McCann, 2000; Neutel *et al.*, 2002; Otto *et al.*, 2007; Olesen *et al.*, 2010; Petchey *et al.*, 2004; Reiss *et al.*, 2009; Woodward *et al.*, 2010). As such, the study of food webs, which can include tens to hundreds of species and thousands of feeding links, has long been a key theme in ecology, especially in recent decades, as highlighted by the exponential increase in publications in this field since the 1970s (Ings *et al.*, 2009). A mainstay of this research has been the collection, compilation, and analysis of food webs from natural ecosystems, although the quality and resolution of these data have not always been appropriate for the tasks to which they have been put (Bersier and Sugihara, 1999; Cohen *et al.*, 1993; Hall and Raffaelli, 1993; Ings *et al.*, 2009; Paine, 1988, 1992; Polis, 1998).

Early food web research highlighted the importance of taxonomic resolution, with seemingly marked changes in network structure being observed depending on whether species, genera, families, or even coarser taxonomic entities were used as the interacting entities (Goldwasser and Roughgarden, 1997; Martinez, 1991). Serious concerns were also raised about systematic biases in taxonomic resolution being confounded with trophic position: large top predators were typically described to species, whereas smaller organisms at the lower trophic levels were often either aggregated far more coarsely (e.g. 'algae') or simply ignored altogether, as pointed out by Cohen *et al.* (1993), Hall and Raffaelli, 1993, and Schmid-Araya *et al.* (2002a,b). In addition to this patchy information on the identity of the nodes and links, the amount of sampling effort used to detect them was, and in many cases still is, highly variable both within and among webs. Some food webs have been constructed from inferential data on interactions from the literature, expert knowledge, or theoretical predictions (e.g. Dunne *et al.*, 2008; Martinez, 1991) and a few have been based solely on direct observation (e.g. Figueroa, 2007; Schmid-Araya *et al.*, 2002a,b; Woodward *et al.*, 2005a), whilst probably most have used some combination of the two (e.g. Layer *et al.*, 2010a,b; Woodward *et al.*, 2008). The former case can risk including links that are not necessarily realised in a particular local food web (although they may be present in other systems), whereas the exclusive reliance on directly observed data can be problematic if no information is provided on the sampling effort used to detect nodes or links, as this can have a huge impact on many web parameters (Martinez *et al.*, 1999).

The recognition of the shortcomings of the early datasets helped to trigger the construction of a new generation of food webs that are more completely sampled and based on more uniform taxonomic data than was the case for many of their precursors (e.g. Banašek-Richter *et al.*, 2009; Benke and Wallace, 1997; Closs and Lake, 1994; Cohen *et al.*, 2003, 2005; de Ruiter *et al.*, 1995; Lafferty *et al.*, 2006; Layer *et al.*, 2010a,b; Martinez, 1992; O'Gorman and Emmerson, 2009; Riede *et al.*, 2010; Schmid-Araya *et al.*, 2002a,b; Tylianakis *et al.*, 2007; van Veen *et al.*, 2008; Warren, 1989; Woodward *et al.*, 2005a,b). These new webs have provided clearer pictures of the structure of trophic networks and have largely superseded the earlier datasets used in the pioneering work carried out in the 1970s and 1980s (e.g. Cohen, 1978; Pimm, 1982).

Many of the data that emerged from about the early 1990s onwards revealed that networks were far more complex than had been previously thought, challenging the theoretical suggestion (May, 1972, 1973; Pimm, 1980, 1982) that complexity was destabilising and should, therefore, be rare. In parallel with the improvements to the empirical data over the last two decades, a range of new dynamical models began to unearth some of the posited 'devious strategies' (May, 1972) by which complex food webs might be stabilised. These included the prevalence of weak versus strong links (McCann, 2000; McCann, *et al.*, 1998), the role of feeding loops (Neutel *et al.*, 2002; Neutel *et al.*, 2007), and the damping of potentially destabilising 'fast' food chains with 'slower' pathways (Rooney *et al.*, 2006). Many of these dynamical models have stressed the importance of body mass as a determinant of interaction strength, and hence stability (e.g. Berlow *et al.*, 2009; Emmerson *et al.*, 2005; Yodzis and Innes, 1992). In addition, a new suite of structural models, building on the initial work of Cohen and colleagues (e.g. Cohen and Newman, 1985; Cohen *et al.*, 1990) have revealed how complex networks might be based on simple rules related to morphological, metabolic, or foraging constraints, many of which are closely correlated with body size (e.g. Beckerman *et al.*, 2006; Cattin *et al.*, 2004; Petchey *et al.*, 2008; Warren, 1996; Williams and Martinez, 2000).

From these multiple lines of evidence, it has become increasingly clear that the body size of species often exerts a powerful influence on ingestion rates ans the dynamical and structural attributes of many food webs (Ings *et al.*, 2009; Montoya *et al.*, 2006, 2009; Perkins *et al.*, 2010; Reiss *et al.*, 2010; Reuman *et al.*, 2009a,b; Woodward and Hildrew, 2002a; Woodward *et al.*, 2005a). A key, but largely overlooked, point of relevance here is that trophic interactions occur between individual organisms, not species *per se* (Stouffer, 2010; Woodward and Warren, 2007). The functional constraints on feeding imposed by size operate at the individual level, and many of the sampling biases in webs are related to the likelihood of observations being made of feeding events. However, the conventional approach to documenting food webs, which has focused primarily on taxonomic entities (species, etc.), may conceal much of this information on size structure, particularly where intraspecific size variation may equal or even exceed that between species averages, and this point has also been largely ignored in both structural and dynamical models (Woodward, 2009; Woodward and Warren, 2007). Although body size is now measured routinely in many food web studies (e.g. Cohen *et al.*, 2003), individual-level variation or the yield–effort data needed to assess potential sampling effects are still rarely considered (but see Goldwasser and Roughgarden, 1997; Ings *et al.*, 2009; Martinez *et al.*, 1999; Woodward and Hildrew, 2001).

We sought to address whether, given that individuals are the relevant interacting entities, constructing networks from this level of organisation could improve upon the traditional process of aggregating via species or size classes and whether considering both size- and species-based perspectives together provide deeper insights than provided by either alone. We addressed these questions using a set of four unusually highly resolved food webs that include information on sampling effort and directly observed data on species identity and the body sizes of the individuals involved in actual feeding interactions.

B. Species Identity and Body Size as Determinants of Food Web Structure

Body size has often been invoked, either implicitly or explicitly, as a potentially key driver behind a range of both dynamical and static food web models. The latter include a group of structural models based on morphological, metabolic, or foraging constraints, and include the cascade (Cohen and Newman, 1985), niche-based (Warren, 1996; Williams and Martinez, 2000), hierarchical (Cattin *et al.*, 2004), and allometric diet breadth models (ADBMs) (Petchey *et al.*, 2008). All the species-based approaches described above effectively deal with ball-and-stick representations of food webs as networks of taxonomic entities (nodes) connected to one another by their trophic interactions (links), following in the footsteps of the first such diagram drawn 130 years ago by Camerano (1880).

Much of the research effort devoted to improving taxonomic resolution and completeness of food webs reflects the field's historical roots in population and community ecology (e.g. *sensu* Elton, 1927; Hutchinson, 1959; MacArthur, 1955), where species identity is often seen as being key (Ings *et al.*, 2009; Raffaelli, 2007). Growing recognition of the role of size in food webs has given rise to more concerted attempts to enrich binary networks of nodes and links by including body size information, whilst retaining the traditional species-based classification. This enrichment has typically involved assigning an average body size to each of the taxonomic entities in a food web; thus allometries describe variation among taxonomic entities. Such taxonomically constructed relationships are the most common types of allometries encountered within the context of community ecology (e.g. Brown *et al.*, 2004; Damuth, 1987; Peters, 1983).

Increasingly, traditional food web networks have been enriched by including data on species' average sizes and their abundance, to produce so-called trivariate food webs (*sensu* Cohen *et al.*, 2003). Here, mean abundance (N) and body mass (M) of each species population are plotted on a log–log scatterplot, with feeding links overlain between the interacting nodes (e.g. Cohen *et al.*, 2003; Jonsson *et al.*, 2005; Layer *et al.*, 2010a,b; McLaughlin *et al.*, 2010; O'Gorman and Emmerson, 2010; Woodward *et al.*, 2005a,b). The allometric MN scaling relationship across the food web as a whole can be remarkably consistent both over time and among systems (e.g. Layer *et al.*, 2010b;

Woodward *et al.*, 2005b), despite considerable species turnover, leading some to suggest that this might point to fundamental ecological constraints that are manifested at this whole-system level of organisation (O'Gorman and Emmerson, 2010). Within trivariate webs, the angles and distances between pairs of connected nodes can also be used to make inferences about the flux of energy and the strength of interactions between species populations, which can have important implications for the functioning and stability of the system as a whole (e.g. Cohen *et al.*, 2003; Emmerson *et al.*, 2005; Layer *et al.*, 2010a; Reuman and Cohen, 2005) and even if abundance data are missing, species-averaged predator–prey mass ratios may be used to infer *per capita* interaction strengths (e.g. Berlow *et al.*, 2009; Emmerson and Raffaelli, 2004). However, these structural and dynamical models effectively ignore the potentially critical variation that occurs within both nodes and links.

A very different approach to documenting trophic networks, body sizes, and abundances has developed in parallel with the species-based approaches described above. Instead of taxonomic entities (such as species), individuals are grouped into size classes (also known as size bins) irrespective of their taxonomy (Petchey and Belgrano, 2010): as such, these size-based approaches are free from the so-called '*curse of the Latin binomial*' (Raffaelli, 2007). The resulting relationships between abundance and size (of a size class) have been termed as size spectra and individual-size distributions. Since they refer to scaling of abundance with body size, one could refer to these relationships as allometries, but care must be taken not to confuse these with 'taxonomically constructed' allometries (Reuman *et al.*, 2008; Yvon-Durocher *et al.*, 2008, 2010).

Size spectra have been used to characterise broad macroecological patterns in energy flux and biomass production for entire ecosystems, with some of the most ambitious studies including 'particles' that range in size from phytoplankton to whales (Kerr and Dickie, 2001). This approach offers a potentially powerful means of simplifying otherwise complex food webs into far more tractable, and hence predictable, forms. Because the slope of a size spectrum describes how quickly abundance declines with size, it can be used, for instance, to gauge the impact of size-selective harvesting of fish assemblages (Pope *et al.*, 1994) and to model the impacts of commercial exploitation on marine ecosystems (e.g. Jennings and Brander, 2010). Size spectra have been most widely used to characterise pelagic marine ecosystems in which gape-limited feeding and indeterminate growth over many orders of magnitude of body mass occur (e.g. for fishes). For instance, as an individual fish grows from a small larva into a large predator weighing tens of kilos (e.g. Atlantic cod), there may be such marked intraspecific ontogenetic shifts in its trophic ecology that the concept of treating each species as a single discrete node often makes little sense (e.g. Hardy, 1924). Consequently, size-based approaches can provide a comparatively straightforward and logistically efficient alternative means of capturing many of the major biologically relevant characteristics within a food web, even in the absence of any information about the taxonomic identity of individuals.

C. Identifying the Relevant Entities and Suitable Measures of Body Size

A critical question that has clouded studies of the size structure of food webs is: 'the size of what?' (Yvon-Durocher *et al.*, 2010). It is essential to define and measure the interacting entities at appropriate levels of organisation and it is important to bear in mind that the individuals that are interacting within a feeding link might be of very different sizes than the averages for their respective species populations. Although species-based approaches have provided invaluable insights into the size structure and dynamics of food webs (e.g. Cohen *et al.*, 2003; Reuman *et al.*, 2005) and macroecological patterns among systems (e.g. Woodward *et al.*, 2005b), they inevitably obscure potentially important individual-level information. Similarly, if the identities of the interacting species are unknown, potentially valuable information is also lost when employing solely size-based approaches. Consequently, individual-level data should not only provide a possible means of better characterising trophic interactions, but could also be used to examine food webs from both species-based and size-based perspectives simultaneously (Figure 1).

Very few food web studies have recorded data about interacting *individuals*, but those that have suggested individual- and species-averaged analyses of size structure can indeed provide very different insights (Ings *et al.*, 2009; Woodward and Warren, 2007). One exception is a recent study carried out in Broadstone Stream, in which several thousands of individuals were identified to species and measured in the environment and in the guts of predators (Woodward and Warren, 2007). When the standard approach of using species-averaged data was used to construct the food web, some predators appeared to feed on prey nearly 100 times larger than themselves (Woodward *et al.*, 2005a), but when exactly the same data were used to construct the food web from an individual-level perspective, predators did not consume prey larger than themselves (Woodward and Warren, 2007). This apparent paradox arose because large individuals of 'small' species fed on small individuals of 'large' species, particularly when generations overlapped. Clearly, the effects of aggregating up to the species level can have large and potentially misleading effects on our perception of the size structure of communities. The current shortage of information about the sizes of interacting individuals within food webs is undoubtedly related to the considerable time and effort required to collect such data, but in recent years several datasets based on

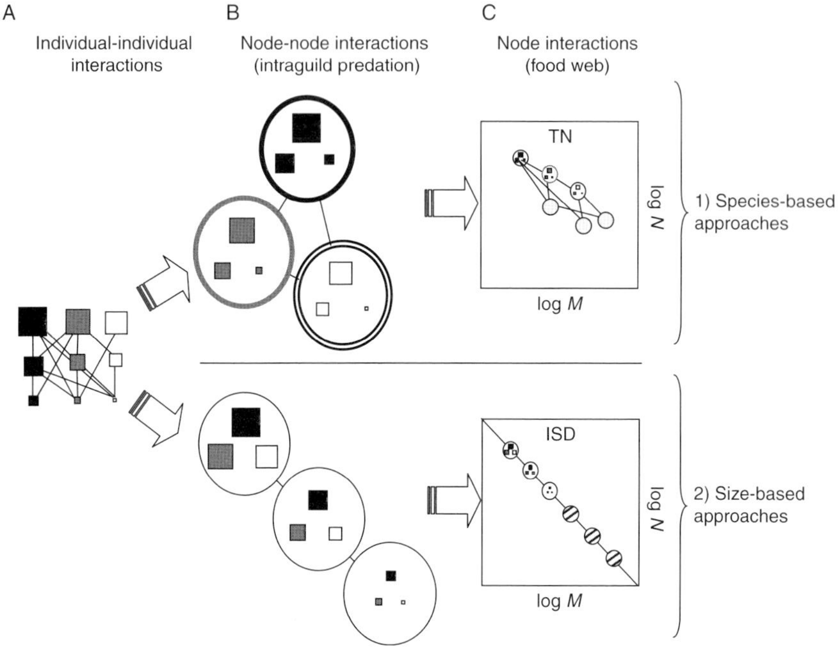

Figure 1 Schematic of (1) species-based and (2) size-based approaches to the construction of food webs, highlighting how both are underpinned by interactions between individuals (depicted as squares, with the three colours each representing a different species). The use of such data therefore enables both perspectives to be viewed simultaneously, across multiple levels of organisation from (A) individual–individual interactions to (B) food web modules (e.g. intraguild predation) to (C) to the entire community as a trophic network (TN), size spectrum or individual-size distribution (ISD).

individual-level interactions have started to emerge (e.g. Barnes *et al.*, 2008; Woodward and Warren, 2007) and the current study is, to the best of our knowledge, the first formal analysis of such data from a range of systems.

D. Coupling Size- and Species-Based Approaches with Individual-Level Data

One of the principal objectives of our study was to assess how switching between taxonomic and size-based views alters the perceived structure of a food web. As well as calculating standard measures of food web structure, we estimated how well a theoretical model could predict the structure of species- versus size-averaged food webs constructed from individual-level data. The structural models we used to compare the two approaches were developed from the ADBM of Petchey *et al.* (2008), which predicts the positioning of links within the feeding matrix using a set of rules based on foraging constraints. The ADBM was used as a relatively simple model of size effects on diet due to the authors' familiarity with it: comparing different types of size-based models of food webs was not the aim of the research presented in this chapter. Although the original incarnation of the ADBM successfully captured much of the variation in network structure from a range of species-averaged food webs, there was considerable variation in its predictive ability and only about 40% of links were predicted correctly for the Broadstone Stream web (Petchey *et al.*, 2008, based on the data of Woodward *et al.*, 2005a). Some of this variability in explanatory power among food webs could be ascribed to variation in the types of trophic interactions (e.g. predatory and herbivorous interactions were predicted more accurately than pathogenic, host–parasitoid, or parasitic ones). It is also possible that mismatches might have arisen from sampling effects and/or species-averaging. In the former case, links that were predicted by the model but 'missing' from the empirical webs might simply reflect incomplete characterisation of consumer's diets, the extent of which can be assessed via yield–effort curves (e.g. Woodward and Hildrew, 2001). In the second case, species-averaging might create 'mirages' due to the crude treatment of body size as a single value for each species. For instance, the relatively poor match between the Broadstone Stream data and the ADBM seemed somewhat surprising, given the strong size-structuring of the web described elsewhere (e.g. Woodward and Hildrew, 2002a; Woodward *et al.*, 2005), but in the Petchey *et al.* (2008) study no information was included on the variation in individual size within either the nodes or links.

We constructed and modelled food webs of predator–prey interactions, using both species-based and size-based approaches derived from individual-level data for four systems: the Broadstone Stream food web (e.g. Hildrew, 2009; Olesen *et al.*, 2010), the Celtic Sea (Blanchard *et al.*, 2005), and two previously undescribed food webs: the Afon Hirnant stream in Wales and the Tadnoll Brook in southern England. For each food web, we collated information on the abundance of species and size classes in the environment and the diets of individuals using directly observed gut contents data, in which the individual body sizes of both consumers and resources were recorded.

We also explored potential sampling effects and how these might influence the attributes of the empirical food webs and the fit of the models to the data: e.g. were predicted links between rare predators and rare prey (as either species or size classes) most likely to be missing from the empirical webs due to undersampling? Finally, we focused on the Broadstone Stream food web as a detailed case study, by using the individual-level data to explore some of the supposed anomalies that are commonly observed in species-averaged food webs (e.g. cannibalistic and mutual predation loops) but which should, according to theory, be rare, due to their perceived destabilising effects (e.g. May, 1973; Pimm, 1982; Tanabe and Namba, 2005; Vandermeer, 2006). This system was selected for more detailed study because predator–prey interactions within its species-averaged food web have been studied intensively for four decades (e.g. Hildrew, 2009; Hildrew and Townsend, 1982; Olesen *et al.*, 2010; Petchey *et al.*, 2008; Schmid-Araya *et al.*, 2002a,b; Townsend and Hildrew, 1977, 1979; Woodward and Warren, 2007; Woodward *et al.*, 2005a), and because it had the most extensive dataset for examining seasonal and ontogenetic effects among our four systems.

II. METHODS AND STUDY SITES

Each of the four study systems and their respective food webs are described in turn below. Only predatory interactions were considered and non-animal basal resources were excluded from all the food webs. Although some of the detailed methods of sample collection differ to some extent between systems, the temporal and spatial scales over which the data were collected, and the extent of sampling effort used, individual level and directly observed data on nodes and links were used to construct the trophic network of each. Within the stream food webs, to standardise taxonomic resolution, and to minimise possible biases arising from the different mesh apertures used for sampling, the very small permanent meiofauna (body length < 500 μm) were excluded. We fitted the ADBM to the data in the same way for each food web, such that the number of entities (i.e. nodes) in the species-based taxonomic web was the same as for its size-based equivalent.

A. Study Sites and Empirical Data Collection

1. The Celtic Sea Food Web

The Celtic Sea is an area of continental shelf bordered by Ireland, the U.K. and the Bay of Biscay. The abundance data by size classes and species were extracted from the annual Cefas Celtic Sea groundfish surveys (after Warnes and Jones, 1995). The data were standardised to account for differences in trawl durations and therefore are indices of relative abundance (numbers of individuals per hour). Only locations consistently sampled through the 1987–2001 time series with the Portuguese high-headline trawl were used (Blanchard *et al.*, 2005). Ideally, abundance data would be derived from exactly the same locations and times as the stomach contents data. However, precise sampling locations and dates were not given in the Barnes *et al.* (2008) dataset, from where the data used in this chapter were extracted.

Consequently, we pooled data over the whole time-period and locations to capture general patterns in the abundance of species and size classes. Due to the size selectivity of fishing gear used in these research surveys, the smaller size classes of fish are not adequately sampled. Species–specific size distributions were extrapolated to smaller classes using linear regression of log abundance versus log body mass class (Jennings *et al.*, 2002).

The feeding links of fishes in the Celtic Sea have been described in a published global dataset of individual predator and prey body sizes and taxonomy (Barnes *et al.*, 2008): in total, 2091 feeding events from 29 predator species were included in the food web presented here. The original stomach contents data were collected from dissections carried out onboard the research vessels Cirolana, Scotia, and Clione during the annual surveys carried out by Cefas (Pinnegar *et al.*, 2003): of all the predators that were dissected, 491 individuals contained ≥ 1 identifiable prey item. Predator and prey length were recorded and converted to body mass by Barnes *et al.* (2008) using established weight to length relationships from regression analysis. Only vertebrate prey were identified and measured, and a vast majority were identified to species (see Appendix 1.1 for the species list).

2. *The Tadnoll Brook Food Web*

The Tadnoll Brook is a nutrient-rich (mean soluble reactive phosphorus and total oxidisable nitrogen 122.9 μg L^{-1} and 7.0 mg L^{-1}, respectively), circumneutral, tributary (mean annual discharge 0.35 $m^3\ s^{-1}$) of the River Frome in southern England (Edwards *et al.*, 2009a,b).

Between February and December 2005, a 240-m reach was sampled (50°41′N, 2°19′W, 30 m a.s.l.) to construct the summary food web. The numerical abundance (numbers m^{-2}) and biomass (mg dry mass m^{-2}) of fish populations were measured every 2 months, using two-pass depletion electrofishing (Seber and LeCren, 1967) or using 30 large-bore Hess samples (0.12 m^2) for *Cottus gobio* L. (after Woodward *et al.*, 2008). Invertebrate abundance was quantified bimonthly using a benthic Surber sampler (0.06 m^2 quadrat; mesh aperture 300 μm), with 20 sample-units taken on each date. Invertebrates were identified to the lowest possible taxonomic level (usually species). Linear dimensions of each individual were measured to the nearest 0.1 mm and used to estimate dry body mass from published length–mass regressions (Benke *et al.*, 1999; Burgherr and Meyer, 1997; Edwards *et al.*, 2008; Edwards *et al.*, 2009a,b; Meyer, 1989; Smock, 1980).

Fishes were anaesthetised using 2-phenoxyethanol, measured and weighed. The guts of brown trout (*Salmo trutta* L.) >70 mm fork length were then flushed and the contents immediately preserved in 4% formalin. For smaller trout and other fish species, samples of specimens were killed and frozen for subsequent dissection. Gut content analysis was carried out for the fish assemblage on each sampling occasion, whereas invertebrate diets were characterised only in May and October. Individuals of the most common invertebrate taxa in the benthos were taken from the Surber samples, and dissected for determination of gut contents. The contents of the foregut were examined for animal prey, which were identified at 400× magnification to species wherever possible and linear body dimensions measured. Dry mass of individual prey items was calculated via the use of published regression equations (Benke *et al.*, 1999; Burgherr and Meyer, 1997; Edwards *et al.*, 2009a,b; Ganihar, 1997; Gonzalez *et al.*, 2002; Meyer, 1989; Sabo *et al.*, 2002). Of all the predators that were dissected, 688 individuals contained ≥ 1 identifiable prey item. All trout captured were fitted with a peritoneal passive integrated transponder (PIT) tag (Prentice *et al.*, 1990) and scanned for subsequent reidentification during each subsequent sampling occasion: this enabled growth rates and diets to be characterised repeatedly for individuals that were recaught and the use of in-river PIT tag readers enabled us to follow life history of the tagged trout (Edwards *et al.*, 2008).

3. *The Afon Hirnant Food Web*

The Afon Hirnant is a tributary of the Welsh Dee in North Wales, U.K. (52° 52′N 03° 34). The study was conducted at three sites (mean annual discharge of 7.26 $m^3 s^{-1}$, 4.93 $m^3 s^{-1}$, and 2.08 $m^3 s^{-1}$, respectively) that were sampled on four dates between 2004 and 2005 (Figueroa, 2007). Within each site, pH ranged from about 5.5 to 7 over the course of the year.

Invertebrates were collected using a benthic Hess sampler (sampling area: 0.028 m^2; mesh aperture 80 μm). Fifteen randomly placed samples were taken at each site on each date, giving a total of 180 samples for the whole sampling period, which were preserved immediately in 100% ethanol and subsequently sorted in the laboratory. All individuals within the benthic samples were counted and identified to the lowest possible taxonomic level (usually species). The foreguts of invertebrates were removed by dissection and examined at 400× magnification. Prey items were identified from a combination of taxonomic keys and reference slides of previously identified species, after Schmid-Araya *et al.* (2002a,b).

The body length of each individual in the benthos (and for prey items in consumers' guts) was measured and used as the dimension of body size in the study (as opposed to mass, as in the other systems). If prey items were too highly digested for body lengths to be measured reliably, previously established regressions based on head capsule width were used as an alternative linear dimension. Of all the predators that were dissected, 253 individuals contained ≥ 1 identifiable prey item.

4. The Broadstone Stream Food Web

Broadstone Stream (51° 05′ N 0° 03′E; 120 m above sea-level) is a second-order tributary of the River Medway in south-east England (see Hildrew, 2009 for a detailed site description). The acidity of the stream (pH 4.7–6.6) excludes fish, resulting in an invertebrate-dominated food web (Woodward and Hildrew, 2001). There are about 25 common invertebrate species (Woodward *et al.*, 2005a). Among the common predators, there are three large species (*Cordulegaster boltonii* Donovan, *Sialis fuliginosa* Pict. and *Plectrocnemia conspersa* [Curtis]) and three small species (the larvae of the tanypod midges *Macropelopia nebulosa* [Meigen], *Trissopelopia longimana* [Staeger], and *Zavrelimyia barbatipes* [Kieffer]), which we focus on in detail here when investigating size-structuring within the intraguild subweb. Detritivorous stoneflies and chironomids dominate the prey assemblage in winter and summer, respectively (Woodward *et al.*, 2005a). Trophic interactions have been studied intensively over four decades, culminating in one of the most completely described food webs for any system (e.g. Hildrew *et al.*, 2004; Hildrew, 2009; Layer *et al.*, 2010b; Schmid-Araya *et al.*, 2002a,b; Woodward *et al.*, 2005a).

Thirty randomly dispersed benthic Surber sample-units (25 cm × 25 cm quadrat; mesh aperture 330 μm) were taken every 2 months between June 1996 and April 1997 and preserved immediately in 5% formalin. All invertebrate taxa were described to species wherever possible, and the few that could not be distinguished with certainty were grouped to the next taxonomic level (usually genus). Very rare taxa (i.e. <0.01% of mean annual standing stock) were excluded from the webs. Most species were univoltine and present only as larvae, so populations were not continually reproducing and recruitment was largely restricted to the summer. Linear body dimensions of all individuals collected from the benthos (or identified in gut contents) were measured and converted to dry mass using published regression equations (listed in Woodward and Hildrew, 2002b).

The foreguts of the predators collected in the Surber samples were dissected and examined at 400× magnification. Gut contents were identified from reference slides: because prey were generally consumed whole, or in large fragments, they could usually be identified to species relatively easily, and their individual body masses were calculated using the same methods as for the benthic samples (Woodward *et al.*, 2005a).

We constructed a summary food web that included all links and species (or size classes) recorded over the six 1996–1997 sampling occasions. In total, 1016 of the dissected predators contained at least one identifiable prey item within their gut, amounting to a total of 1818 prey individuals. Each predator and prey individual involved in a feeding link was measured and transformed into their respective body masses, which resulted in 2909 unique measurements. Sampling effort (the number of predator guts containing at least one recognisable prey items) and predator size were fitted as independent variables in an OLS multiple regression model used to predict total diet width (i.e. the asymptotic number of prey species eaten) for each predator species. This regression model was used to gauge the joint effects of body size and sampling effort on diet characterisation.

B. Construction of the Food Webs

All individuals were grouped into their respective species to create a taxonomic food web from the feeding interaction data. A trophic link assigned between two species if an individual of one was ever found in the gut of at least one individual of the other species. To create a size-based food web, the size range of all individuals in the web was divided into equal logarithmic size classes, with as many size classes as there were species in the same web (for comparability), and individuals were allocated to those size classes irrespective of their taxonomy. Logarithm size classes were used so that the distribution of individuals per size classes was more even than would have been the case if arithmetic sizes were used. A trophic link was assigned between two size classes if any individual from one size class was ever found in the gut of any individual from another size class. Estimates of abundance of individuals within species, and of individuals within size classes were compiled from samples not associated with the interaction data. Suctorial predators were omitted from the food webs, because their feeding links could not be described (after Closs and Lake, 1994).

C. Modelling and Analyses

Predation matrices are frequently used to visualise aspects of food web structure, such as size structure and diet contiguity. As well as visual inspection, we calculated a range of food web statistics for each study system, including: maximum chain length (the number of trophic elements in the longest food chain); directed connectance was calculated as $C=L/S^2$ (Martinez, 1992), where L is the number of realised trophic links observed and S is the number of trophic elements in the web; complexity was calculated as $SC=S(L/(S[S-1]/2))$ (after Polis, 1991). The mean number of links per species, d, was calculated as L/S. Mean shortest path lengths and clustering coefficients were calculated using the functions shortest.paths() and transitivity() in the igraph R package (Csardi and Nepusz, 2006).

In addition, each food web was separately modelled using the ADBM, following Petchey *et al.* (2008). This model has two main components: First, a set of allometric relationships that relate body size to variables that influence and represent individual foraging, namely, attack rates, handling times, energy contents, and densities. Second, and separately for each consumer, the contingency model of optimal foraging (Emlen, 1966; MacArthur and Pianka, 1966) translates these variables into predicted diets. Since available empirical data about the allometries of foraging variables was insufficient and or unsuitable for parameterising the model [Supplementary Information in Petchey *et al.* (2008)], we selected allometries that resulted in predicted connectance equal to observed connectance and also the maximum number of links predicted correctly (Petchey *et al.*, 2008). In all the food web modelling, we assumed that individuals could be *a priori* assigned as predators or non-predators. Therefore, the models attempted only to explain the diets of consumer taxa/size classes. For comparison, we also calculated the proportion of observed links that were correctly predicted if we simply assumed that all the predatory nodes consumed all other species.

We also analysed characteristics of the species that were involved in links that were either observed but not predicted, or predicted but not observed. Specifically, we used logistic regression to assess if a false-positive prediction, whereby the model predicted a links that were not observed in the data, was related to the abundances of the prey and predator involved in the link (i.e. a potential sampling effect). Also, we used randomisations to test if false-negatives, whereby empirical data described a link that was not predicted by the model, could be explained by coincidental (rather than intentional) ingestion of prey by predators; for instance, we tested if unusually small prey were disproportionately over-represented in predator guts that contained another more typically sized prey item, and potential instances of 'Russian dolls' in which the gut contents of a large predator also included the gut contents of a smaller predator it has ingested.

III. RESULTS

A. General Properties of Empirical Food Webs and Comparisons Between Models and Data

The food webs contained 28–59 nodes and 112–170 links in the taxonomic versions, and 190–484 links in the size-based versions. In all cases, the size-based webs were more densely connected than their taxonomic equivalents: on average, the number of links, linkage density, and connectance were about twice as high; clustering coefficients and maximum and mean trophic height were higher; and path lengths were shorter (Table 1). Predator–prey mass ratios were also higher than in the taxonomic webs. Predator diet width (W, $\log_{10}$ number of links) in the taxonomic webs was well predicted by both consumer body mass (M, $\log_{10}$ mg) and sample size (NG, $\log_{10}$ number of

Table 1 Parameters for the four study systems for their respective species- and size-based food webs

	Celtic Sea		Tadnoll Brook		Afon Hirnant		Broadstone Stream	
	Species	Size	Species	Size	Species	Size	Species	Size
Number of guts dissected that contained prey	491	491	688	688	253	253	1016	1016
Nodes feedings on smaller nodes (%)	0.98	0.98	0.98	0.98	0.92	0.92	0.82	0.82
Individuals feeding on smaller individuals (%)	1	1	1	1	0.90	0.90	0.99	0.99
Number of nodes	48	48 (2)	59	59 (6)	33	33 (1)	28	28 (1)
Number of links	169	376	170	484	112	190	124	185
Linkage density	3.52	7.83 (8.17)	2.88	8.2 (9.13)	3.39	5.76 (5.94)	4.43	6.61 (6.85)
Directed connectance	0.07	0.16 (0.17)	0.05	0.14 (0.17)	0.10	0.17 (0.19)	0.16	0.24 (0.25)
Predatory nodes (%)	0.6	0.56	0.32	0.66	0.39	0.58	0.32	0.57
Top predators (%)	0.38	0.19	0.1	0.24	0.24	0.18	0	0.14
Basal species (%)	0.4	0.33	0.68	0.24	0.61	0.42	0.68	0.39
Intermediate species (%)	0.23	0.38	0.22	0.42	0.15	0.39	0.32	0.43
Cannibalistic species (%)	0.12	0	0.02	0	0.06	0.15	0.25	0.32
Mean shortest path length	2.3	1.59	2.04	1.69	1.84	1.74	1.69	1.50
Clustering coefficient	0.3	0.5	0.2	0.39	0.28	0.56	0.50	0.70
Maximum trophic height of a node	4	3.67	2.7	4.08	3.13	3.57	2.60	4.26
Mean trophic height of a node	2	1.99	1.42	2.61	1.52	1.95	1.45	2.36
Mean predator prey size ratio (by links between nodes)	1.3	2.2	1.4	3.7	1.1	1.3	1.48	3.72
Mean predator prey size ratio (individual interactions)	42.7	42.7	10715.2	10715.2	3.8	3.8	117.49	117.49
ADBM predictive success (%) (allometric abundances)	0.3	0.73	0.43	0.73	0.42	0.5	0.63	0.83
ADBM predictive success (%) (actual abundances)	0.3	0.74	0.45	0.74	0.43	0.53	0.62	0.83

guts that contained prey) in three of the webs (Celtic Sea: $W = -0.254 + 0.116M + 0.627NG$; $r^2_{(adj)} = 0.92$; $F_{2,27} = 158.33$; $p < 0.001$; Tadnoll Brook: $W = 0.218 + 0.053M + 0.531NG$; $r^2_{(adj)} = 0.87$; $F_{2,18} = 58.31$; $p < 0.001$; Broadstone Stream: $W = 0.216 + 0.156M + 0.389NG$; $r^2_{(adj)} = 0.96$; $F_{2,8} = 88.94$; $p < 0.001$), and by sample size alone in the Afon Hirnant ($W = 0.133 + 0.584NG$; $r^2 = 0.92$; $F_{1,12} = 117.69$; $p < 0.001$). In terms of predicting the positioning of the links within the feeding matrix, in each case the ADBM showed a marked improvement in the fit between the model and the data in the size-based webs (30–61% correctly predicted) relative to their species-based (50–84% correctly predicted) equivalents. The Broadstone Stream food web provided the best fit between the models and data in both cases (62% and 83% in the species- and size-based webs, respectively), with the most marked improvement between the species-based (30%) and size-based (74%) web being evident in the Celtic Sea web (Table 1).

B. Specific Properties of Each Empirical Food Web and Comparisons Between Models and Data

1. Celtic Sea Food Web

a. Properties of the Species- and Size-Based Webs. The diet breadth of the predators within the food web was well described by their mean individual mass and the number of guts examined that contained prey (Figure 2). The species-based food web contained 48 taxa and 169 links (Figure 3). Nineteen of these consumed no other taxa; the 29 other taxa were both predators and prey, and six of these were cannibals (Figure 3A). This food web contained the highest proportion of top predators (0.38) among all eight networks across the four systems, and the longest single food chain (4) (Table 1).

The size-based food web contained 376 links (Figure 3B). Thirty-two size classes were consumers, of which nine, of the predominantly larger classes, were not themselves consumed. There were 16 (predominantly smaller) 'basal' size classes that did not consume any other, but were themselves consumed. Eighteen were intermediate, and none were cannibals. Five size classes (mostly smaller ones) were neither consumers nor prey and were therefore unconnected because no individuals occurred in these size classes. Up to 21 size classes were present in the diets of the predatory size classes. The individuals in each size class represented up to 18 taxa, and the number of species displayed a hump-shaped relationship with size (Figure 3B) as did the number of prey species and predator species in each size class.

In both versions of the food web 98% of nodes fed on smaller prey: on average, predators were 1.3 and 2.2 times larger than their prey in the species and size-based webs, respectively, and across all individuals predators were

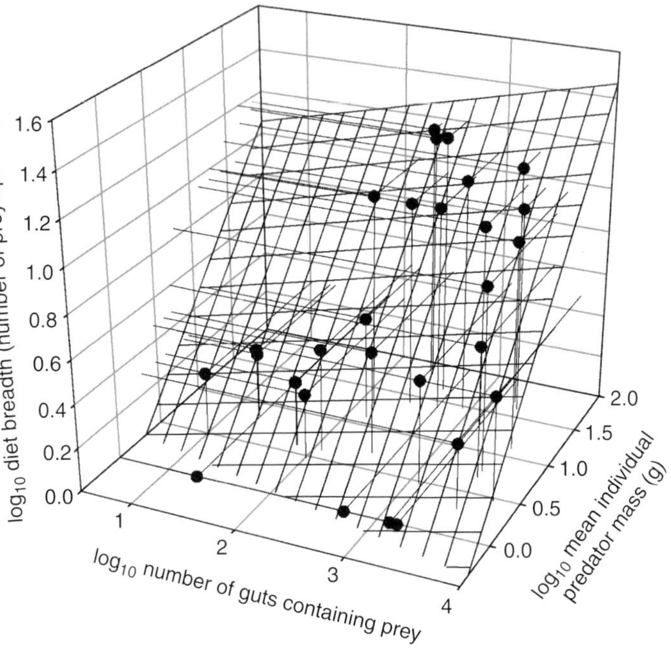

Figure 2 *Celtic Sea food web.* Diet breadth of as a function of sampling effort and mean individual predator body mass ($\log_{10}$ number of links $= -0.254 + 0.116$ (mean individual predator mass, $\log_{10}$ mg) $+ 0.627$($\log_{10}$ number of guts); $r^2_{(adj)} = 0.92$; $F_{2,27} = 158.33$; $p < 0.001$).

43 times larger than their prey (Table 1). The size-based web was more reticulate than the species-based version, with higher connectance, linkage density, and clustering coefficients and lower mean shortest path length (Table 1)

b. Explaining the Arrangement of Feeding Interactions. A simple 'model' that assumed that the 29 predator species could consume all 48 species (including themselves) predicted 29×48 links (1392), of which 169 are correctly assigned ($169/1392 = 12\%$). The apparent vulnerability of prey (the number of species the prey is consumed by) was significantly positively associated with their abundance (generalised linear model, Poisson errors, z value of slope $= 4.7$, 27 df for error, $p < 0.0001$). The apparent generality of predator taxa (the number of taxa they consumed) was not associated with their abundance (generalised linear model, Poisson errors, z value of slope $= 0.1$, 27 df for error, $p = 0.89$).

The ADBM correctly predicted 30% of the links in the taxonomic food web; consequently the patterning of predicted links was clearly different from the observed links (Figure 3A): the diets were predicted to include a much smaller range of prey taxa, and much more diet contiguity, than was observed. When given the identity of the predatory size classes, however, the ADBM predicted 74% of the links correctly (Figure 3B; Table 1). Most of the deviations occurred due to the lack of predicted links from consumers to relatively small size classes. The chance of the ADBM predicting a link incorrectly was higher when the prey taxa or either prey or predator size class involved in link was rare (binomial generalised linear models, $p < 0.05$ for the relevant model terms) (Figure 4).

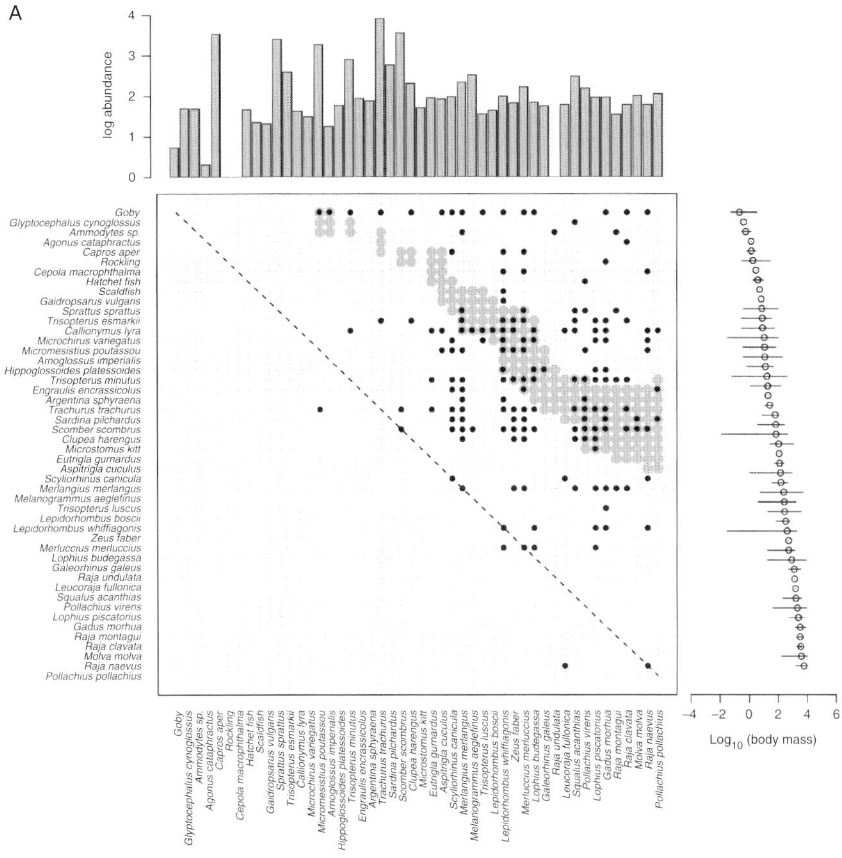

Figure 3 (Continued)

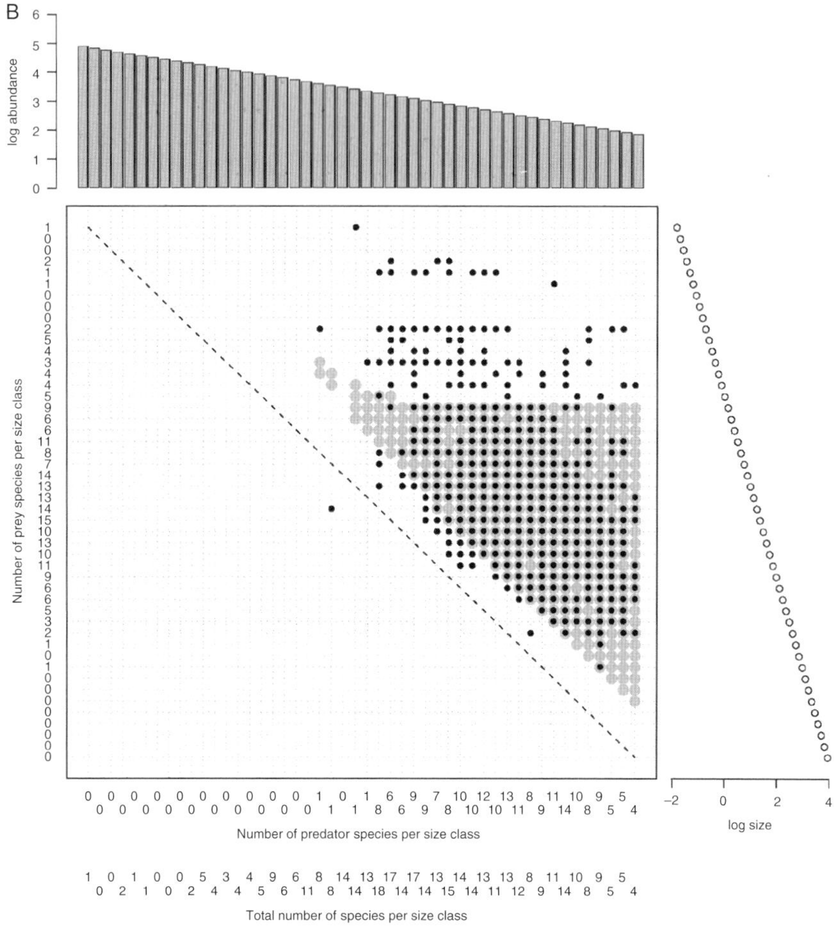

Figure 3 *The Celtic Sea food web, abundance, and body size information.* The central predation matrix indicates an observed feeding interaction between a consumer (column) and resource (row) with a black dot. Thus, columns can be viewed as the diet composition of consumer. The larger grey dots indicate the feeding interactions predicted by the ADBM. Consumer columns and resource rows are ordered by body size, with the smallest in the top left corner and largest in the bottom right. Thus feeding interactions in which a predator is larger than its prey occur in the triangle above the diagonal dashed line. The upper bar chart shows the abundances in the environment of the species in the predation matrix. The plot on the right shows the mean and range of body sizes of individuals in each of the species in the predation matrix. (A) A version of the Celtic Sea food web in which individuals are grouped according to taxonomy. (B) A version of the Celtic Sea food web in which individuals are grouped according to size. In (B) a number of species represented in each size class are displayed to the left and below the predation matrix. All information is as in Figures 6, 9, and 12, except that the abundances in (B) are allometric estimates (see Methods, in Section III), since estimates of actual abundance were constrained by sampling methods and gear, so could not be used.

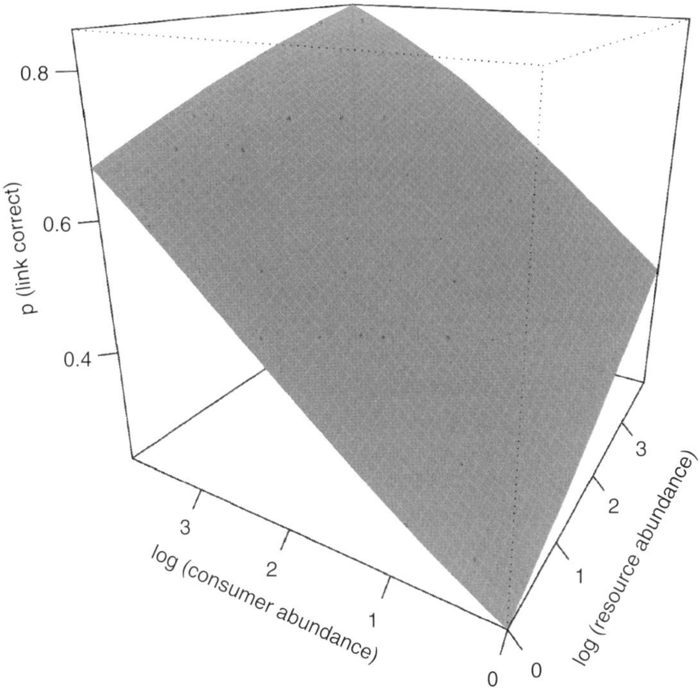

Figure 4 In the Celtic Sea food web, the relationship between the probability that the link predicted by the ADBM is present in the observed food web and the abundance of the resource and consumer involved in that link.

2. *Tadnoll Brook Food Web*

a. Properties of the Species- and Size-Based Webs. The diet breadth of the predators was well described by their mean individual mass in the benthos and the number of guts examined that contained prey (Figure 5). The taxonomic food web contained 59 taxa and 170 links (Figure 6A; Table 1). Forty of these consumed no other taxa; 13 other taxa were both predators and prey, six were predators only, and one of these was cannibalistic (Figure 6A). The predatory taxa consumed up to 50 other species. The size-based food web contained 484 links (Figure 6B). Fourteen of the mainly smaller size classes did not consume any other size classes, but were themselves consumed. Thirty-nine were consumers, of which the 14 largest were consumers but were not themselves consumed. Twenty-five were both consumed and consumers, and none of these were cannibals. Six size classes were

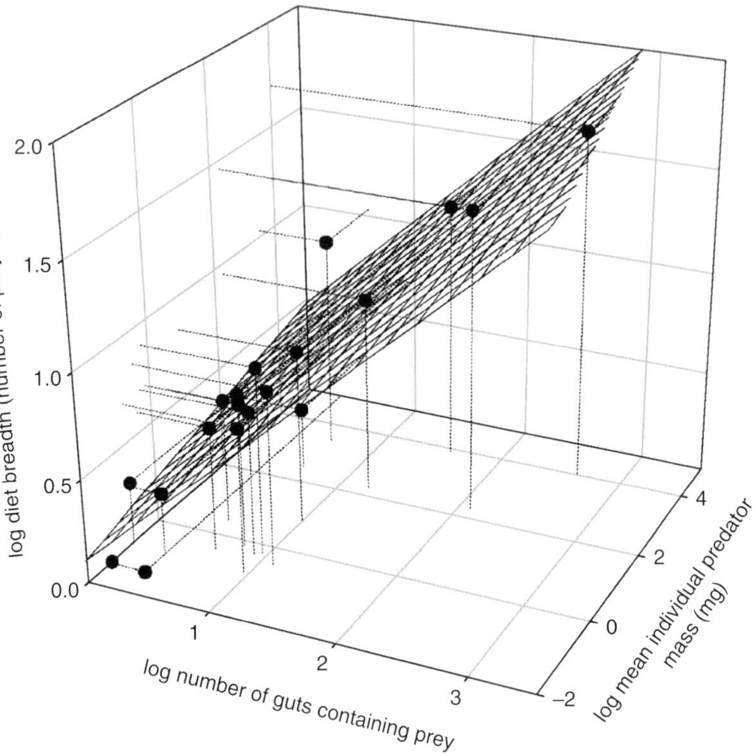

Figure 5 *Tadnoll Brook food web.* Diet breadth of as a function of sampling effort and mean individual predator body mass.

unconnected in the matrix because they contained no individuals. From 8 to 18 size classes were present in the diets of the predatory size classes. The individuals in each size class represented up to 25 taxa, and the number of species displayed a hump-shaped relationship with size (Figure 6B), as did the number of prey species and predator species in each size class (Figure 6B).

In both versions of the food web, 98% of nodes fed on prey that were smaller than themselves. On average, predators were 1.4 and 3.7 times larger than their prey in the species and size-based webs, respectively, but across all individual predators were 10,715 times larger than their prey (Table 1). As in the other webs, the size-based web had higher connectance, linkage density, and clustering coefficients and reduced mean shortest path length relative to its species-based equivalent (Table 1).

b. Explaining the Arrangement of Feeding Interactions. A simple 'model' which assumed that the 19 predator species could consume all 59 species (including themselves) predicted 19×59 links (1121), of which 170 were correctly assigned ($170/1121 = 15\%$). The vulnerability of species was significantly positively associated with their abundance (generalised linear model, Poisson errors, z value of slope $= 4.7$, 27 df for error, $p < 0.0001$), but this was not the case for their apparent generality (generalised linear model, Poisson errors, z value of slope $= 0.8$, 7 df for error, $p = 0.43$).

The ADBM predicted 45% of the links in the taxonomic food web correctly, with some consumers predicted to have very narrow diets, when in reality they preyed on many species. When given the identity of the predatory

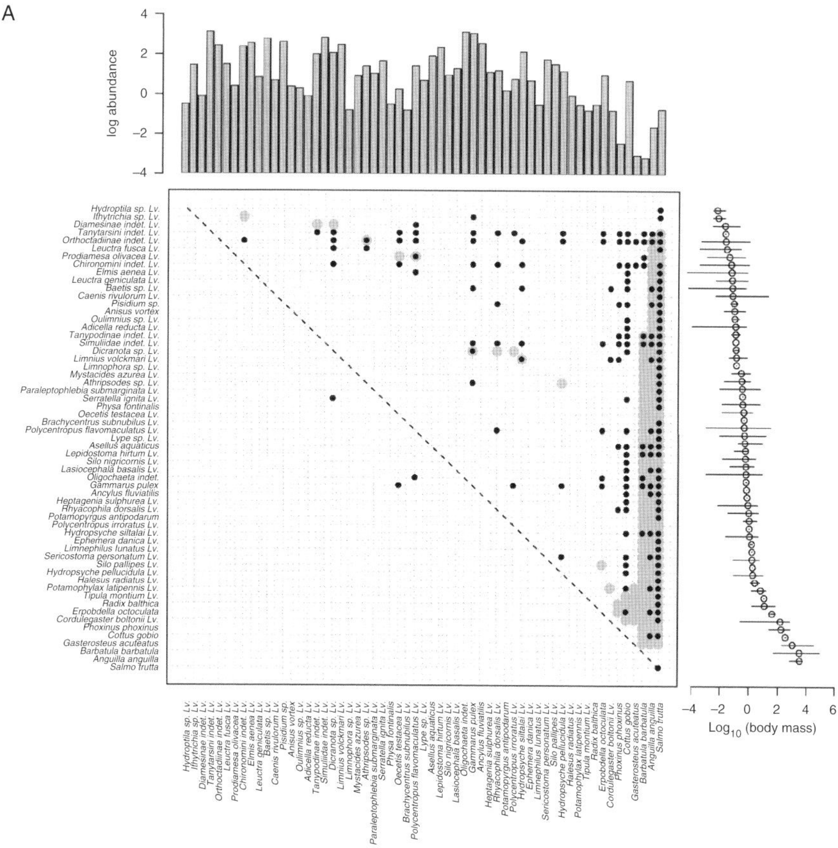

Figure 6 (Continued)

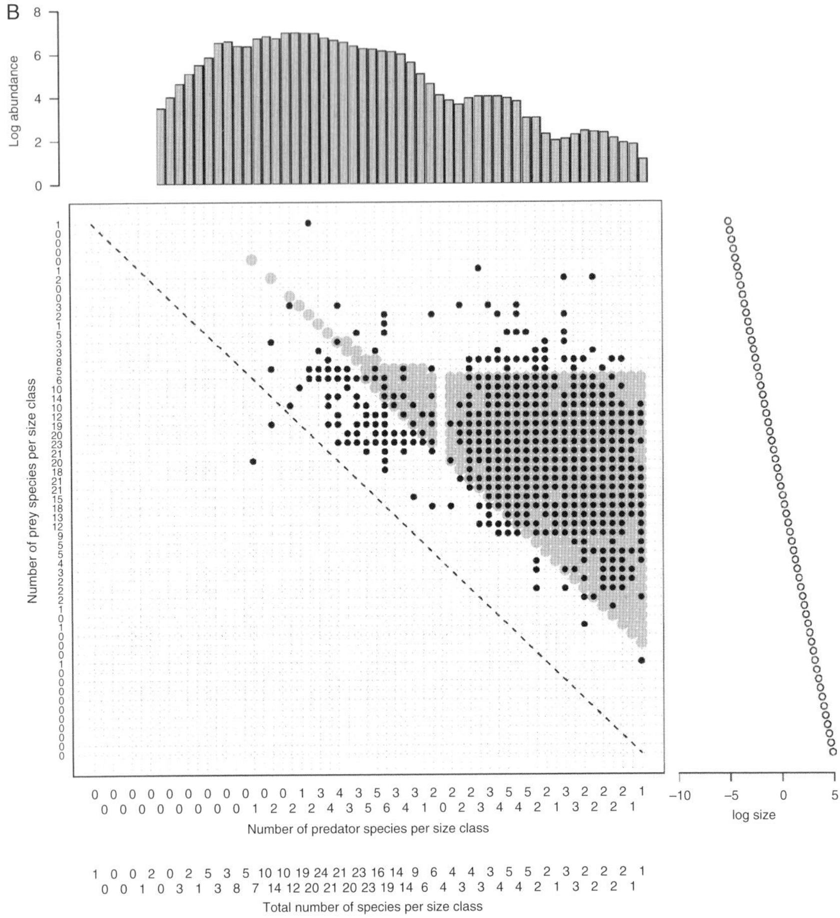

Figure 6 Tadnoll Brook food web, abundance, and body size information. See Figure 3 legend for details. The upper bar chart shows the abundances in the environment of the species in the predation matrix. The plot on the right shows the mean and range of body sizes of individuals in each of the species in the predation matrix. (A) A version in which individuals are grouped according to taxonomy. (B) A version in which individuals are grouped according to size. In (B) a number of species represented in each size class are displayed to the left and below the predation matrix.

size classes, the ADBM predicted 74% of the links correctly in the size-based web (Figure 6B). The chance of the ADBM predicting a link incorrectly was higher when the prey or predator species were common, but was highest when the size class involved in link was rare (binomial generalised linear models, $p < 0.03$ for all main effects and interaction term) (Figure 7): it is possible that this reversal of patterns between the taxonomic and size-based webs is due to the more exhaustive sampling of the diets of the larger fish in this community.

c. Ontogenetic Dietary Shifts Within Individual Predators. The most frequently recaptured individual of the top predator species, brown trout, *S. trutta* L, was caught on eight occasions, over which period it more than doubled its body mass (from 5.2 to 13.2 g) and exhibited marked shifts in its diet, from feeding primarily within the invertebrate-dominated portion of the size spectrum to becoming markedly more piscivorous as it moved upwards through the food web (Figure 8). This shift from invertivory to piscivory was manifested when the fish exceeded 10 g, and was reflected by a sudden increase in mean ingested prey mass of about two orders of magnitude and, ultimately, a marked reduction in the size range of prey being taken.

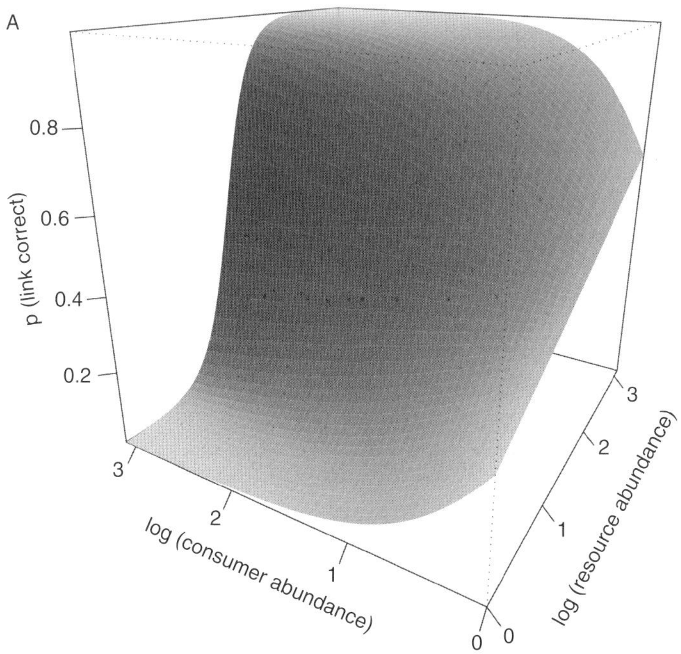

Figure 7 (Continued)

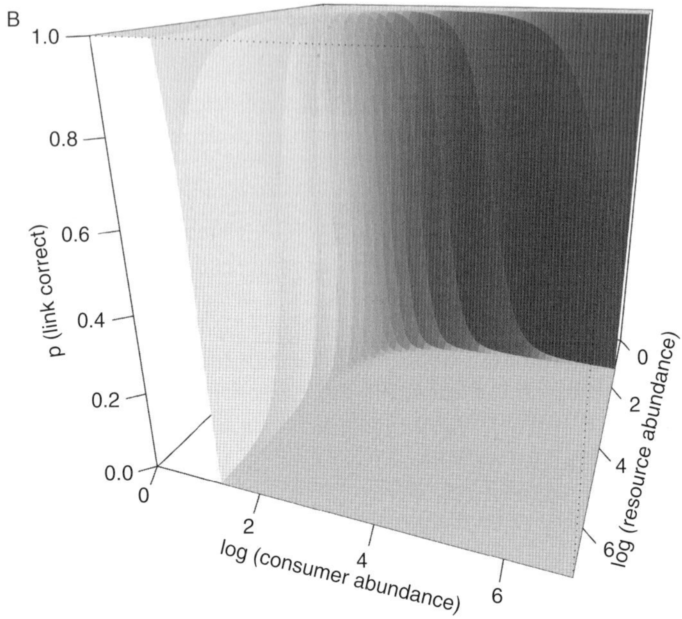

Figure 7 In the Tadnoll Brook food web, the relationship between the probability that the link predicted by the ADBM is present in the observed food web, and the abundance of the resource and consumer involved in that link. The surface is that predicted by a binomial regression fitted to the data. (A) Model of the taxonomic version of Tadnoll Brook food web. (B) Size-based version of the Tadnoll Brook food web. In both cases, the resource and consumer abundance were both significantly associated with probability of correct prediction.

3. *Afon Hirnant Food Web*

a. Properties of the Species- and Size-Based Webs. The diet breadth of the predators was well described by the number of guts examined that contained prey (Figure 9). The taxonomic food web contained 33 taxa and 112 links (Figure 10A). Twenty of these consumed no other taxa, 13 other taxa were both predators and prey, eight were predators only, and two of these were cannibalistic (Figure 10A). The predatory taxa consumed up to 19 other species.

The size-based food web contained 190 links (Figure 10B). Fourteen size classes were 'basal' nodes and these were mostly located in the smaller size classes. Nineteen size classes were consumers, of which six were 'top predators' that were not eaten by other size classes, and 13 size classes were intermediate. Up to 23 size classes were preyed upon. The individuals in each

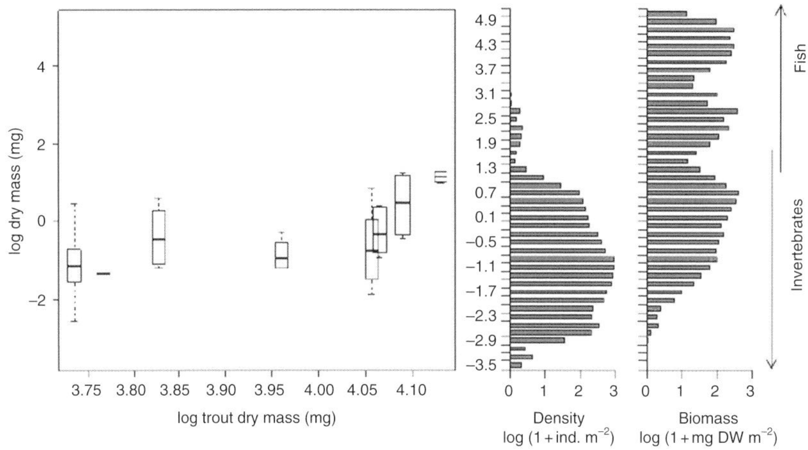

Figure 8 Ontogenetic changes in the size range of prey in the diet of an individual trout with increasing consumer size. The trout in the Tadnoll Brook were tagged on first capture enabling us to follow changes in the diet of individuals (by repeated gut-flushing). The box plot represents data for one trout repeatedly sampled over eight occasions in its first year of growth. Box plots of the size distributions of prey items consumed by this individual trout are plotted against its mass. The two histograms in the right-hand panels represent the density and biomass of size categories of potential prey in the environment. The arrows on the far right indicate size range of invertebrates and fish.

size class represented from 1 to 15 taxa, and the total number of species per class displayed a unimodal relationship with size, as did the number of prey and predator species (Figure 10B).

In both versions of the food web, 92% of the consumer nodes fed on smaller prey: on average, predators were 1.1 and 1.3 times larger than their prey in the species and size-based webs, respectively, and across all individual predators were 3.8 times larger than their prey (Table 1). The size-based web was again more densely connected than was the case for the species-based version: connectance, linkage density, and clustering coefficients were all higher and mean shortest path length was reduced (Table 1).

b. Explaining the Arrangement of Feeding Interactions. Assuming that the 13 predator species could consume all 33 species (including themselves) predicted 13×33 links (429), of which 112 were correctly assigned ($112/429 = 26\%$). The ADBM predicted 43% of the links in the taxonomic food web correctly but when supplied with the identity of the predatory size classes this increased to 53% in the size-based webs (Figure 10B; Table 1). The chance of the ADBM predicting a link incorrectly was higher when the

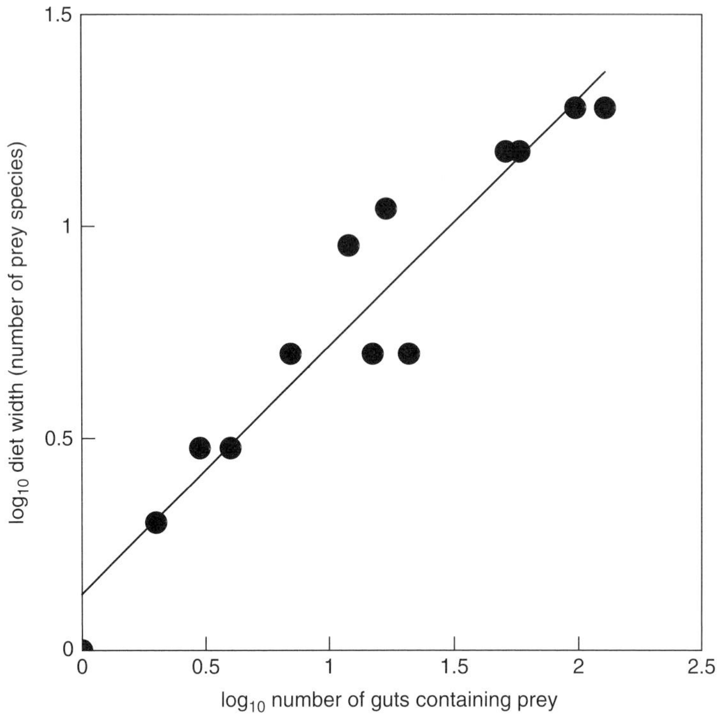

Figure 9 *Afon Hirnant food web*. Diet breadth of as a function of sampling effort and mean individual predator body mass ($\log_{10}$ number of links $= 0.133 + 0.584(\log_{10}$ number of guts); $r^2_{(adj)} = 0.92$.; $F_{1,12} = 117.69$; $p < 0.001$).

predator species or size class involved in link was rare (binomial generalised linear models, $p < 0.01$ for all main effects; Figure 11).

4. *Broadstone Stream Food Web*

a. Properties of the Species- and Size-Based Webs. The diet breadth of the predators in the Broadstone Stream food web was well described by mean individual predator mass in the benthos and the number of guts examined that contained at last one prey item (Figure 12). The taxonomic food web contained 28 taxa and 124 links (Figure 13A). Nineteen of these consumed no other taxa; the nine other taxa were both predators and prey, and seven these were cannibals. The predatory taxa consumed 6–25 other species, and some (e.g. *Zavrelimyia* feeding on *Trissopelopia*) appeared to feed on species that were nearly 20 times larger than themselves (Figure 13A).

The size-based web contained many more (185) links (Figure 13B). Eleven (mostly small) size classes were 'basal' nodes and the four largest size classes represented 'top predators' that were not preyed upon. Twelve size classes were intermediate consumers and nine of these were 'cannibals'. One size class (the second smallest) was empty and unconnected within the matrix because no individuals occurred within it. Between 8 and 19 size classes were present in the diets of the predators. There were 11 instances (i.e. <5% of links) of predators eating prey that were larger than themselves, and 55% of these were represented by size classes adjacent to one another on the ordinal body size scale (Figure 13B). Each size class represented 0–23 species, and the total number of species per node displayed a hump-shaped relationship with size, as did both the number of prey species and predator species (Figure 13B).

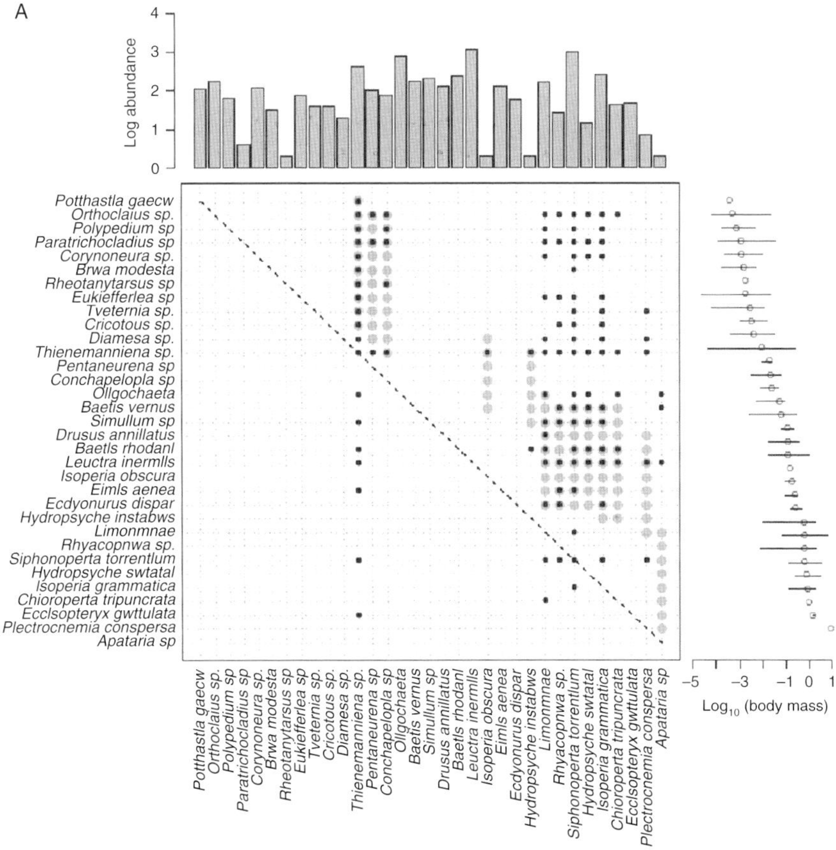

Figure 10 (Continued)

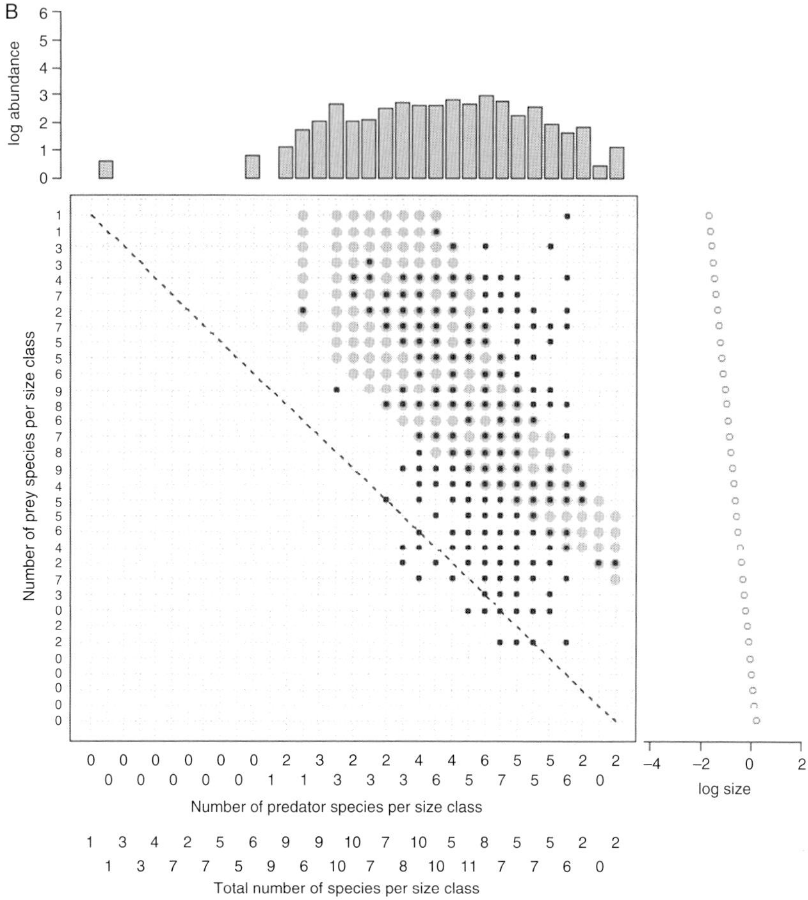

Figure 10 *Afon Hirnant food web, abundance, and body size information.* The central predation matrix indicates an observed feeding interaction between a consumer (column) and resource (row) with a black dot. See Figures 3 and 6 for details. (A) A version in which individuals are grouped according to taxonomy. (B) A version in which individuals are grouped according to size. In (B) the number of species represented in each size class are displayed to the left and below the predation matrix.

Eighty-four percent of the nodes in both versions of the food web preyed on nodes were smaller than themselves. On average, predators were 1.5 and 3.7 times larger than their prey in the species and size-based webs, respectively, and across all individuals, predators were 117 times larger than their prey (Table 1). The size-based web was again markedly more

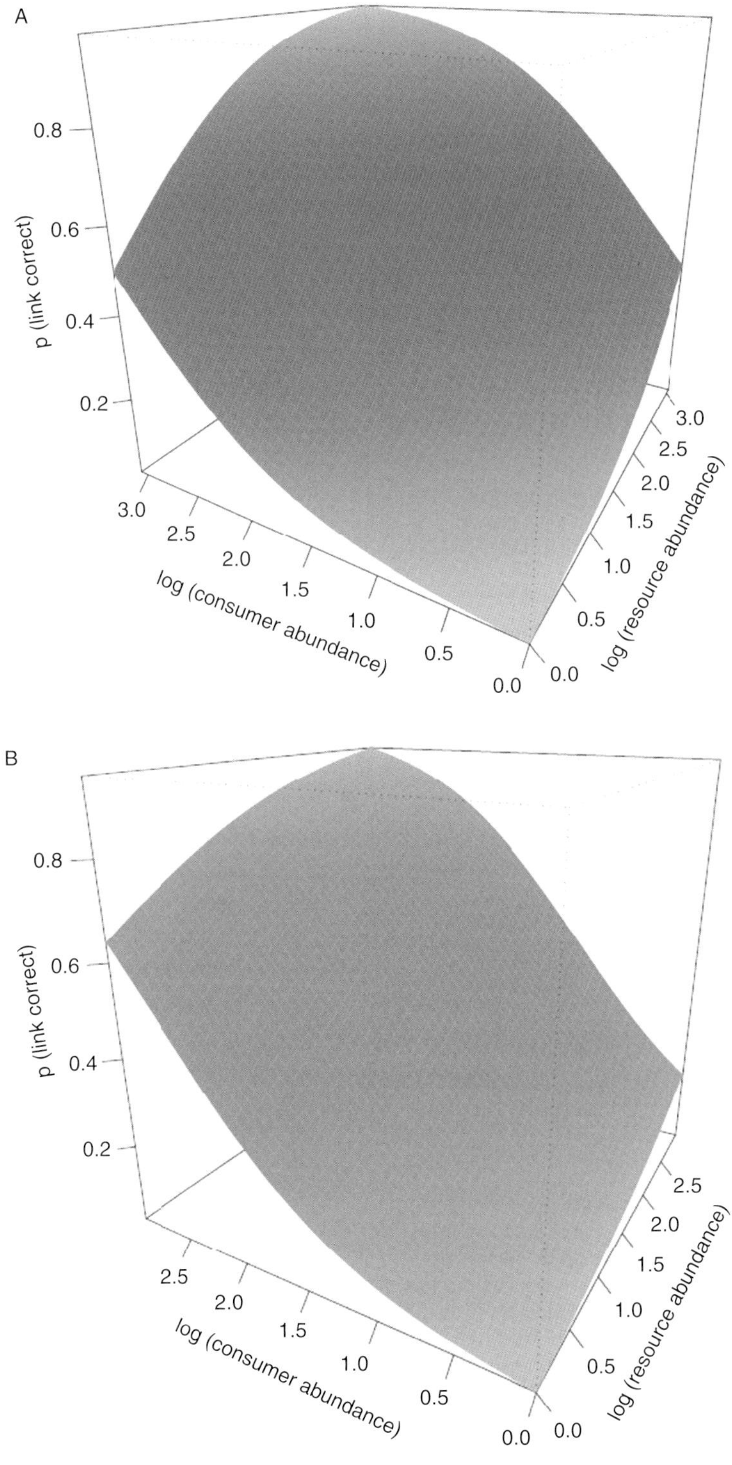

Figure 11 (legend on following page)

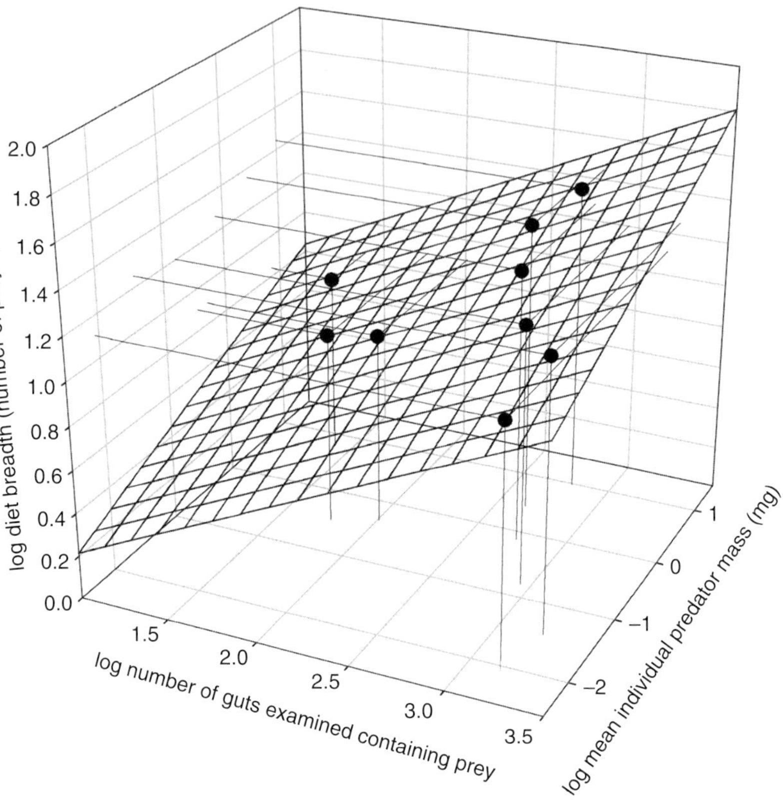

Figure 12 Estimating diet breadth in the species-averaged Broadstone Stream food web, as a function of sampling effort (the number of predator guts examined that contained at least one prey item) and mean predator body mass ($r^2_{(adj)}=0.96$; $F_{2,8}=88.94$; $p<0.001$). Note all axes are $\log_{10}$ transformed.

interconnected than the species-based version, with higher connectance, linkage density, and clustering coefficients and reduced mean shortest path length (Table 1).

Figure 11 In the Afon Hirnant food web, the relationship between the probability that the link predicted by the ADBM is present in the observed food web, and the abundance of the resource and consumer involved in that link. The surface is that predicted by a binomial regression fitted to the data. (A) Model of the taxonomic version of Afon Hirnant food web. (B) Size-based version of Afon Hirnant food web. In both cases, the resource and consumer abundance were both significantly associated with probability of correct prediction.

b. Explaining the Arrangement of Feeding Interactions. If we assume that the nine predator species can consume all 28 species (including themselves) 9×28 links (252) are predicted, of which 132 were correctly assigned (132/252 = 52%). The vulnerability of species was significantly positively associated with their abundance (generalised linear model, Poisson errors, z value of slope = 4.7, 27 df for error, $p < 0.0001$), but this was not the case for their apparent generality (generalised linear model, Poisson errors, z value of slope = 0.8, 7 df for error, $p = 0.43$).

The ADBM correctly predicted 62% of the links in the taxonomic food web, but the patterning of predicted links was clearly different to the observed links (Figure 13A). For example, the diets of some consumers were predicted to be nearly empty, when in reality they preyed on many of the other species. In addition, none of the taxa were predicted to feed on taxa much larger than themselves, whereas this was not the case in the empirical

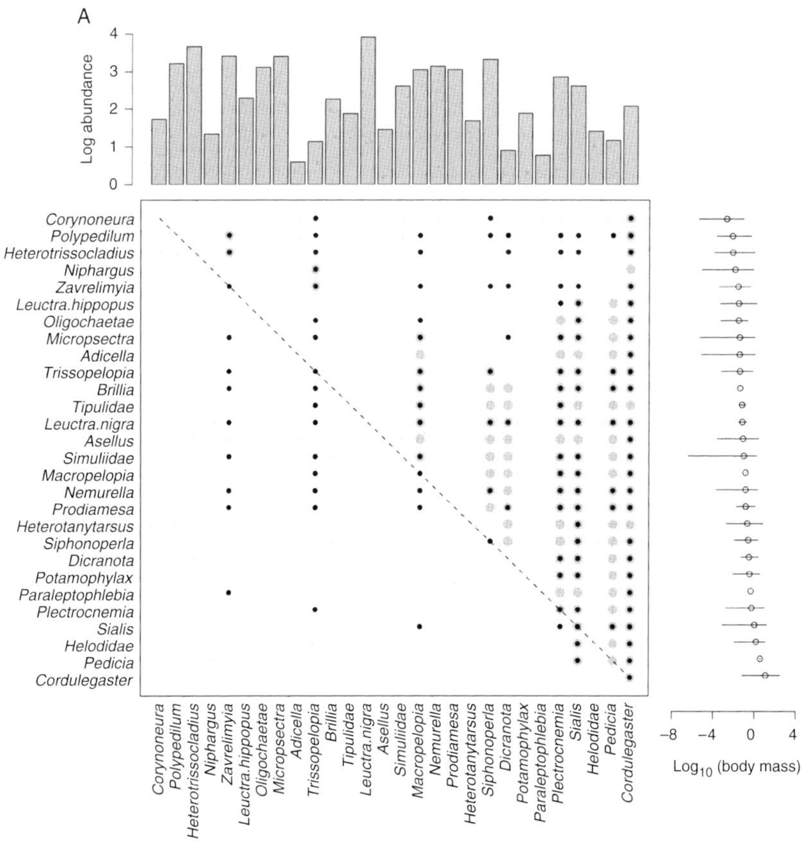

Figure 13 (Continued)

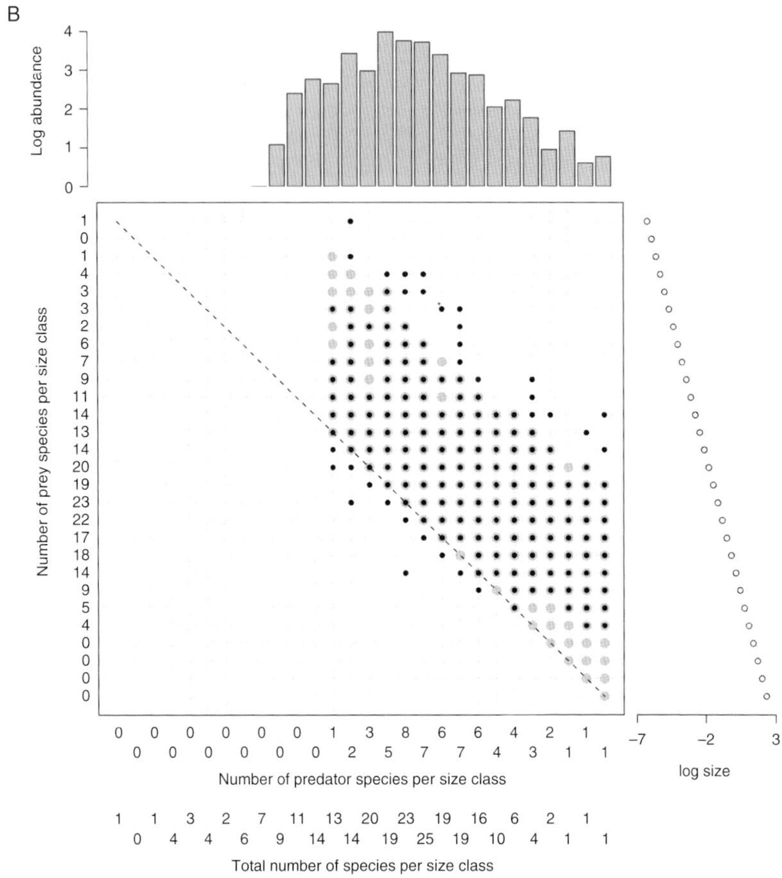

Figure 13 *Broadstone stream food web, abundance, and body size information.* The central predation matrix indicates an observed feeding interaction between a consumer (column) and resource (row) with a black dot. See Figures 3 and 6 for details. (A) A version of Broadstone stream in which individuals are grouped according to taxonomy. (B) A version of Broadstone stream in which individuals are grouped according to size. In (B) a number of species represented in each size class are displayed to the left and below the predation matrix.

data. Note that this value of 62% is different from the ~40% given in Petchey *et al.* (2008), due to differences in the empirical data: in Petchey *et al.* (2008) data were extracted from Brose *et al.* (2005), whereas the current analyses are based on the raw individual diet data. For the size-based web the ADBM predicted 83% of the observed links correctly (Figure 13B). The chance of the ADBM predicting a link incorrectly was higher when the predator species or size class involved in link was rare (binomial generalised linear models, $p < 0.01$ for all main effects; Figure 14). These findings tallied with the yield–effort curves, which indicated incomplete sampling of the diets of *Pedicia* (7 prey), *Dicranota* (6 prey), *Siphonoperla* (7 prey) (cf. Woodward *et al.*, 2005a), which were all at least one order of magnitude rarer than the six other predator species.

C. Seasonal and Ontogenetic Shifts Within the Broadstone Stream Food Web

When feeding links within the trivariate species-based food web were decomposed into their associated size–frequency distributions for nodes and links, the previously hidden variation of species-averaging became more readily apparent (Figure 15). This was highlighted for one of the most frequently observed interactions in the web, that of the top predator *C. boltonii* preying upon the stonefly *N. pictetii*. The frequency distribution of individual predator–prey mass ratios within this link was log-normal, and only a subset

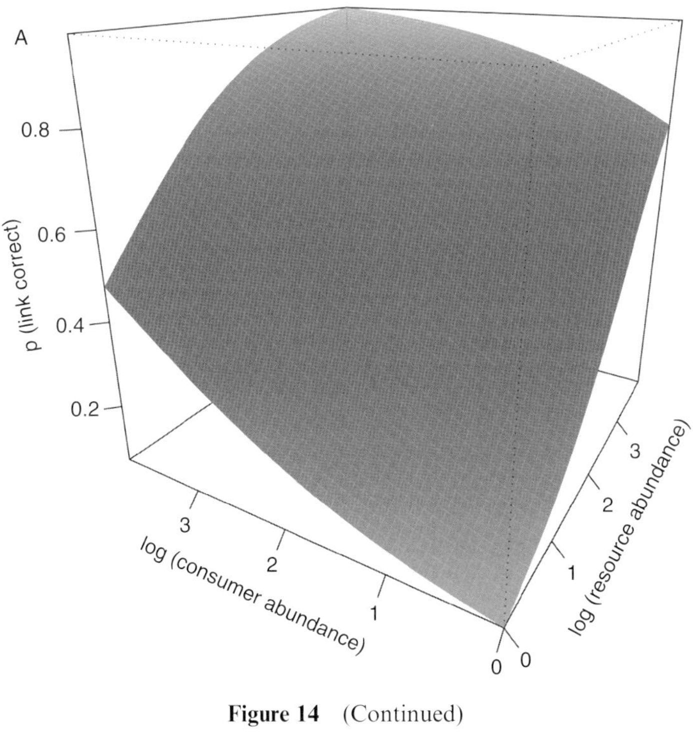

Figure 14 (Continued)

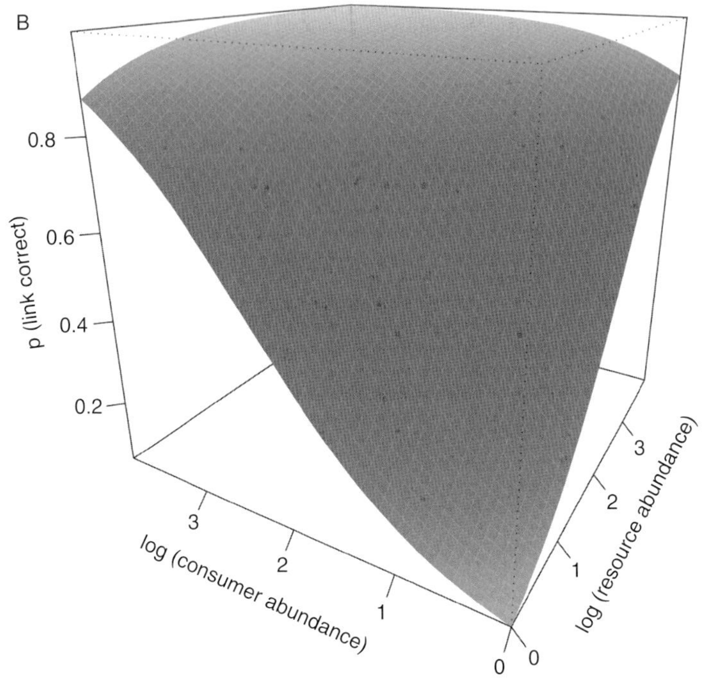

Figure 14 In the Broadstone stream food web, the relationship between the probability that a link predicted by the ADBM is present in the observed food web, and the abundance of the resource and consumer involved in that link. The surface is that predicted by a binomial regression fitted to the data. (A) Model of the taxonomic version of Broadstone stream food web. (B) Size-based version of Broadstone stream. In both cases, the resource and consumer abundance were both significantly associated with probability of correct prediction.

of the predator population interacted with a subset of the prey population (Figure 15). For the highlighted *C. boltonii–N. pictetii* feeding link, the predators were represented by the larger individuals and the prey by the smaller individuals within their respective populations: thus the individual-based ratio was about 1.7 times higher than that of the species-averaged data due to the implicit inclusion of the larger (invulnerable) prey in the calculations of the latter ratio. When resolved temporally (Figure 16), marked ontogenetic and seasonal shifts in the occurrence of this link were revealed, with apparent mismatches occurring at certain times of the year (e.g. in June) when interactions between particular size classes were not observed. These contrasted with other periods of more intense linkages when the size disparity

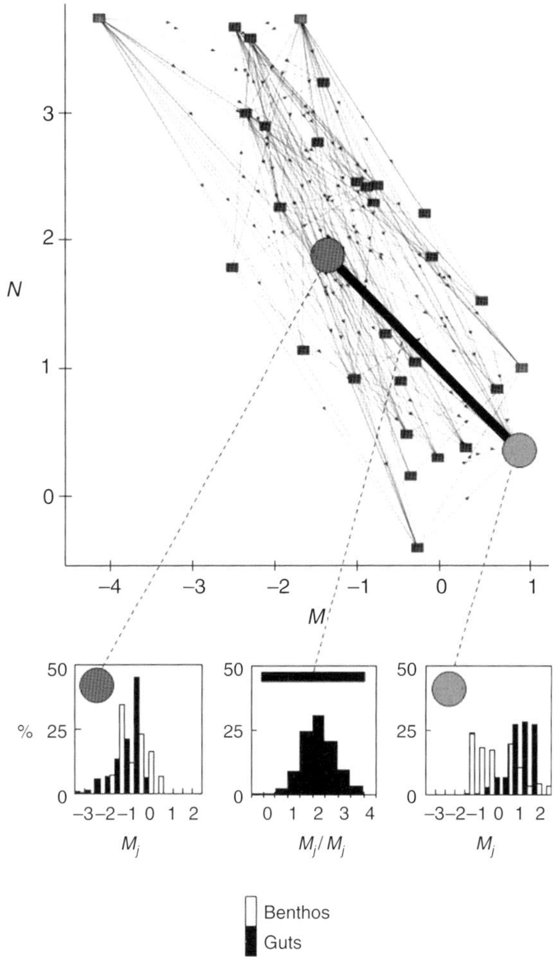

Figure 15 Species-averaged *MN* scaling in the Broadstone Stream food web. Trivariate patterns in the Broadstone Stream food web, in which log abundance (*N*) is plotted as a function of log mean individual body mass (*M*) with feeding links overlain. The species-averaged graph obscures the considerable variation in body masses of individual protagonists within each predator–prey interaction, as revealed in the lower panels.

was sufficient to enable a greater proportion of the predators to exploit this common prey species more extensively (e.g. in August).

Within the guild of the six common predators, there were clear differences in the sizes of the individuals that interacted with one another as either predators or prey relative to their respective populations as a whole. This was evident from the distinct size–frequency distributions for the IGP subweb

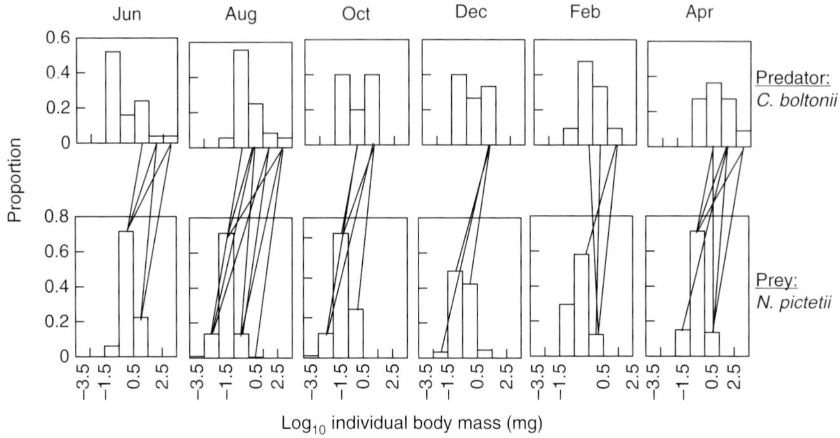

Figure 16 Size distributions of predators and prey in the environment for the *C. boltonii–N. pictetii* feeding link highlighted in Figure 15, showing seasonal patterns in the manifestation of the link and the size classes involved.

as a whole versus those for the individuals actually involved in trophic interactions as either 'predators' or 'prey' (Figure 17). Although mutual predation and cannibalism occurred in many permutations within the IGP subweb, when these predator species were subdivided into size classes (Figure 18), all interactions were accounted for by small predators being eaten by larger individuals, irrespective of species identity. Size refugia were also evident within the IGP subweb for both very large and very small size classes: that is, there was a 'window of vulnerability' at intermediate size ratios. IGP was rarely manifested by the smallest predator size classes, which tended to feed instead on smaller primary consumers. In contrast, *C. boltonii* fed on all other predator species and exploited a wider range of progressively larger 'prey' as it grew, until it eventually became the largest predator at the top of the food web. Consequently, feeding loops both between and within species were accounted for by life-history omnivory when individual-based predator–prey mass ratios > 1, rather than species identity *per se*.

IV. DISCUSSION

A. Size-Based and Species-Based Food Webs

The taxonomic and size-based food webs presented and analysed above highlight the marked differences in our perception of network structure that the two views can create. Both the structure of the predation matrices

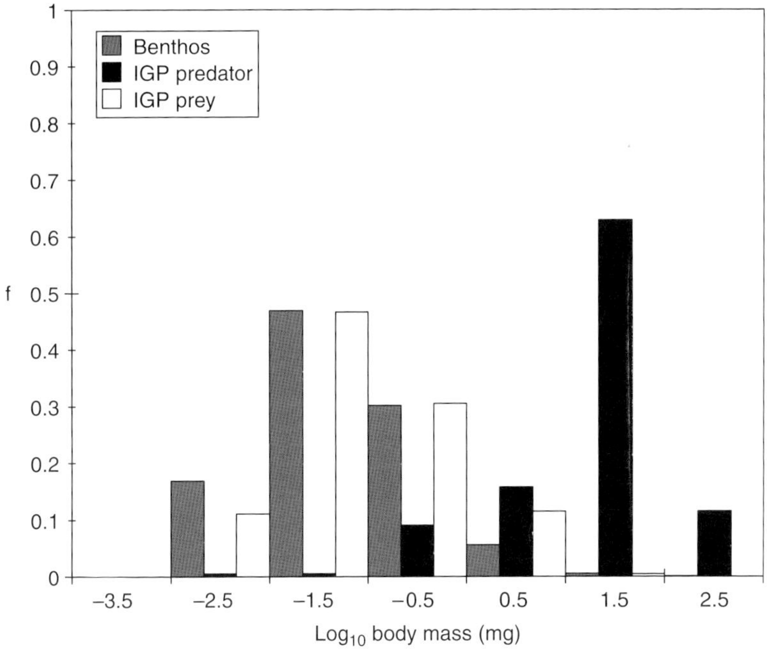

Figure 17 The individual-based body size distribution of nodes and links for the six dominant species within the intraguild predator assemblage of Broadstone Stream. Grey bars represent the body size distribution of predators in the environment; black bars represent the body size distribution of predators acting as consumers in IGP interactions; white bars represent the body size distribution of predators acting as prey in IGP interactions.

and the ability of a model to explain the structure were affected markedly by the choice of how to group individuals. For all four systems, the two versions of each food web differed markedly in many key attributes. The number of links, linkage density, connectance, and the match between the empirical data and the models were, on average, almost twice as high in size-based webs relative to their species-based equivalents. They also exhibited more pronounced ‘small world’ properties (e.g. Dunne *et al.*, 2002; Dunne, 2005; Montoya *et al.*, 2006), with higher clustering coefficients and reduced mean shortest path lengths. Taxonomic grouping of individuals has, to date, dominated food web research, raising the possibility that the dominance of this approach might be limiting our ability to understand more fully the drivers of food web structure. Both body size and sampling effort played important roles in determining the number of links recorded in each web, and also their patterning within the web, which was generally well predicted by the ADBM, especially for the size-based webs.

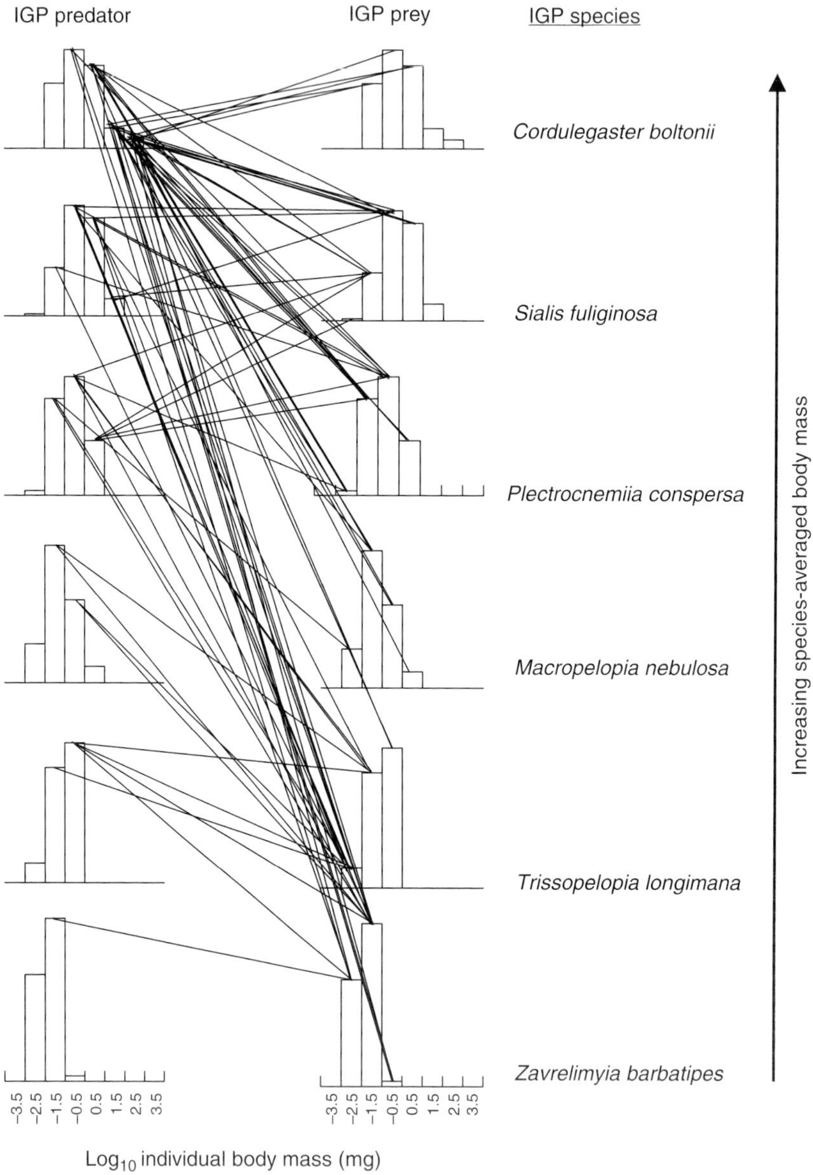

Figure 18 Decomposing the species-averaged food web into size-based interactions within the predator guild in Broadstone Stream. The six common predator species are arranged vertically, with the largest (*Cordulegaster boltonii*) at the top. Each row represents one predator species, as either predators or prey (thus links within rows represent cannibalism). In no instance did an individual predator consume a larger individual, either within or between species.

We stress that neither the taxonomic nor the size-based approach is necessarily the 'correct' way to represent a food web, as both have their limitations. Taxonomic food webs are unable to capture the size structure of feeding interactions completely, whilst size-based food web cannot characterise the role of taxonomy in determining the presence and patterning of feeding interactions. For instance, unlikely or seemingly impossible predator–prey ratios (e.g. prey that were more than an order of magnitude bigger than their consumers) occurred frequently in our taxonomic webs. Similarly, in the size-based web non-predatory species occurred in size classes that contained predatory individuals. Thus, in both taxonomic and size-based webs a link between two entities in the matrix did not mean that all individuals in that group could potentially feed on all individuals in the other.

Species-averaging masked the fact that predators and prey interacted via subsets of their respective size distributions, such that prey that survived to a large size could become invulnerable to the smaller predators, whilst at the other end of the spectrum very small (and hence energetically unrewarding) prey were rarely taken by the larger predators. In extreme cases, size disparities within the predator guild could even be sufficient to reverse trophic status, as occurred when large individuals of 'small' species were able to prey upon small individuals of 'large' species (Figure 18), which occurred when generations overlapped (Woodward and Hildrew, 2002a,b). This is suggestive of foraging constraints on the ability to detect and handle prey effectively determining the presence (and strength) of feeding links, and emphasises the connection between individual behaviour and the more complex system-level properties of the food web as a whole (Woodward and Warren, 2007).

Foraging theory based on individuals' decisions about what resource types to include in their diets appeared less effective in describing the patterns observed in the taxonomic food webs (30%, 45%, 43%, 62% of links correctly predicted) than in the size-based food webs (74%, 74%, 53%, 83% of the links correctly predicted). Thus, ignoring species identities, and constructing a food web based only on the size of individuals produced an often extremely good fit between the observed data and the model, with up to 83% links being predicted correctly in the size-based Broadstone Stream web. At one level, this is perhaps not surprising: if we construct a food web in such a way as to unambiguously reflect the size relationships of consumers and their resources, then a model based on mechanisms that are also size dependent should provide a better fit to those data than the same food web expressed in taxonomic form. This is not a test of the model *per se*, but an examination of the extent to which understanding the potential role of particular mechanisms, or processes in determining food web structure, is a process of reflection between model and data, considering the processes each encapsulates. It seems likely that other models incorporating rules based on size dependent foraging would also fit better to size-based than taxonomic webs.

Nonetheless, we would observe that the improved fit of the model to the size-based webs, is consistent with the view that foraging theory coupled with body size allometries in individual foraging behaviour could provide plausible explanations of food web structure (but see Allesina 2008; Rohr *et al.*, 2010). This raises the often neglected question of how food web models should be evaluated. The scarcity of food web data, and its limitations have tended to mean that the data are taken as a given, and models, even rather differently constructed or motivated models, are compared to each other on the strength of their performance against that one representation of the food webs. Whilst the concept of a food web is simple, their empirical representation is neither straightforward nor singular, and perhaps evaluation of increasingly sophisticated food web models will require more attention to the nature of the data representation, as well as the details of the models themselves.

B. Understanding the High Explanatory Power of the ADBM

The ADBM predicted some of the important features of the size-based food webs very effectively, such as absence of predation by larger size classes on the smallest size classes (lower size-refugia). This occurs because the relationship between profitability (energetic value of a resource divided by handling time) and resource size is hump-shaped: thus intermediate sized resources should have the highest profitability, whereas more extreme size disparities reduce profitability. Large consumers could, in theory, consume the smallest resources, but this appears not to be the case as doing so would decrease the overall rate at which they gain energy.

The unimodal relationship between resource profitability and size results from assumptions made about how handling time scales with resource size. As resources become smaller, handling time approaches a minimum value: thus very small resource items, which necessarily have very low energetic value, take a disproportionately long time to handle. The handling time scaling function also assumes that as resource size increases and approaches a critical fraction of consumer size (b) handling times approach infinity. That is, a consumer cannot handle a resource item larger than this fraction of its own size. Any resources above this size ratio have infinite handling time, zero profitability, and are never included in a consumer's diet. The ratio that produced the best fit of the model to the observed pattern of interactions among species was 0.78 in the Broadstone Stream analysis. In this sense, handling time as used in the model represents a measure of the mechanical (upper and lower) constraints on feeding that consumers face. In support of these arguments, there is evidence that upper- and lower-size refugia, as assumed by the ADBM, do exist in natural food webs (Schmid-Araya *et al.*, 2002a,b; Woodward *et al.*, 2005a,b).

C. Explaining 'Missing' and 'Unlikely' Links

The fact that the presence of 'missing links' were often associated with pairwise combinations of rare predators and rare prey is interesting as it suggests that the gaps could be artefacts of undersampling—that is, the model could be correctly predicting where a link is expected, but the data are not sufficiently complete to detect it. Some recently published analyses of empirical food webs concern the contiguity of diets (Allesina *et al.*, 2008; Cattin *et al.*, 2004; Stouffer *et al.*, 2006; Williams and Martinez, 2000): that is, if the available resources are ordered along a single axis, for example, body size, do consumers feed on a set of resources that occupy a contiguous set along the axis, or are there 'gaps'? One explanation for dietary gaps relates to multiple dimensions of resource variation (i.e. body size being one dimension, and perhaps defensive structures and or chemical being a second dimension) (Allesina *et al.*, 2008; Petchey *et al.*, 2008): that is, the fact that the vulnerability of a prey to a particular consumer may be determined by a combination of traits, such as size and colouration (Allesina *et al.*, 2008; Petchey *et al.*, 2008; Warren, 1996).

It is also possible that apparent dietary gaps might be due to sampling effects, if for instance, low abundance of prey and/or predators results in incomplete characterisation of feeding links. Other anomalies related to the consumption of prey types that lie well outside the 'core' dietary range of a consumer, might reflect accidental ingestion or 'Russian dolls', whereby the gut contents of an intermediate predator are recorded in the guts of a larger predator. There are several lines of evidence that lend support to these suggestions. For instance, incomplete sampling resulted in underestimation of diet width in the empirical data and instances when the models predicted links that were missing from the empirical data were most likely for combinations of rare predators and rare prey. The individual-based size distribution was far more robust to these effects than were the species-based size distribution, as the former tended to smoothe over the effects of undersampling rare species by allocating these individuals more evenly among nodes, rather than treating them as fixed entities in the latter. The Broadstone data provide an illustrative example of these joint effects of sampling and body size, which were especially marked in the species-averaged web. Here, the three rarest predators, whose yield–effort curves for links were not saturated (see also Woodward *et al.*, 2005a,b), had observed diets that were far narrower than would be predicted on the basis of their body mass alone (Figure 12). Because they overlapped with the six dominant predators, in terms of their body mass distributions, in the size-based web these three rare (and undersampled) species were mixed within size classes with the common species whose diet was more completely sampled. Intriguingly, in the species-averaged Broadstone Stream web these scarce predators with few observed links were also most similar in body mass to the dominant predators, with the latter being evenly spaced on a log-scale body mass gradient (cf. Figure 13 in Olesen *et al.*, 2010). This might imply that the coexistence of predators within the food web is related to the apportioning of the body mass (and hence prey availability) distribution, unless other dimensions come into play that might create contiguous dietary niches.

False-negative model predictions occurred when the empirical data described a link that was not predicted by the model. Typically, these arose where the data exceeded the 1:1 line between predator and prey body mass and also where predators ate very small prey relative to the rest of their diet. These might represent measurement errors in some cases, but they could also represent real phenomena that enable certain species to somehow bend or break the general rules of size-structuring. In many cases, these exceptions to the rule related to interactions involving the caddisfly *P. conspersa*, a species that uses a net to capture its prey and which can therefore take very small prey and also (occasionally) prey items that are larger than itself (Townsend and Hildrew, 1977, 1979). Many other instances of predators feeding on unusually small prey appeared to represent accidental ingestion or the presence of 'Russian dolls', whereby the prey items eaten by an intermediate predator are visible in the guts of a larger predator. For instance, of the 1087 individual predators, only 51 contained a prey individual less than four orders of magnitude smaller than themselves and 41% of these contained two or more prey individuals, a significantly higher proportion than would be expected for 51 randomly chosen predator individuals ($p=0.013$).

The overall results of the current study highlight the need for models of food web structure that can include the sampling processes that result in the observed data. Two of the key sampling processes in operation that require consideration here are the sampling of prey by predators, and the sampling of predators by researchers. Typically, both of these have been ignored in food web research, despite the powerful influences they can have on many food web properties and our ability to marry data to models in a meaningful way. An analogy may be drawn here with state space models of population dynamics (e.g. Clark and Bjornstad, 2004), which explicitly include biological and observation processes in quantitative models fit to empirical data: we are unaware of similar models of food web structure. Such models would include the potentially stochastic sampling of the environment by consumers and the stochastic nature of sample of the consumers by ecologists, both of which couple with the biological processes to produce the observed data.

D. Seasonal and Ontogenetic Effects

When combined abundances were low (e.g. rare predators and rare prey) feeding links in the summary web were often missing from seasonal versions of the Broadstone Stream web, despite being within the range of feasible links based on the respective body masses of the predators and prey (e.g. Figure 12 in the current study and Figure 13 in Olesen *et al.*, 2010). The seasonal patterns in the expression of links across the six sampling occasions reported here are also reflected in shifts in the extent of dietary niche overlap within the predator guild (cf. Woodward and Hildrew, 2002b), the structure of the wider food web, and the frequency of predation (cf. Woodward *et al.*, 2005a). Essentially, niche overlap, network complexity and *per capita* ingestion rates peaked during the summer (i.e. when predator and prey body masses spanned the largest intraspecific range as generations overlapped during recruitment periods) and declined progressively over the winter through to spring (Woodward *et al.*, 2005a). Much of this seasonal variability within the food web was related to a combination of imperfect synchrony of reproductive pulses, generational overlap, and differential growth rates of predators and their prey (cf. Woodward and Hildrew, 2002a,b; Woodward *et al.*, 2005a). Such situations, where only subsets of species populations interact intensely at certain times and at others little, or not at all, could potentially produce complex dynamics that differ considerably from those expected if all individuals are equivalent at all times (de Roos and Persson, 2002).

Mutual predation loops and omnivory often produce dynamics that are highly unstable in mathematical models and should, in theory, be rare (e.g. May, 1973). This is clearly not the case, however, in most of the well-described empirical webs, where such interactions are common (e.g. Petchey *et al.*, 2008; Williams and Martinez, 2000; Woodward and Hildrew, 2001). Our data were no exception but, unlike most previous studies, our individual-based approach enabled us to ascribe many of these apparent anomalies to ontogenetic shifts in trophic position, which would have otherwise inevitably been obscured by species-averaging. In Broadstone Stream, for instance, the predators moved progressively upwards through the food web and became more generalist as they grew, both within and among species. Thus, the biggest individuals of the dragonfly *C. boltonii* had the broadest diets within the web, and this species ultimately became an apex predator when it entered its largest size classes (Figure 18; see also Woodward and Hildrew, 2002a,b). Such shifts in diet and trophic status along a continuum of body mass are likely to be prevalent in food webs that contain predator species that overlap in body size and also display indeterminate growth, as is true for many aquatic ecosystems (e.g. Hardy, 1924).

Unfortunately, true ontogenetic shifts are often very difficult to detect in natural systems, because in most food webs the collection of gut contents involves destructive sampling. Even if this can be avoided, repeated mark–recapture of individuals over a protracted time series is often logistically unfeasible. We were unusually fortunate, therefore, to be able to obtain repeated measures of diet and body mass data for individual fish within the Tadnoll Brook food web. At this finest scale of individual-level data that we explored, body size was again a key determinant of diet and trophic position, as highlighted in the example of the dietary data from a single brown trout that was tracked and gut-flushed on eight sampling occasions as it more than doubled its body mass. When small, this individual preyed primarily on the portion of the prey assemblage that contained the most abundant size classes of invertebrates, but as it grew it began to feed higher in the food web by exploiting the larger but rarer size classes, which were predominantly fishes (Figure 8). This sudden change in diet is suggestive of a behavioural switch that occurred at a particular size threshold, and within the context of salmonid life-cycle biology, data such as these could help resolve the long-standing question as to whether there is a particular size threshold that determines whether an individual remains in freshwater as a resident brown trout or migrates to the ocean to develop into the much larger sea trout form. This exceptionally high data resolution, which is rarely available in food web studies, thus enabled us to characterise 'true' individual-level ontogenetic shifts in feeding directly, rather than relying on the use of proxy snapshots of a number of individuals of different size. Future studies that can embed this level of detail within empirical food webs could offer potentially much deeper insight than is currently possible: for instance, they could be used to ascertain whether the feeding niches of apparently 'generalist species' are comprised of a wide range of specialist individuals or a set of similarly generalist individuals.

E. Implications, Caveats, and Future Directions

There are several caveats associated with both the models and data presented here that merit further consideration. For instance, the algorithm we used to create the size classes of the size-based food webs simply designated evenly log spaced size classes within the range defined by the smallest to largest individuals observed in each system. It was therefore possible for a size class to be created when in fact no organisms fell in that class, as was the case for three of the four size-based webs. We considered both the likely consequences of the creation of empty size classes, and possible alternatives. The size-based webs had a greater number of links and connectance than their taxonomic counterparts: removing empty nodes would decrease the number of nodes (but not links), thereby increasing connectance still further and amplifying the contrasts with the taxonomic webs. Therefore, at least for differences in connectance, the presence of empty nodes has not affected the qualitative effects seen. Simply removing empty nodes would leave the size-based food webs with fewer nodes than taxonomic ones, and we were striving to keep the number of nodes (albeit potentially empty ones) equal. Another option would be to create unevenly spaced size classes, and ones that therefore could all contain individuals: this could potentially lead to additional interesting changes in the perception of size-based food web structure and could be addressed in further research.

Our study is necessarily restricted to the four highly-resolved aquatic food webs that have been described using individual-based data for both nodes and links. Since individual-level information, especially in populations of commercially exploited fish stocks, can reveal important information on the ecological effects of fishing on their dynamics (Stenseth and Rouyer, 2008), such highly-resolved data seem certain to emerge far more extensively, at least from aquatic systems, in the near future. Given the prospect of these projected increases in the number of available webs, this raises the question as to what extent this approach might be applied more generally to include other ecosystem types, including terrestrial food webs. It also highlights the need to try to standardise data collection in terms of both what measure of body size is used, how exhaustive sampling is, and how those samples are collected, since all these variables had strong effects on food web structure and the fit between theoretical predictions and empirical observations. It would also be instructive to explore how sampling effort influences the structure of both species- and sized-based webs: for instance, we assigned links based on the presence of a single feeding interaction between nodes, but other criteria (e.g. excluding very rare links and/or nodes) might alter the two types of web in different ways, with the taxonomic webs being potentially far more sensitive to sampling effects than their size-based equivalents.

There is a general perception that terrestrial and aquatic ecosystems differ fundamentally in their size structure, with the former typically being strongly size-structured and the latter seemingly less so (Ings *et al.*, 2009; Yvon-Durocher *et al.*, 2010). It is not inconceivable, however, that at least some of the perceived differences between the importance of size in aquatic and terrestrial ecosystems might be due to the respective disciplines measuring and analysing different things in different ways (Raffaelli, 2007; Yvon-Durocher *et al.*, 2010). Increasing the catalogue of available individual-based data will ultimately allow this to be tested in a more rigorous and standardised manner, and to reassess food webs that have previously been described from either species-based or size-based approaches, but not both.

Size-structuring is expected to be particularly strong where organisms are 'unitary particles' and gape-limited consumption predominates (Cyr and Curtis, 1999; Jennings and Brander, 2010; Woodward and Warren, 2007), but such conditions are not necessarily restricted to aquatic systems and they might be broadly applicable to predator–prey interactions in general (e.g. Brose *et al.*, 2006). Conversely, body size might not be expected to be a major player for different types of organism (e.g. modular species) or for other types of interaction (e.g. in mutualistic networks), where it may have a more subsidiary role to other species traits (Ings *et al.*, 2009; but see Stang *et al.*, 2006, 2009). Similarly, it might not necessarily be the case for interactions between primary consumers and basal resources: one obvious general case of a lack of size structure is the coexistence of both very small (insect) and large (mammalian) herbivores that both feed on many terrestrial plants that also vary greatly in size. This would clearly make it difficult, or even impossible, to characterise and predict the structure and dynamics of such systems based on simple size-based rules, and it is probably meaningless to attempt to do so in such cases.

There have been very few systematic and quantitative analyses of the importance of size for community structure across ecosystem types, but one exception is the extensive compilation and analysis by Brose *et al.* (2005, 2006) of resource–consumer body size information from 20 studies from marine, freshwater, and terrestrial systems. These data include four types of trophic interaction (predator, parasitoid, herbivore, and detritivore) and three consumer types (invertebrates, ectothermic vertebrates, and endothermic vertebrates), although not all possible combinations were evident within each study. Resource–consumer body size ratios differed between ecosystem types, feeding types, and consumer types, but nonetheless the vast majority of feeding interactions involved consumers that were larger than their resources (Brose *et al.*, 2006). These data and some more recent analyses (e.g. Berlow *et al.*, 2009; Yvon-Durocher *et al.*, 2010) suggest that resource–consumer interactions exhibit general size structure, but that its importance varies among systems: an individual-based approach could potentially shed further light on these apparent patterns.

The potentially universal importance of individual body size, as emphasised by its deep connections with many physiological rates (Gillooly *et al.*, 2001), provides the opportunity for it to be applied as a first principle, based on formal mathematical theory, for understanding patterns and processes in complex ecosystems (Petchey and Belgrano, 2010). Using individual size as a metric for thinking about food webs provides scope to seek commonalities and to explain unifying patterns, but it is also important for exploring deviations from these in systems where it is not the primary axis of network structure or dynamics. Despite the apparent potential for adding evermore complexity to what are already very complex systems, at any one time consumers and resources within a food web can always be ordered by body size, and underlying size constraints on feeding can then be explored, as we have done here. Clearly, if we are to obtain more meaningful insight into food webs we need to view them as more than simply coarse-grained agglomerations of species populations or size classes, and to integrate both taxonomic and size-based approaches (Petchey and Belgrano, 2010). In reality, both taxonomy and size are likely to play important roles, as revealed by some population dynamic models (Andersen and Pedersen, 2010; Blanchard *et al.*, 2009; De Roos *et al.*, 2008; Persson *et al.*, 1998) which could potentially be extended to multispecies systems. The explicit collection of new individual-level data from natural food webs would go a long way to providing a simultaneous understanding of both the general rules (e.g. allometries) and the specific exceptions within ecology that can be related to other drivers, such as, for instance, those associated with biodiversity and biogeochemical fluxes (Belgrano *et al.*, 2002; Naeem, 2006; Reiss *et al.*, 2009).

Future research using an individual-based approach could also address a range of additional questions relating to the relative importance of intraspecific and interspecific variation in body size within food webs, and the implications of this for species coexistence: for instance, if predator diets are nested subsets based largely on body size, do predators partition resources by staggering their phenologies? This approach could therefore help to identify whether niche assembly mechanisms are driven by frequency disruptive selection or neutral processes, and it could improve our ability to develop and test a range of new mechanistic models of individual-based food webs.

ACKNOWLEDGEMENTS

We would like to thank Carlos Melian and Jens Olesen, whose insightful comments greatly improved this manuscript, and all the numerous people who provided assistance in the field and laboratory. This project was supported by two Natural Environment Research Council grants awarded to JI and GW (NE/C511905/1 and NE/E012175/2), an NERC Centre for Population Biology grant awarded to GW and two ESF-funded SIZEMIC Working Groups led by OP, Julia Reiss, and Ute Jacob. We are also grateful to Sarah Pottinger and Jacques Deere for helping to organise the CPB workshops.

ELECTRONIC APPENDICES

E1 Food web data
E2 Yield–effort curves

REFERENCES

Allesina, S., Alonso, D., and Pascual, M. (2008). A general model for food web structure. *Science* **320**, 658–661.

Andersen, K.H., and Pedersen, M. (2010). Damped trophic cascades driven by fishing in model marine ecosystems. *Proc. R. Soc. B-Biol. Sci.* **277**, 795–802.

Banašek-Richter, C., Bersier, L.F., Cattin, M.F., Baltensperger, R., Gabriel, J.-P., Merz, Y., *et al.* (2009). Complexity in quantitative food webs. *Ecology* **90**, 1470–1477.

Barnes, C., Bethea, D.M., Brodeur, R.D., Spitz, J., Ridoux, V., Pusineri, C., *et al.* (2008). Predator and prey body sizes in marine food webs. *Ecology* **89**, 881.

Beckerman, A.P., Petchey, O.L., and Warren, P.H. (2006). Foraging biology predicts food web complexity. *Proc. Natl. Acad. Sci. USA* **103**, 13745–13749.

Belgrano, A., Allen, A.P., Enquist, B.J., and Gillooly, J.F. (2002). Allometric scaling of maximum population density: A common rule for marine phytoplankton and terrestrial plants. *Ecol. Lett.* **5**, 611–613.

Benke, A.C., and Wallace, J.B. (1997). Trophic basis of production among riverine caddisflies: Implications for food web analysis. *Ecology* **78**, 1132–1145.

Benke, A.C., Huryn, A.D., *et al.* (1999). Length–mass relationships for freshwater macroinvertebrates in North America with particular reference to the southeastern United States. *J. North Am. Benthol. Soc.* **18**, 308–343.

Berlow, E.A., Dunne, J.A., Martinez, N.D., Starke, P.B., Williams, R.J., and Brose, U. (2009). Simple prediction of interaction strengths in complex food webs. *Proc. Natl. Acad. Sci. USA* **106**, 187–191.

Bersier, L.F., and Sugihara, G. (1999). Scale-invariant or scale dependent behavior of the link density property in food webs. *Am. Nat.* **153**, 676–682.

Blanchard, J.L., Dulvy, N.K., Jennings, S., Ellis, J.E., Pinnegar, J.K., Tidd, A., *et al.* (2005). Do climate and fishing influence size-based indicators of Celtic Sea fish community structure? *ICES J. Mar. Sci.* **62**, 405–411.

Blanchard, J.L., Jennings, S., Law, R., Castle, M.D., McCloghrie, P., Rochet, M.J., *et al.* (2009). How does abundance scale with body size in coupled size-structured food webs? *J. Anim. Ecol.* **78**, 270–280.

Brose, U., Cushing, L., Banasek-Richter, C., Berlow, E., Bersier, L.F., Blanchard, J., *et al.* (2005). Empirical consumer–resource body size ratios. *Ecology* **86**, 2545.

Brose, U., Jonsson, T., Berlow, E.L., Warren, P., Banasek-Richter, C., Bersier, L.-F., *et al.* (2006). Consumer-resource body–size relationships in natural food webs. *Ecology* **57**, 2411–2417.

Brown, J.H., Gillooly, J.F., Allen, A.P., Savage, V.M., and West, G.B. (2004). Toward a metabolic theory of ecology. *Ecology* **85**, 1771–1789.

Burgherr, P., and Meyer, E.I. (1997). Regression analysis of linear body dimensions vs. dry mass in stream macroinvertebrates. *Arch. Hydrobiol.* **139**, 101–112.

Camerano, L. (1880). Dell' equilibrio dei viventi mercé la reciproca distruzione. *Atti della Reale Accademia delle Scienze di Torino* **15**, 393–414.

Cattin, M.-F., Bersier, L.-F., Banašek-Richter, C., Baltensperger, R., and Gabriel, J.-P. (2004). Phylogenetic constraints and adaptation explain food–web structure. *Nature* **427**, 835–839.

Clark, J.S., and Bjornstad, C.N. (2004). Population time series: Process variability, observation errors, missing values, lags, and hidden states. *Ecology* **85**, 3140–3150.

Closs, G.P., and Lake, P.S. (1994). Spatial and temporal variation in the structure of an intermittent-stream food-web. *Ecol. Monogr.* **64**, 1–21.

Cohen, J.E. (1978). *Food Webs and Niche Space*. Princeton (now) University Press, Princeton, New Jersey, USA.
Cohen, J.E., and Newman, C.M. (1985). A stochastic theory of community food webs. I. Models and aggregated data. *Proc. R. Soc. Lond. B* **224**, 421–448.
Cohen, J.E., Briand, F., and Newman, C.M. (1990). Community Food Webs: Data and Theory. Springer-Verlag, New York, USA.
Cohen, J.E., Beaver, R.A., Cousins, S.H., DeAngelis, D.L., Goldwasser, L., Heong, K.L., *et al.* (1993). Improving food webs. *Ecology* **74**, 252–258.
Cohen, J.E., Jonsson, T., and Carpenter, S.R. (2003). Ecological community description using the food web, species abundance, and body size. *Proc. Natl. Acad. Sci. USA* **100**, 1781–1786.
Cohen, J.E., Jonsson, T., Muller, C.B., Godfray, H.C.J., and Savage, V.M. (2005). Body sizes of hosts and parasitoids in individual feeding relationships. *Proc. Natl. Acad. Sci. USA* **102**, 684–689.
Csardi, G., and Nepusz, T. (1695). The igraph software package for complex network research. *InterJournal*.
Cyr, H., and Curtis, J.M. (1999). Zooplankton community size structure and taxonomic composition affects size-selective grazing in natural communities. *Oecologia* **118**, 306–315.
Damuth, J. (1987). Interspecific allometry of population density in mammals and other animals: The independence of body mass and population energy-use. *Biol. J. Linn. Soc.* **31**, 193–246.
De Roos, A.M., Schellekens, T., Van Kooten, T., van de Wolfshaar, K.E., Claessen, D., and Persson, L. (2008). Simplifying a physiologically structured population model to a stage-structured biomass model. *Theor. Popul. Biol.* **73**, 47–62.
De Roos, A.M., Leonardsson, K., Persson, L., and Mittelbach, G.G. (2002). Ontogenetic niche shifts and flexible behavior in size-structured populations. *Ecol. Monogr.* **72**, 271–292.
de Ruiter, P.C., Neutel, A.M., and Moore, J.C. (1995). Energetics, patterns of interaction strengths, and stability in real ecosystems. *Science* **269**, 1257–1260.
Dunne, J. (2005). The network structure of food webs. In: *Ecological Networks: Linking Structure to Dynamics in Food Webs* (Ed. by M.P.a.J.A. Dunne), pp. 27–86. Oxford University Press, New York.
Dunne, J.A., Williams, R.J., and Martinez, N.D. (2002). Food–web structure and network theory: The role of connectance and size. *Proc. Natl. Acad. Sci. USA* **99**, 12917–12922.
Dunne, J.A., Williams, R.J., Martinez, N.D., Wood, R.A., and Erwin (2008). Compilation and network analyses of Cambrian food webs. *PLoS Biol.* **6**, 693–708.
Edwards, F.K., Lauridsen, R.B., Duerdoth, C.R., and Jones, J.I. (2008). Dry and ash-free dry mass to length relationships of bullhead (*Cottus gobio* L.). *Fundam. Appl. Limnol.* **172**, 315–316.
Edwards, F.K., Lauridsen, R.B., Armand, L., Vincent, H.M., and Jones, J.I. (2009a). The relationship between length, mass and preservation time for three species of freshwater leeches (Hirudinea). *Fundam. Appl. Limnol.* **173**, 321–327.
Edwards, F.K., Lauridsen, R.B., *et al.* (2009b). Re-introduction of Atlantic salmon, *Salmo salar* L. to the Tadnoll Brook, Dorset. *Proc. Dorset Nat. Hist. Archaeol. Soc.* **130**, 1–8.
Elton, C.S. (1927). Animal Ecology. Sedgewick and Jackson, London.
Emlen, J.M. (1966). The role of time and energy in food preference. *Am. Nat.* **100**, 611–617.
Emmerson, M., and Raffaelli, D. (2004). Predator–prey body size, interaction strength and the stability of a real food web. *J. Anim. Ecol.* **73**, 399–409.

Emmerson, M.C., Montoya, J.M., and Woodward, G. (2005). Body size, interaction strength and food web dynamics. In: *Dynamic Food Webs: Multispecies Assemblages, Ecosystem Development and Environmental Change* (Ed. by P. de Ruiter, V. Wolters and J.C. Moore), pp. 167–178. Academic Press, New York.

Figueroa, D. (2007). *Food Web Dynamics: New Patterns from Southern South America and North Wales UK, and the Role of Basal Species Structuring Food Webs*. PhD, University of London, London.

Ganihar, S.R. (1997). Biomass estimates of terrestrial arthropods based on body length. *J. Biosci.* **22**, 219–224.

Gillooly, J., Brown, J.H., West, G.B., Savage, V.M., and Charnov, E.L. (2001). Effects of size and temperature on metabolic rate. *Science* **293**, 2248–2251.

Goldwasser, L., and Roughgarden, J. (1997). Sampling effects and the estimation of food-web properties. *Ecology* **78**, 41–54.

Gonzalez, J.M., Basaguren, A., *et al.* (2002). Size–mass relationships of stream invertebrates in a northern Spain stream. *Hydrobiologia* **489**, 131–137.

Hall, S.J., and Raffaelli, D.G. (1993). Food webs—theory and reality. *Adv. Ecol. Res.* **24**, 187–239.

Hardy, A.C. (1924). The Herring in Relation to its Animate Environment, Part 1. Ministry of Agriculture and Fisheries. *Fishery Investigations series 2*, **7**, 3 UK.

Hildrew, A.G. (2009). Sustained research on stream communities: A model system and the comparative approach. *Adv. Ecol. Res.* **41**, 175–312.

Hildrew, A.G., and Townsend, C.R. (1982). Predators and prey in a patchy environment: A freshwater study. *J. Anim. Ecol.* **51**, 797–815.

Hutchinson, G.E. (1959). Homage to Santa Rosalia or why are there so many kinds of animals? *Am. Nat.* **93**, 145–159.

Ings, T.C., Montoya, J.M., Bascompte, J., Bluthgen, N., Brown, L., Dormann, C.F., *et al.* (2009). Ecological networks—Beyond food webs. *J. Anim. Ecol.* **78**, 253–269.

Jennings, S.J., and Brander, K. (2010). Predicting the effects of climate change on marine communities and the consequences for fisheries. *J. Mar. Syst.* **79**, 418–426.

Jennings, S., Warr, K.J., and Mackinson, S. (2002). Use of size-based production and stable isotope analyses to predict trophic transfer efficiencies and predator–prey body mass ratios in food webs. *Mar. Ecol. Prog. Ser.* **240**, 11–20.

Jonsson, T., Cohen, J.E., and Carpenter, S.R. (2005). Food webs, body size, and species abundance in ecological community description. *Adv. Ecol. Res.* **36**, 1–84.

Kerr, S.R., and Dickie, L.M. (2001). *The Biomass Spectrum: A Predator–Prey Theory of Aquatic Production*. Columbia University Press, New York, 320 pp.

Lafferty, K.D., Dobson, A.P., and Kuris, A.M. (2006). Parasites dominate food web links. *Proc. Natl. Acad. Sci. USA* **103**, 11211–11216.

Layer, K., Hildrew, A.G., Monteith, D., and Woodward, G. (2010a). Long-term variation in the littoral food web of an acidified mountain lake. *Glob. Change Biol.* **16** (11), 133–3143.

Layer, K., Riede, J.O., Hildrew, A.G., and Woodward, G. (2010b). Food web structure and stability in 20 streams across a wide pH gradient. *Adv. Ecol. Res.* **42**, 265–299.

MacArthur, R. (1955). Fluctuations of animal populations and a measure of community stability. *Ecology* **36**, 533–536.

MacArthur, R.H., and Pianka, E.R. (1966). On optimal use of a patchy environment. *Am. Nat.* **100**, 603–609.

Martinez, N.D. (1991). Artifacts of attributes? Effects of resolution on the Little Rock lake food web. *Ecol. Monogr.* **61**, 367–392.

Martinez, N.D., Hawkins, B.A., Dawah, H.A., and Feifarek, B.P. (1999). Effects of sampling effort on characterization of food–web structure. *Ecology* **80**, 1044–1055.

May, R.M. (1972). Will a large complex system be stable? *Nature* **238**, 413–414.

May, R.M. (1973). *Stability and complexity in model ecosystems*. Princeton Univ. Press, Princeton.

McCann, K.S. (2000). The diversity–stability debate. *Nature* **405**, 228–233.

McCann, K., Hastings, A., and Huxel, G.R. (1998). Weak trophic interactions and the balance of nature. *Nature* **395**, 794–798.

McLaughlin, O.B., Jonsson, T., and Emmerson, M.C. (2010). Temporal variability in predator–prey relationships of a forest floor food web. *Adv. Ecol. Res.* **42**, 171–264.

Meyer, E. (1989). The relationship between body length parameters and dry mass in running water invertebrates. *Arch. Hydrobiol.* **117**, 191–203.

Montoya, J.M., Pimm, S.L., and Solé, R.V. (2006). Ecological networks and their fragility. *Nature* **442**, 259–264.

Montoya, J.M., Woodward, G., Emmerson, M.C., and Sole, R.C. (2009). Indirect effects propagate disturbances in real food webs. *Ecology* **90**, 2426–2433.

Naeem, S. (2006). Expanding scales in biodiversity-based research: challenges and solutions for marine systems. *Mar. Ecol. Prog. Ser.* **311**, 273–283.

Neutel, A.-M., Heesterbeek, J.A.P., and de Ruiter, P.C. (2002). Stability in real food webs: Weak links in long loops. *Science* **296**, 1120–1123.

Neutel, A.M., Heesterbeek, J.A.P., van de Koppel, J., Hoenderboom, G., Vos, A., Kaldeway, C., *et al.* (2007). Reconciling complexity with stability in naturally assembling food webs. *Nature* **449**, 599–602.

O'Gorman, E.J., and Emmerson, M.C. (2009). Perturbations to trophic interactions and the stability of complex food webs. *Proc. Natl. Acad. Sci. USA* **106**, 13393–13398.

O'Gorman, E.J., and Emmerson, M.C. (2010). Manipulating interaction strengths and the consequences for trivariate patterns in a marine food web. *Adv. Ecol. Res.* **42**, 301–419.

Olesen, J.M., Dupont, Y.L., O'Gorman, E.J., Ings, T.C., Layer, K., Melián, C.J., *et al.* (2010). From Broadstone to Zackenberg: Space, time and hierarchies in ecological networks. *Adv. Ecol. Res.* **42**, 1–69.

Otto, S.B., Rall, B.C., and Brose, U. (2007). Allometric degree distributions facilitate food–web stability. *Nature* **450**, 1226–1229.

Paine, R.T. (1988). Food webs—Road maps of interactions or grist for theoretical development. *Ecology* **69**, 1648–1654.

Paine, R.T. (1992). Food–web analysis through field measurement of per capita interaction strength. *Nature* **355**, 73–75.

Perkins, D.M., McKie, B.G., Malmqvist, B., Gilmour, S.G., Reiss, J., and Woodward, G. (2010). Environmental warming and biodiversity-ecosystem functioning in freshwater microcosms: partitioning the effects of species identity, richness and metabolism. *Adv. Ecol. Res.* **43**, 177–209.

Persson, L., Leonardsson, K., de Roos, A.M., Gyllenberg, M., and Christensen, B. (1998). Ontogenetic scaling of foraging rates and the dynamics of a size-structured consumer–resource model. *Theor. Popul. Biol.* **54**, 270–293.

Petchey, O.L., and Belgrano, A. (2010). Body-size distributions and size-spectra: universal indicators of ecological status? *Biol. Lett.* **6**, 434–437.

Petchey, O.L., Beckerman, A.P., and Warren, P.H. (2008). Size, foraging, and food web structure. *Proc. Natl. Acad. Sci. USA* **105**, 4191–4196.

Petchey, O.L., Downing, A.L., Mittelbach, G.G., Persson, L., Steiner, C.F., Warren, P.H., and Woodward, G. (2004). Species loss and the structure and functioning of multitrophic aquatic ecosystems. *Oikos* **104**, 467–478.

Peters, R.H. (1983). In: *The Ecological Implications of Body Size* (Ed. by E. Beck, H.J.B. Birks and E.F. Connor), pp. 24–44. Cambridge University Press, Cambridge, UK.

Pimm, S.L. (1980). Food web design and the effects of species deletion. *Oikos* **35**, 139–149.

Pimm, S.L. (1982). *Food Webs*. The University of Chicago Press, Chicago.

Pinnegar, J.K., Trenkel, V.M., Tidd, A.N., Dawson, W.A., and Du Buit, M.H. (2003). Does diet in Celtic Sea fishes reflect prey availability? *J. Fish Biol.* **63**, 197–212.

Polis, G.A. (1991). Complex trophic interactions in deserts: An emprirical critique of food web theory. *Am. Nat.* **138**, 123–155.

Polis, G.A. (1998). Stability is woven by complex webs. *Nature* **395**, 744–745.

Pope, J.G., Shepherd, J.G., and Webb, J. (1994). Successful surf-riding on size spectra: The secret of survival in the sea. *Phil. Trans. Roy. Soc. B.* **343**, 41–49.

Prentice, E.F., Flagg, T.A., and McCutcheon, S. (1990). Feasibility of using implantable passive integrated transponder (PIT) tags in salmonids. In: *Fish Marking Techniques* (Ed. by N.C. Parker, A.E. Giorgi, R.C. Heidinger, D.B. Jester, Jr., E.D. Prince and G.A. Winans), *American Fisheries Society Symposium*, **Vol. 7**, pp. 317–322.

Raffaelli, D. (2007). Food webs, body size and the curse of the latin binomial. In: *From Energetics to Ecosystems: The Dynamics and Structure of Ecological Systems* (Ed. by N. Rooney, K.S. McCann and D.L.G. Noakes), pp. 53–64. Springer, Netherlands.

Reiss, J., Bridle, J.R., Montoya, J.M., and Woodward, G. (2009). Emerging horizons in biodiversity and ecosystem functioning research. *Trends Ecol. Evol.* **24**, 505–514.

Reiss, J., Bailey, R.A., Cássio, F., Woodward, G., and Pascoal, C. (2010). Assessing the contribution of micro-organisms and macrofauna to biodiversity-ecosystem functioning relationships in freshwater microcosms. *Adv. Ecol. Res.* **43**, 151–176.

Reuman, D.C., and Cohen, J.E. (2005). Estimating relative energy fluxes using the food web, species abundance, and body size. *Adv. Ecol. Res.* **36**, 137–182, Elsevier Academic Press Inc, San Diego.

Reuman, D.C., Mulder, C., Raffaelli, D., and Cohen, J.E. (2008). Three allometric relations of population density to body mass: Theoretical integration and empirical tests in 149 food webs. *Ecol. Lett.* **11**, 1216–1228.

Reuman, D.C., Mulder, C., Banašek-Richter, C., Cattin Blandenier, M.-F., Breure, A.M., Den Hollander, H.A., *et al.* (2009a). Allometry of body size and abundance in 166 food webs. *Adv. Ecol. Res.* **41**, 1–44.

Reuman, D.C., Cohen, J.E., and Mulder, C. (2009b). Human and environmental factors influence soil food webs' abundance–mass allometry and structure. *Adv. Ecol. Res.* **41**, 45–85.

Riede, J.O., Rall, B.C., Banasek-Richter, C., Navarrete, S.A., Wieters, E.A., and Brose, U. (2010). Scaling of food–web properties with diversity and complexity across ecosystems. *Adv. Ecol. Res.* **42**, 139–170.

Rohr, R.P., Scherer, H., Kehrli, P., Mazza, C., and Bersier, L.-F. (2010). Modeling food webs: Exploring unexplained structure using latent traits. *Am. Nat.* **176**, 170–177.

Rooney, N., McCann, K., Gellner, G., and Moore, J.C. (2006). Structural asymmetry and the stability of diverse food webs. *Nature* **442**, 265–269.

Sabo, J.L., Bastow, J.L., *et al.* (2002). Length–mass relationships for adult aquatic and terrestrial invertebrates in a California watershed. *J. North Am. Benthol. Soc.* **21**, 336–343.
Schmid-Araya, J.M., Hildrew, A.G., Robertson, A., Schmid, P.E., and Winterbottom, J. (2002a). The importance of meiofauna in food webs: Evidence from an acid stream. *Ecology* **83**, 1271–1285.
Schmid-Araya, J.M., Schmid, P.E., Robertson, A., Winterbottom, J.H., Gjerløv, C., and Hildrew, A.G. (2002b). Connectance in stream food webs. *J. Anim. Ecol.* **71**, 1062.
Seber, G.A.F., and LeCren, E.D. (1967). Estimating population parameters from catches large to the relative population. *J. Anim. Ecol.* **36**, 631–643.
Smock, L.A. (1980). Relationships between body size and biomass of aquatic insects. *Freshw. Biol.* **10**, 375–383.
Stang, M., Klinkhamer, P.G.L., and van der Meijden, E. (2006). Size constraints and flower abundance determine the number of interactions in a plant–flower visitor web. *Oikos* **112**, 111–121.
Stang, M., Klinkhamer, P.G.L., Waser, N.M., Stang, I., and van der Mejden, E. (2009). Size-specific interaction patterns and size matching in a plant–pollinator interaction web. *Ann. Bot. Lond.* **103**, 1459–1469.
Stenseth, N.C., and Rouyer, T. (2008). Ecology: Destabilized fish stocks. *Nature* **452**, 825–826.
Stouffer, D.B. (2010). Scaling from individuals to networks in food webs. *Funct. Ecol.* **24**(1), 44–51.
Stouffer, D.B., Camacho, J., and Amaral, L.A.N. (2006). A robust measure of food web intervality. *Proc. Natl. Acad. Sci. USA* **103**, 19015–19020.
Tanabe, K., and Namba, T. (2005). Omnivory creates chaos in simple food web models. *Ecology* **86**, 3411–3414.
Townsend, C.R., and Hildrew, A.G. (1977). Predation strategy and resource utilization by *Plectrocnemia conspersa* (Curtis) (Trichoptera: Polycentropodidae). In: *Proceedings of the 2nd International Symposium on Trichoptera* Junk, The Hague.
Townsend, C.R., and Hildrew, A.G. (1979). Resource partitioning by two freshwater invertebrate predators with contrasting foraging strategies. *J. Anim. Ecol.* **48**, 909–920.
Tylianakis, J.M., Tscharntke, T., and Lewis, O.T. (2007). Habitat modification alters the structure of tropical host-parasitoid food webs. *Nature* **445**, 202–205.
van Veen, F.J.F., Muller, C.B., Pell, J.K., and Godfray, H.C.J. (2008). Food web structure of three guilds of natural enemies: Predators, parasitoids and pathogens of aphids. *J. Anim. Ecol.* **77**, 191–200.
Vandermeer, J. (2006). Omnivory and the stability of food webs. *J. Theor. Biol.* **238**, 497–504.
Warnes, S., and Jones, B.W. (1995). *Species Distributions from English Celtic Sea Groundfish Surveys, 1984 to 1991.* MAFF Directorate of Fisheries Research, LowestoftReport 98.
Warren, P.H. (1989). Spatial and temporal variation in the structure of a freshwater food web. *Oikos* **55**, 299–311.
Warren, P.H. (1996). Structural constraints on food web assembly. In: *Aspects of the Genesis and Maintenance of Biological Diversity* (Ed. by M.E. Hochberg, J. Clobert and R. Barbault), pp. 142–161. Oxford University Press, Oxford, U.K.
Williams, R.J., and Martinez, N.D. (2000). Simple rules yield complex food webs. *Nature* **404**, 180–183.

Woodward, G. (2009). Biodiversity, ecosystem functioning and food webs in fresh waters: Assembling the jigsaw puzzle. *Freshw. Biol.* **54**, 2171–2187.
Woodward, G., and Hildrew, A.G. (2001). Invasion of a stream food web by a new top predator. *J. Anim. Ecol.* **70**, 273–288.
Woodward, G., and Hildrew, A.G. (2002a). Differential vulnerability of prey to an invading top predator: Integrating field surveys and laboratory experiments. *Ecol. Entomol.* **27**, 732–744.
Woodward, G., and Hildrew, A.G. (2002b). Body-size determinants of niche overlap and intraguild predation within a complex food web. *J. Anim. Ecol.* **71**, 1063–1074.
Woodward, G., and Warren, P.H. (2007). Body size and predatory interactions in freshwaters: Scaling from individuals to communities. In: *Body size: The Structure and Function of Aquatic Ecosystems* (Ed. by A.G. Hildrew, D. Raffaelli and R. Edmonds-Brown), pp. 98–117. Cambridge University Press.
Woodward, G., Speirs, D.C., and Hildrew, A.G. (2005a). Quantification and resolution of a complex, size-structured food web. *Adv. Ecol. Res.* **36**, 85–135.
Woodward, G., Ebenman, B., Emmerson, M., Montoya, J.M., Olesen, J.M., Valido, A., *et al.* (2005b). Body-size in ecological networks. *Trends Ecol. Evol.* **20**, 402–409.
Woodward, G., Papantoniou, G., Edwards, F., and Lauridsen, R.B. (2008). Trophic trickles and cascades in a complex food web: Impacts of a keystone predator on stream community structure and ecosystem processes. *Oikos* **117**, 683–692.
Woodward, G., Benstead, J.P., Beveridge, O.S., Blanchard, J., Brey, T., Brown, L., *et al.* (2010). Ecological networks in a changing climate. *Adv. Ecol. Res.* **42**, 71–138.
Yodzis, P., and Innes, S. (1992). Body size and consumer–resource dynamics. *Am. Nat.* **139**, 1151–1175.
Yvon-Durocher, G., Montoya, J.M., Emmerson, M.C., and Woodward, G. (2008). Macroecological patterns and niche structure in a new marine food web. *Cent. Eur. J. Biol.* **3**, 91–103.
Yvon-Durocher, G., Reiss, J., Blanchard, Ebenman, B., Perkins, D.M., Reuman, D.C., *et al.* (2010). Across ecosystem comparisons of size structure: Methods, approaches, and prospects. *Oikos* (in press).

The Temperature Dependence of the Carbon Cycle in Aquatic Ecosystems

GABRIEL YVON-DUROCHER, ANDREW P. ALLEN, JOSÉ M. MONTOYA, MARK TRIMMER AND GUY WOODWARD

ADVANCES IN ECOLOGICAL RESEARCH VOL. 43
© 2010 Elsevier Ltd. All rights reserved
0065-2504/10 $35.00
DOI: 10.1016/S0065-2504(10)43007-X

SUMMARY

The carbon cycle modulates climate change via the regulation of atmospheric CO_2, and represents one of the most important ecosystem services of value to humans. However, considerable uncertainties remain concerning potential feedbacks between the biota and the climate. We developed theoretical models derived from the metabolic theory of ecology (MTE), and tested them in an ecosystem-level manipulative experiment in freshwater mesocosms. The year-long experiment simulated a warming scenario (A1B; [IPCC, 2007]) expected by the end of the century. The key components of the carbon cycle – that is gross primary production (GPP), ecosystem respiration (ER) and CH_4 efflux (ME) – measured in our experiment were all strongly related to temperature. Their temperature dependence was typically constrained by the average activation energy of their particular metabolic pathway, and as predicted by our models, this increased progressively for GPP, ER and ME. Warming of 4 °C decreased the sequestration of CO_2 by 13%, increased the fraction of primary production effluxing as methane by 20% and the fraction of ER as methane by 9%, in line with the offset in their respective activation energies. Because methane has 21 times the greenhouse gas radiative potential of CO_2, these results suggest aquatic ecosystems could drive a previously unknown positive feedback between warming and the carbon cycle.

We then used a series of global data compilations of measurements of rates of primary production and respiration to better understand the temperature dependence of the carbon cycle in other aquatic ecosystems and to compare them with data from terrestrial systems. Our experimental results were mirrored by our global data compilations, with the effective activation energy for marine and freshwater primary production identical to GPP measured in our experiment. Similarly, the temperature dependences of respiration in estuaries, lakes and the ocean were indistinguishable from that of ER in our experiment. Finally, our study suggests that the temperature dependence of primary production and respiration in aquatic ecosystems might differ from those in terrestrial ecosystems, and this could be crucial in predicting the future response of the carbon cycle in these different systems to global warming.

I. INTRODUCTION

A. The Ecological Consequences of Global Warming

The biosphere is in the midst of a pronounced warming trend: global surface temperatures have already risen by an average of ~0.74 °C over the last century and are projected to increase by a further 3–5 °C over the next century (IPCC, 2007). Future warming has important consequences for all levels of ecological organisation, from individual organisms to entire ecosystems (Montoya and Raffaelli, 2010; Walther *et al.*, 2002). At the population and species levels, ecological responses to global warming have been demonstrated for a large and growing list of taxa; these responses include extirpation of local populations, latitudinal and altitudinal shifts in the distribution of species populations, and altered phenology (Parmesan and Yohe, 2003; Woodward *et al.*, 2010a,b). At the ecosystem level, however, there is still no firm consensus regarding the magnitudes, types and directions of biotic responses to warming. This uncertainty is often ascribed to the complexity of ecological networks (e.g. food webs) that are embedded within ecosystems, and which is often perceived as confounding in attempts to derive predictions at these higher levels of organisation (Montoya *et al.*, 2006; Perkins *et al.*, 2010; Ptacnik *et al.*, 2010; Purdy *et al.*, 2010; Reiss *et al.*, 2010a,b; Woodward *et al.*, 2010a,b). The composition of the interaction networks in the new communities that will emerge as temperatures rise will be determined by differential rates of range shifts, with certain species moving faster and further than others; however, since the direction and magnitude of a particular species' response to climate change are often unexpected, such community-level responses are often difficult to predict (Montoya and Raffaelli, 2010; Walther *et al.*, 2002). In contrast, predicting the effects of environmental change on ecosystem-level processes (i.e. primary production, ecosystem respiration, nutrient cycling) might be substantially easier because comparable rates of ecosystem processes can result from vastly different community compositions (Manning *et al.*, 2006; Perkins *et al.*, 2010; Reiss *et al.*, 2010a).

In this chapter we focus on the effects of warming – a key component of future climate change – on ecosystem processes in aquatic systems. Global warming has the potential to affect ecosystem processes and the provisioning of their associated services to humanity (e.g. carbon sequestration, crop production) profoundly, via a variety of direct and indirect mechanisms (Montoya and Raffaelli, 2010; Schroter *et al.*, 2005). Temperature may, for example, directly alter the carbon sequestration capacity and the greenhouse gas efflux potential of ecosystems through its effects on the balance between plant growth and decomposition (Allen *et al.*, 2005; Yvon-Durocher *et al.*, 2010b). Both of these rates exhibit predictable temperature dependencies at the organismal level, which will ultimately determine responses at the whole ecosystem level (Allen *et al.*, 2005; Gillooly *et al.*, 2001; Lopez-Urrutia *et al.*, 2006; Yvon-Durocher *et al.*, 2010a). In terrestrial ecosystems, the potential for strong positive feedbacks between warming and carbon sequestration through temperature-induced enhancement of microbial respiration is well established (Davidson and Janssens, 2006; Knorr *et al.*, 2005). However, its magnitude remains unclear because the fraction of soil carbon susceptible to enhanced decomposition in response to warming is highly variable between ecosystems (Bellamy *et al.*, 2005; Luo *et al.*, 2001). Given that turnover rates of some soil carbon fractions span many years (Trumbore, 2000), and that a 10% change in total soil organic carbon would be equivalent to all the anthropogenic CO_2 emitted over a 30-year period (Kirschbaum, 2000), temperature-induced enhancement of soil respiration has the potential to affect atmospheric carbon chemistry for decades (Bellamy *et al.*, 2005). Continued warming may also help facilitate decomposition of large quantities of carbon currently stored in permafrost, thereby substantially increasing efflux of carbon dioxide and methane to the atmosphere (Mastepanov *et al.*, 2008).

Recent work in aquatic ecosystems has also highlighted the potential for significant enhancement of respiration rates (Lopez-Urrutia *et al.*, 2006) and associated declines in carbon sequestration (Gudasz *et al.*, 2010), again due to the strong temperature dependence of metabolism. Furthermore, concerns have recently been raised regarding the colossal potential source of methane which lies beneath the sea bed in the form of clathrate methane hydrates which, if they were to become unstable due to warming, could have a catastrophic impact on our climate (Westbrook *et al.*, 2009).

Overall, the potential importance of positive feedbacks between biogeochemical cycles and global warming highlights the need to link biotic responses explicitly to the components of the carbon cycle that are likely to be affected by climate change (Cox *et al.*, 2000; Friedlingstein *et al.*, 2006). This paper outlines several approaches, via the use of bioenergetic models, whole-ecosystem experiments and global meta-analyses, to address the potential for feedbacks between warming and the biogeochemcial cycling of carbon in aquatic ecosystems. The models and their experimental verification are primarily derived from, and represent extensions of, the metabolic theory of ecology (MTE) (Allen *et al.*, 2005; Brown *et al.*, 2004; Enquist *et al.*, 2003; Lopez-Urrutia *et al.*, 2006; Yvon-Durocher *et al.*, 2010a,b), which uses fundamental physical and chemical principles to understand and predict fluxes of energy and matter through individuals and ecosystems. We focus on the biotic controls that determine the dynamics of carbon dioxide and methane – the dominant end products of the remineralisation of organic carbon in aquatic ecosystems and the two most important greenhouse gases with potential to cause positive feedbacks with global warming. Our paper is therefore structured around two central questions. First: how and by how much will CO_2 sequestration in aquatic ecosystems be affected by the 4 °C rise in temperature expected by the end of the century (IPCC, 2007)? Second: how and by how much will CH_4 emissions change relative carbon sequestration under the warming scenario expected by the end of the century? We begin by presenting the predictions of our theoretical models, which attempt to delineate the temperature dependence of the key components of the carbon cycle in aquatic ecosystems – that is, gross primary production (GPP), ecosystem respiration (ER) and methane efflux. We then discuss the degree of concordance between our theoretical predictions and two direct tests of their assumptions. The first of these uses a large-scale freshwater mesocosm experiment as a model system, and the second uses global compilations of data of measurements of primary production and respiration in natural marine and freshwater ecosystems. We conclude each section, which is delineated by a particular metabolic flux, by emphasising the similarities and differences between its temperature dependence in aquatic and terrestrial ecosystems. We then attempt to predict how the balance between GPP, ER, and methane efflux will respond to future warming and test our predictions in our mesocosm experiment. Finally, we conclude by discussing the prospects and challenges for extending these approaches in the future and in a broader context.

B. Carbon Cycling Within an Ecosystem: A Tractable Model

Given that carbon is *the* universal currency used by biota to store and expend energy (Baird and Ulanowicz, 1989), delineating the factors governing its transformations and fluxes is key to understanding biotic interactions within ecosystems and how they might be affected by environmental warming. We therefore begin by considering the cycling of carbon between the organic (i.e. biotic) and inorganic (i.e. abiotic) pools with a generic and hypothetical standing freshwater ecosystem (Figure 1). In this simplified view, biomass is synthesised either by capturing photons (light energy) via photosynthesis, or by harnessing exergonic redox reactions of inorganic chemical compounds via chemosynthesis. This biomass, which is predominantly composed of organic carbon compounds, is transferred through the food web from autotrophs (i.e. the photo- and chemo-synthesisers) to heterotrophic primary consumers, secondary consumers, and so on via consumption, and also partially to the microbial consortium via senescence and biomass excretion. This detrital biomass is then remineralised by microbes via aerobic and anaerobic metabolism, yielding inorganic constituents. At each step in the transfer of energy ($\approx$biomass), a substantial proportion is lost due to metabolism and heat production (Lindeman, 1942), such that energy availability declines towards the top of trophic pyramids, and this is one of the key reasons why large predators are typically rare (Hutchinson, 1959).

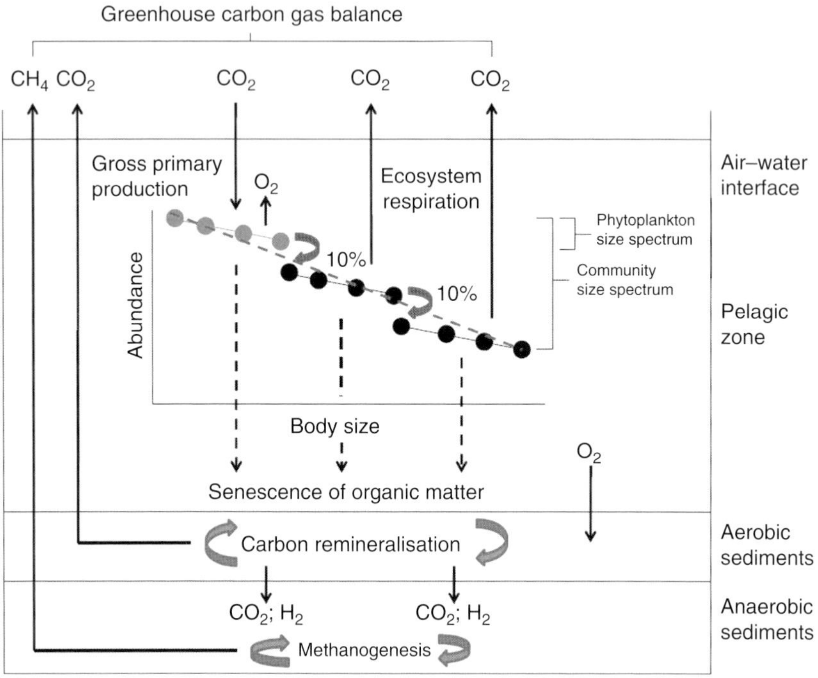

Figure 1 Simplified view of the carbon cycle in a model aquatic ecosystem. The structure of the biotic community is illustrated by the size spectrum. Here the community size spectrum is denoted by the dashed line, and the phytoplankton size spectrum is denoted by the grey circles and line, while the heterotrophic size spectrum is given by the black circles and lines. According to energetic equivalence and food web theory the slope of the community size spectrum is expected to be steeper than the slope of the size spectra within trophic levels because approximately only 10% of production is transferred between trophic levels. The biogeochemical processes are denoted by the bold arrows.

The net carbon balance of an ecosystem is defined as the difference between the gross fixation of CO_2 from the atmosphere through photosynthesis (GPP) and the total metabolism of fixed carbon, which is released back to the atmosphere primarily as CO_2 (ER) and/or methane (CH_4 efflux; ME) (e.g. Yvon-Durocher *et al.*, 2010a,b). In an idealised closed system, where the sizes of the autotrophic and heterotrophic carbon pools and fluxes between them remain constant, the net carbon balance is zero at steady state. Because CO_2 and methane are both greenhouse gases, and therefore have radiative forcing potential, the net carbon balance of ecosystems is crucial to the regulation of global temperature as it determines whether ecosystems act primarily as either sources or sinks of atmospheric CO_2 (Lovelock, 1972;

Whiting and Chanton, 2001; Woodwell *et al.*, 1998). A related concept, the greenhouse gas efflux potential of an ecosystem, concerns the overall effect of a non-zero net carbon balance on potential radiative forcing in the atmosphere (Whiting and Chanton, 2001). Consequently, in the context of biotic feedbacks with climate change, the two most important gaseous end products of the remineralisation of organic carbon are CO_2 and CH_4. Understanding how the carbon balance and greenhouse gas efflux potential of aquatic ecosystems are related to environmental temperature is thus crucial for generating robust predictions from coupled climate–carbon models (Cox *et al.*, 2000; Friedlingstein *et al.*, 2006). Achieving this goal will require a detailed understanding of the mechanisms that control the temperature dependence of the key metabolic process in the carbon cycle at the ecosystem level (i.e. GPP, ER, and CH_4 efflux [ME]). Although recent progress has been made in this area within the last decade, particularly in terrestrial systems (Allen *et al.*, 2005; Davidson and Janssens, 2006; Enquist *et al.*, 2003, 2007; Gudasz *et al.*, 2010; Knorr *et al.*, 2005; Laws *et al.*, 2000; Lopez-Urrutia *et al.*, 2006; Luo *et al.*, 2001; Melillo *et al.*, 2002; Yvon-Durocher *et al.*, 2010a,b), there remains much uncertainty, especially in aquatic ecosystems.

In the terrestrial carbon cycle, the temperature dependence of GPP has been demonstrated to be constrained by the rate of Rubisco carboxylation, which is determined primarily by the concentration of CO_2 at the active site of the enzyme. Therefore, potential differences between aquatic and terrestrial plants (e.g. temperature-dependent changes in aqueous concentration gradient of CO_2 at the site of photosynthesis due to Henry's Law, or differences in Rubisco kinetics between aquatic and terrestrial plants) may result in a divergence in the temperature dependence of photosynthesis between these systems. In terrestrial ecosystems the temperature dependence of ER is constrained to be equal to that of GPP at steady state because fixed carbon from GPP provides the substrate that fuels ER in these systems, which receive little if any carbon subsidies from adjacent ecosystems. This acclimation of the temperature dependence of ER to GPP over temporal scales relevant to the potential feedbacks between warming and the carbon cycle has led some authors to suggest that the potential positive feedbacks between warming and terrestrial respiration may be weaker than expected from short-term experimental studies (Gifford, 2003; Luo *et al.*, 2001). The long-term temperature sensitivity of respiration in aquatic ecosystems may be complicated by the fact that many aquatic ecosystems (e.g. lakes, streams, rivers, estuaries and coastal seas) receive considerable subsidies of carbon from adjacent ecosystems – for example coastal seas receive carbon from estuaries, estuaries receive carbon from rivers, while lakes, streams and rivers receive carbon from the terrestrial catchments they drain (Cole and

Caraco, 2001; Cole *et al.*, 2000, 2002; del Giorgio and Peters, 1994; del Giorgio *et al.*, 1997; Pace *et al.*, 2004; Ram *et al.*, 2007). Therefore, in many aquatic ecosystems, ER is not necessarily at steady state with respect to GPP and might not be constrained by autochthonous production (Cole *et al.*, 2000; del Giorgio and Peters, 1994; del Giorgio *et al.*, 1997). This simple, yet fundamental, difference may have far reaching consequences for the propagation of positive feedbacks between warming and the carbon cycle in aquatic ecosystems.

C. Predicting the Effects of Warming: Combining Theory, Experiments and Empirical Data

Metabolic rate is the speed at which an organism uptakes energetic and material resources from its environment, transforms them into useable forms, and provisions them to the biochemical processes necessary for growth, survival, and reproduction (Brown *et al.*, 2004). Thus, it is a fundamental determinant of the contribution an individual organism makes to the overall flux of energy and materials in an ecosystem (Brown *et al.*, 2004). The MTE describes how two key variables, body size and temperature, influence the metabolic rates of cells, individual organisms, communities and ecosystems (e.g. Allen *et al.*, 2005). Therefore, despite concerns raised by its critics (Clarke, 2004; Kozlowski and Konarzewski, 2005; Makarieva *et al.*, 2005) the MTE provides a potentially useful framework to build a predictive understanding of the feedbacks between warming, and the biogeochemical cycling of carbon within ecosystems (Allen *et al.*, 2005; Lopez-Urrutia *et al.*, 2006; Yvon-Durocher *et al.*, 2010a,b), and forms the basis of much of the theoretical component of this chapter.

The experimental component of this study involved simulating the potential effects of future warming on aquatic ecosystems using a replicated, ecosystem-level manipulation in freshwater mesocosms (Figure 2). The experimental design is described in detail in Section III (Yvon-Durocher *et al.*, 2010a,b). Briefly, the experiment was conducted using 20 freshwater mesocosms, 10 of which were maintained at 3–5 °C above ambient temperature (mean 4 °C), in line with the A1B warming scenario predicted for temperate latitudes by the end of the twenty-first century (IPCC, 2007). The remaining 10 were maintained at ambient temperature, but were otherwise identical, and served as unheated controls (Yvon-Durocher *et al.*, 2010a,b).

Mesocosm experiments represent an inevitable compromise between the control and replication of laboratory studies and the realism of

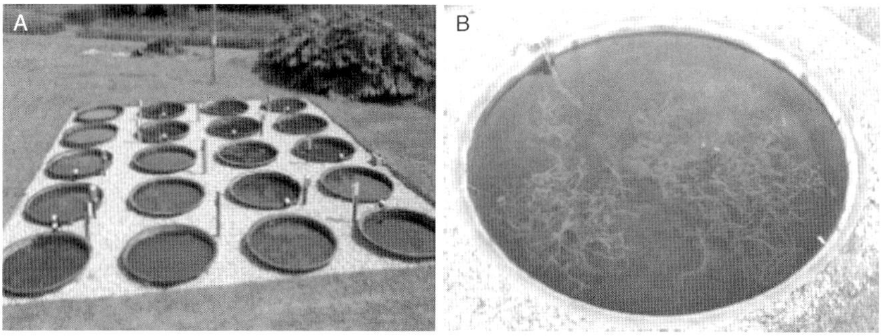

Figure 2 (A) Aerial view of the global warming mesocosm experiment in April 2008. The experimental plot consisted of 20 mesocosms: 10 heated and 10 unheated. (B) Close-up of mesocosm 1 (heated) in April 2008, highlighting the presence of diverse floral and faunal assemblages. Figure from Yvon-Durocher *et al.* (2010a)

descriptive field surveys. Despite their potential limitations, because they allow cause-and-effect relationships to be identified in relatively realistic settings they represent a useful tool for predicting how global change scenarios might affect ecosystem-level processes that cannot be achieved using laboratory studies or field surveys (Benton *et al.*, 2007; Woodward *et al.*, 2010a,b). In particular, they afford the opportunity to isolate the effects of temperature from other potentially confounding variables (e.g. latitude, altitude and nutrient availability) and permit direct comparisons to be made between ecosystems under ambient and warmed conditions from the perspective of both applied and basic science. Such results could provide much-needed insight into the potential future consequences of warming on aquatic ecosystems and from a basic science perspective, when interpreted in light of theoretical predictions, they can aid in developing more mechanistic, predictive capabilities for addressing future climate change and for understanding fundamental relationships between community structure and ecosystem functioning (Woodward *et al.*, 2010a,b).

Because mesocosm experiments are limited in their spatial and temporal extent, caution must be exercised when extrapolating their findings to natural systems and other ecosystem types (e.g. lakes, rivers and the coastal and open ocean). To address these potential caveats related to generality and ecological scales of investigation, we also compiled a broad database that combined many previously published global data compilations of measurements of respiration and primary production across a range of aquatic ecosystems (e.g. lakes, streams, estuaries and the open ocean) to further scrutinise our theoretical models and the application of our experimental findings to natural ecosystems.

D. Aims of the Study

The overarching goal of this study was to develop a more predictive understanding of how temperature regulates carbon cycling in aquatic ecosystems and to provide at least an initial glimpse of potential responses to future warming. More specifically, our objectives were:

1. To understand the mechanisms controlling the temperature dependence of GPP, ER and the efflux of CH_4 to the atmosphere (i.e. the three key metabolic fluxes in the aquatic carbon cycle).
2. To understand the mechanisms that control the temperature dependence of balance between GPP and ER, which determines the carbon sequestration capacity of aquatic ecosystems and to gain predictive capabilities to assess how this balance might respond to future global warming scenarios.
3. To develop a mechanistic understanding of how the rate of methane efflux in relation to GPP (i.e. CO_2 fixation) and ER (CO_2 efflux) will be affected by warming. Together, this balance determines the greenhouse gas efflux potential of aquatic ecosystems.
4. To compare potential differences between these processes in aquatic and terrestrial ecosystems.

II. THEORETICAL FRAMEWORK

A. Carbon Fluxes at the Individual-Level

At the level of an individual organism, carbon fluxes occur during photosynthesis, respiration and methanogenesis, all of which have predictable mass M_i and temperature T (K) dependencies, which can, in theory, ultimately be scaled up to the entire ecosystem (Brown *et al.*, 2004; Gillooly *et al.*, 2001):

$$B_i = b_0 \mathrm{e}^{-Ei/kT} M_i^{\alpha} \quad (1)$$

In this expression, B_i is the basal metabolic rate of individual i (g carbon s^{-1}), α is a scaling exponent that typically takes a value near 0.75 (Savage *et al.*, 2004), and b_0 is a normalization constant independent of mass and temperature (g carbon $g^{-\alpha}$ s^{-1}). For what follows, we will replace b_0 with metabolic pathway-specific normalization constants of r_0, p_0, and m_0, which correspond to respiratory, photosynthetic, and methanogenic fluxes of heterotrophs, autotrophs, and methanogens, respectively. The Boltzmann–Arrhenius factor, $\mathrm{e}^{-Ei/kT}$, describes the exponential effect of temperature on metabolic rate, where k is Boltzmann's constant (8.62×10^{-5} eV K^{-1}) and E_i is the average activation energy of the ith metabolic pathway. The Boltzmann–Arrhenius factor quantifies how the speed of a reaction (e.g. ATP turnover in respiration) is governed by the average kinetic energy (i.e. temperature) of molecules, which must collide with a force exceeding the activation energy barrier, E, required for a given reaction to proceed.

The magnitude of the activation energy, E, varies depending on the metabolic pathway. Therefore, we will use E_r, E_m and E_p to characterize to the temperature dependence of aerobic respiration, methanogenesis and photosynthesis, respectively. The average activation energy of heterotrophic respiratory metabolism is $E_r \approx 0.65$ eV (Gillooly *et al.*, 2001), which corresponds to a 15-fold increase in rates from 0 to 30 °C, whereas methanogenesis increases by about 35-fold (activation energy $E_m \approx 0.85$ eV, Yvon-Durocher *et al.*, 2010b). In contrast, photosynthesis increases only about four-fold over this temperature range (effective activation energy of $E_p \approx 0.32$ eV, Allen *et al.*, 2005). We refer to E_p as an "effective" activation energy because the temperature dependence of C3 photosynthesis is hyperbolic (Medlyn *et al.*, 2002), rather than exponential, due primarily to the process of photorespiration leading to the presence of temperature optima (Badger and Collatz, 1977). Using E_p as an approximation allows for direct comparison between the temperature dependences of photosynthesis with those of respiration and methanogenesis (Allen *et al.*, 2005; Lopez-Urrutia *et al.*, 2006). This approximation was derived by Allen *et al.* (2005) using a well-established model of C3 photosynthesis (Farquhar *et al.*, 1980) based on assumptions that are reasonable for terrestrial plants (internal CO_2 concentrations are about 70% of ambient, co-limitation of photosynthesis by Rubisco, similar kinetic properties for Rubisco across species).

B. Relating Individual-Level Fluxes to Ecosystem Processes: The Temperature Dependence of the Aquatic Carbon Cycle

Metabolic flux at the ecosystem level is equal to the sum of the metabolic fluxes of all its constituent individual organisms, and this knowldege may help in understanding important aspects of the structure and functioning of ecosystems (Allen *et al.*, 2005; Brown *et al.*, 2004; Enquist *et al.*, 2003; Lopez-Urrutia *et al.*, 2006; Yvon-Durocher *et al.*, 2010a). Here we consider how each of the three fundamental components of the carbon cycle (GPP, ER and ME) could potentially respond to warming.

The rate of GPP for a whole ecosystem can be estimated from the sum of the individual photosynthetic rates of all of its autotrophic organisms (Allen *et al.*, 2005; Lopez-Urrutia *et al.*, 2006; Yvon-Durocher *et al.*, 2010a), following Eq. 1:

$$\mathrm{GPP} = \frac{1}{V}\sum_{i=1}^{N_\mathrm{p}} B_i = \frac{1}{V} p_0 \mathrm{e}^{-E_\mathrm{p}/kT} \sum_{i=1}^{N_\mathrm{p}} M_i^{\alpha} = p_0 \mathrm{e}^{-E_\mathrm{p}/kT} M_{\mathrm{TOT}}^{\mathrm{p}} \langle M^{1-\alpha} \rangle_\mathrm{p} \quad (2)$$

In this expression, N_p is the number of photosynthetic organisms in an ecosystem of volume V, $M_{\mathrm{TOT}}^{\mathrm{p}} = \frac{1}{V}\sum_{i=1}^{N_\mathrm{p}} M_i$ is the total biomass of photosynthetic organisms per unit volume, and $\langle M^{1-\alpha} \rangle_\mathrm{p} = \sum_{i=1}^{N_\mathrm{p}} M_i^{3/4} / \sum_{i=1}^{N_\mathrm{p}} M_i$ is an average for body size (Allen *et al.*, 2005; Lopez-Urrutia *et al.*, 2006). By taking logs of both sides of Eq. 2 we can derive a general expression for the temperature dependence of GPP:

$$\ln(\mathrm{GPP}) = -E_\mathrm{p}\left(\frac{1}{kT}\right) + \ln\left[p_0 M_{\mathrm{TOT}}^{\mathrm{p}} \langle M^{1-\alpha} \rangle_\mathrm{p}\right] \quad (3)$$

GPP is the gross fixation of CO_2, and thus is equal to the sum of net primary production

$$\mathrm{NPP} = \varepsilon \mathrm{GPP} = \varepsilon p_0 \mathrm{e}^{-E_\mathrm{p}/kT} M_{\mathrm{TOT}}^{\mathrm{p}} \langle M^{1-\alpha} \rangle_\mathrm{p} \quad (4)$$

and autotroph respiration

$$\mathrm{AR} = (1-\varepsilon)\mathrm{GPP} = (1-\varepsilon) p_0 \mathrm{e}^{-E_\mathrm{p}/kT} M_{\mathrm{TOT}}^{\mathrm{p}} \langle M^{1-\alpha} \rangle_\mathrm{p} \quad (5)$$

where GPP = NPP + AR, and ε is the fraction of photosynthate allocated to net primary production. Equation 5 is consistent with the observation that autotrophic respiration is ultimately limited by, and therefore tightly coupled to, photosynthesis (Atkin and Tjoelker, 2003; Dewar *et al.*, 1999). Consequently, whilst the temperature dependence of autotrophic respiration has an activation energy of E_r over the short term, it is ultimately constrained to equal to that of photosynthesis, E_p, over longer time scales (Allen *et al.*, 2005). This process, which is referred to as type I respiratory acclimation, has been observed empirically (Atkin and Tjoelker, 2003) and experimentally (Dewar *et al.*, 1999). We therefore assume that autotroph respiration has an activation energy of E_p for the derivation that follows.

In a similar way, we can readily obtain ecosystem-level expressions for heterotroph respiration

$$\mathrm{HR} = r_0 \mathrm{e}^{-E_\mathrm{r}/kT} M_{\mathrm{TOT}}^{\mathrm{r}} \langle M^{1-\alpha} \rangle_\mathrm{r} \quad (6)$$

by summing the respiratory contributions of the heterotrophs (Allen *et al.*, 2005; Enquist *et al.*, 2003; Lopez-Urrutia *et al.*, 2006; Yvon-Durocher *et al.*, 2010a) where $M_{\mathrm{TOT}}^{\mathrm{r}}$ is the total biomass of heterotrophs and $\langle M^{1-\alpha} \rangle_\mathrm{r}$ is calculated based on the size distribution of heterotrophs. The total rate of ER is readily obtained by combining Eqs. 5 and 6:

$$\mathrm{ER} = \mathrm{AR} + \mathrm{HR} = (1-\varepsilon) p_0 \mathrm{e}^{-E_\mathrm{p}/kT} M_{\mathrm{TOT}}^{\mathrm{p}} \langle M^{1-\alpha} \rangle_\mathrm{p} + r_0 \mathrm{e}^{-E_\mathrm{r}/kT} M_{\mathrm{TOT}}^{\mathrm{r}} \langle M^{1-\alpha} \rangle_\mathrm{r} \quad (7)$$

Note that in contrast to the previous equations, the temperature dependence of ER is not governed by a single activation energy because it includes contributions from both heterotrophic and autotrophic organisms. At steady state, ER is limited by substrate availability, and must therefore equal GPP. Here, we define steady state as the tendency for ER to equal GPP over time periods of 1 year or greater. During the short term - that is over diurnal or seasonal cycles - the carbon balance frequently deviates from steady state (e.g. during the spring bloom in aquatic ecosystems) and GPP can exceed ER (or vice versa), though over the year at steady state this is no net gain or loss of carbon. However, when an ecosystem deviates from steady state (i.e. ER < GPP or ER > GPP), ER is no longer constrained by GPP. During long term (i.e. a year or greater) non-steady state dynamics, such as those that may potentially occur in our experiment (i.e. because warming is expected to act as a perturbation), heterotrophic respiration may exceed NPP (i.e. the potential contemporary carbon substrate) over temporal scales dependent on the turnover time of the carbon stores in the ecosystem. Further, this situation of non-steady state dynamics might exist more generally in aquatic ecosystems provided that they hold sufficient stored carbon or receive significant allochthonous subsidies to fuel heterotrophic metabolism (Cole *et al.*, 2000, 2002; del Giorgio and Peters, 1994; del Giorgio *et al.*, 1997; Pace *et al.*, 2004; Ram *et al.*, 2007). Under such conditions heterotrophic metabolism tends towards maximum capacity. Therefore, during non-steady state dynamics, because $E_r > E_p$, ER is expected to have a temperature dependence approaching that of heterotrophic metabolism, E_r – that is greater than the activation energy for GPP. Thus, assuming that ER and GPP are not at steady state, we can make the simplifying assumption that the term for AR in Eq. 7 is insignificant relative to HR in ER and remove it. We then have:

$$\ln(\mathrm{ER}) = -E_{\mathrm{r}}\left(\frac{1}{kT}\right) + \ln\left[r_0 M_{\mathrm{TOT}}^{\mathrm{r}} \langle M^{1-\alpha} \rangle_{\mathrm{r}}\right] \tag{8}$$

which is a general expression for the temperature dependence of ER at non-steady state.

An ecosystem-level expression for methane production can be obtained from summing the respiratory contributions of the methanogens in the ecosystem

$$\mathrm{MP} = m_0 \mathrm{e}^{-E_{\mathrm{m}}/kT} M_{\mathrm{TOT}}^{\mathrm{m}} \langle M^{1-\alpha} \rangle_{\mathrm{m}}. \tag{9}$$

where $M_{\mathrm{TOT}}^{\mathrm{m}}$ is the total biomass of methanogens, and $\langle M^{1-\alpha} \rangle_{\mathrm{m}}$ is calculated based on the size distribution of the methanogens. Here, we are attempting to understand how biogeochemical feedbacks will respond to global warming and to predict how the net efflux of CH_4 to the atmosphere behaves in response to temperature. There is, however, an important distinction between CH_4 efflux and production (MP). CH_4 efflux (ME) to the atmosphere is the net result of the difference between CH_4 production in the anaerobic zone of the sediment and the oxidation of CH_4 by methanotrophs in the oxic layers of the sediment and the overlying water column (Segers, 1998). CH_4 oxidation has the potential to oxidize ~90% of CH_4 produced in the sediments of lakes and wetlands (Schutz *et al.*, 1989) and substantially reduce net CH_4 efflux. In our model, we make the simplifying assumption that although CH_4 oxidation can considerably reduce net CH_4 efflux, it has little effect on its temperature dependence (after Yvon-Durocher *et al.*, 2010b). As such, we expect CH_4 efflux to be a constant proportion of CH_4 production with temperature, therefore the temperature dependence of CH_4 efflux should be equivalent to the activation energy for methanogenesis. Given the assumption that CH_4 oxidation has little effect on the temperature dependence of CH_4 efflux (ME), we can derive a general expression for the temperature dependence of ME by taking logs of both sides of Eq. 9, after Yvon-Durocher *et al.* (2010b)

$$\ln(\mathrm{ME}) = -E_{\mathrm{m}}\left(\frac{1}{kT}\right) + \ln\left[m_0 M_{\mathrm{TOT}}^{\mathrm{m}} \left\langle M^{1-\alpha}\right\rangle_{\mathrm{m}}\right] \tag{10}$$

Equations 2–10 predict the temperature dependence of GPP, ER and MP (and therefore CH_4 efflux [ME]) and highlight the importance of the activation energies of sub-cellular metabolism in controlling the temperature response of ecosystem-level flux rates. The theory above provides a platform from which a mechanistic understanding of the potential consequences of global warming on the carbon balance of ecosystems can be drawn and leads to a number of important predictions, which we tested experimentally. First, the temperature dependence of GPP, ER and ME will be governed by their respective activation energies at the cellular level. Therefore, the relationship between ln(GPP) and $1/kT$ should approximate a slope of $E_p \sim 0.32$ eV, ln(ER) (after Allen *et al.*, 2005). Assuming non-steady state dynamics, the temperature dependence of ER should be greater than that of GPP and the slope of the relationship between ln(ER) and $1/kT$ should approach the activation energy of heterotrophic metabolism, $E_r \sim 0.65$ eV (after Gillooly *et al.*, 2001). The temperature dependence of CH_4 efflux should be governed by the activation energy of methanogenesis. Thus, the slope of the relationship between ln(ME) and the reciprocal of absolute temperature $1/kT$ should approximate a slope equal to $E_m \sim 0.88$ eV (after Yvon-Durocher *et al.*, 2010b). Finally, and most importantly, because of the differential temperature dependences of these three processes, warming is expected to shift the overall carbon balance of ecosystems, thereby altering rates of carbon sequestration, greenhouse gas emission and thus potential biotic–abiotic feedbacks. We test these theoretical predictions using data from our mesocosm experiment and global data compilations of primary production and respiration in natural ecosystems.

C. Relating Individual-Level Fluxes to Ecosystem Processes: The Carbon Balance

With an understanding of the mechanisms controlling the temperature dependence of GPP, ER and ME, using the above theory it is possible to predict the temperature dependence of the carbon balance of aquatic ecosystems. Here we will focus on the metabolic balance (ER/GPP), which is the ability of an ecosystem to sequester carbon, the balance between ME and GPP, and that of ME and ER, which determines the greenhouse gas efflux potential of aquatic ecosystems. The metabolic balance is defined as ER/GPP and combining Eqs. 3 and 8, we have:

$$\mathrm{ER/GPP} = \frac{r_0 \mathrm{e}^{-E_\mathrm{r}/kT} M_\mathrm{TOT}^\mathrm{r} \langle M^{1-\alpha} \rangle_\mathrm{r}}{p_0 \mathrm{e}^{-E_\mathrm{p}/kT} M_\mathrm{TOT}^\mathrm{p} \langle M^{1-\alpha} \rangle_\mathrm{p}} \tag{11}$$

This can be simplified to

$$\mathrm{ER/GPP} = \frac{r_0}{p_0}(E_\mathrm{r} - E_\mathrm{p}) \frac{M_\mathrm{TOT}^\mathrm{r} \langle M^{1-\alpha} \rangle_\mathrm{r}}{M_\mathrm{TOT}^\mathrm{p} \langle M^{1-\alpha} \rangle_\mathrm{p}} \tag{12}$$

this, by taking logs, can be rearranged to

$$\ln(\mathrm{ER/GPP}) = E_\mathrm{r} - E_\mathrm{p}\left(\frac{1}{kT}\right) + \ln\left[\frac{r_0}{p_0} \times \frac{M_\mathrm{TOT}^\mathrm{r} \langle M^{1-\alpha} \rangle_\mathrm{r}}{M_\mathrm{TOT}^\mathrm{p} \langle M^{1-\alpha} \rangle_\mathrm{p}}\right] \tag{13}$$

which yields a general expression for the temperature dependence of the ER/GPP ratio during non-steady state dynamics where ER is not constrained by GPP. Similarly we can derive non-steady state solutions for the temperature dependence of the ME/GPP ratio by combining Eqs. 3 and 10:

$$\ln(\mathrm{ME/GPP}) = E_\mathrm{m} - E_\mathrm{p}\left(\frac{1}{kT}\right) + \ln\left[\frac{m_0}{p_0} \times \frac{M_\mathrm{TOT}^\mathrm{m} \langle M^{1-\alpha} \rangle_\mathrm{m}}{M_\mathrm{TOT}^\mathrm{p} \langle M^{1-\alpha} \rangle_\mathrm{p}}\right] \tag{14}$$

and the ME/ER ratio by combining Eqs. 8 and 10:

$$\ln(\mathrm{ME/ER}) = E_\mathrm{m} - E_\mathrm{r}\left(\frac{1}{kT}\right) + \ln\left[\frac{m_0}{r_0} \times \frac{M_\mathrm{TOT}^\mathrm{m} \langle M^{1-\alpha} \rangle_\mathrm{m}}{M_\mathrm{TOT}^\mathrm{r} \langle M^{1-\alpha} \rangle_\mathrm{r}}\right] \tag{15}$$

Equations 13–15 predict that in ecosystems where autochthonous carbon production (i.e. GPP) and consumption (i.e. ER and ME) are uncoupled, either through a perturbation (e.g. long-term changes in temperature) or by allochthonous carbon subsidies, as is frequently prevalent in freshwater, estuarine and coastal ecosystems (Cole *et al.*, 2000, 2002; del Giorgio and Peters, 1994; del Giorgio *et al.*, 1997; Pace *et al.*, 2004; Ram *et al.*, 2007), the temperature dependence of the carbon balance of aquatic ecosystems will be governed by the differences in the activation energies of the key carbon fluxes. Therefore, our models predict that ln(ER/GPP) should approximate a linear function of $1/kT$ with a slope of $E_r - E_p$. While ln(ME/GPP) versus $1/kT$ should have a slope equal to $E_m - E_p$ and ln(ME/ER) versus $1/kT$ should have a slope of approximately $E_m - E_r$. We test these theoretical predictions using our mesocosm experiment.

III. MATERIALS AND METHODS

A. Study Site and Experimental Design

The outdoor mesocosm experiment was based at the Freshwater Biological Association River Laboratory (2°10'W, 50°13'N) in East Stoke, Dorset, UK. Twenty artificial ponds, each holding 1 m^3 of water were set up to mimic shallow lake ecosystems (Figure 2): this scale of mesocosm is designed to reproduce many of the key elements of community structure (e.g. diversity, trophic complexity) and functioning (e.g. nutrient cycling) of shallow lake ecosystems (Jones *et al.*, 2002; McKee *et al.*, 2003; Ventura *et al.*, 2008). Ten of the 20 ponds were warmed 3–5 °C above ambient temperature, in accordance with the IPCC A1B global warming projections for the next 100 years for temperate areas in the northern hemisphere (IPCC, 2007). Experimental warming was achieved by an electronic heating element connected to a thermocouple which monitored the temperature in a given heated and unheated treatment pair of mesocosms. Temperatures were logged every 5 min over the entire year using HOBO temperature-irradiance data loggers and regular adjustments were made to ensure that temperature differences between treatments were ~4 °C. The mean annual temperature difference between treatments was 4.1 °C ± SE 0.01(Yvon-Durocher *et al.*, 2010a).

The mesocosms were seeded in December 2005 with organic substrates and a suite of organisms that included representative species from primary producers (phytoplankton, macrophytes) to top predators (Roach, *Rutilus rutilus*), and a suite of intermediate invertebrate consumers (zooplankton, including cladocerans, copepods and rotifers, and benthic macroinvertebrates, including Mollusca, Malacostraca, Trichoptera, Ephemeroptera and Odonata) to mimic, as far as possible, the general organismal composition, trophic complexity and physical structure of shallow lake ecosystems. The biota was left to establish for 10 months prior to the onset of experimental warming, which commenced in September 2006, thereby allowing time for further natural colonisation before the start of the annual sampling period in April 2007. Populations of the introduced top predator, *R. rutilus* were maintained at constant densities [two individuals (age 1+) per mesocosm ($\sim$12 g carbon m^{-3})] in all mesocosoms and monitored via regular electrofishing surveys. Because the fish were maintained at predetermined biomass-densities they merely served to 'complete' the food webs to mimic natural shallow lakes and are not considered further here.

B. Measuring Primary Production and Respiration

GPP and ER were measured over a 24 h diel cycle for each replicate of each treatment on alternate months over the course of 1 year (April 2007 to April 2008) using the dissolved oxygen (DO) change technique (Marzolf *et al.*, 1994; Mulholland *et al.*, 2001), resulting in a total of 140 measurements of each. This technique assumes that changes in DO concentration over a diel cycle represent the metabolic activity (photosynthetic and respiratory) of an aquatic ecosystem. To measure the concentration of DO in the mesocosms, YSI 600XLM multiparameter Sondes equipped with 6562 rapid pulse™ DO sensors were deployed in each heated and unheated treatment pair on each of the seven sampling occasions over the year. Measurements of DO, temperature and pH were taken every 15 min for 24 h at mid depth (0.25 m) in the water column of each pond. At the beginning of each sampling occasion the calibration of each Sonde was tested by deploying both Sondes in the same pond for 1 h to ensure equivalence in DO readings, and re-calibration was carried out when necessary. Subsequently, prior to deployment in each treatment pair, the Sondes were calibrated in water-saturated air with a correction for barometric pressure. Calibration accuracy was verified by monitoring the DO concentration of water-saturated air for 10 min and checking against 100% O_2 saturation for the measured temperature and pressure (Yvon-Durocher *et al.*, 2010a).

The record of continuous DO measurements was used to calculate the GPP and ER for each mesocosm on each sampling occasion, after Yvon-Durocher *et al.* (2010a). The dissolved oxygen change (ΔDO) for each 15 min time interval was calculated as the difference in O_2 concentration between t_1 and t_2 (i.e. $t_2 - t_1$). The daylight and night-time analysis periods were delimited as follows: the total analysis period was defined from the minimum O_2 concentration on the first night and extended for 24 h to include the minimum O_2 concentration on the second night. Photosynthetic dawn was identified as the minimum O_2 concentration after which all subsequent values were greater than it. Photosynthetic dusk was defined as the maximum O_2 concentration after which all subsequent values were lower (Bales and Nardi, 2007; Yvon-Durocher *et al.*, 2010a). Each O_2 change value was then assigned to a day or night-time category. Subsequently the metabolic parameters were calculated by numerical integration. GPP was calculated as:

$$\mathrm{GPP} = \sum \Delta O_{2\mathrm{day}} + R_{\mathrm{day}} \tag{16}$$

where R_{day} is day-time respiration. Since it is impossible to directly measure R_{day}, it was estimated, in keeping with the literature, by extrapolating the mean night time respiration value across the hours of daylight (Bales, 2007; Marzolf *et al.*, 1994; Mulholland *et al.*, 2001). ER was calculated as (see appendix details):

$$\mathrm{ER} = R_{\mathrm{day}} + \sum \Delta O_{2\mathrm{night}} \tag{17}$$

The metabolic balance of each replicate of each treatment was then determined as the ratio of ER/GPP. In the rare event of significant instrument drift or failure, the entire replicate was removed from the final analysis (9 measurements were removed from a total of 140; $n = 131$). Current biogeochemical techniques presently preclude the disentanglement of autotrophic and heterotrophic respiration at the ecosystem level (Mulholland *et al.*, 2001) and rule out the estimation of photorespiration (Marzolf *et al.*, 1994). Consequently, measures of GPP using the DO change technique may be slightly overestimated given the inclusion of heterotrophic respiration in calculation of R_{day}.

Benthic respiration (oxygen uptake) was measured using dark *in situ* benthic chambers which enclosed a sample of 500 mL at the sediment–water interface. A magnetic stirrer in the chamber ensured that the sample was evenly mixed. Benthic respiration was measured by the removal of 25 mL samples at the beginning and the end of the 6 h incubations. The samples were gently discharged into gas-tight vials (12 mL, Exetainers, Labco Ltd, High Wycombe, UK) and allowed to overflow twice (to minimize atmospheric gas exchange), and fixed for later Winkler analysis. The samples were immediately fixed and stored in a fridge at 5 °C to minimize light and temperature fluctuations until they could be titrated in the laboratory (<5 days). To ensure linearity of oxygen uptake an initial timed series of samples was taken, subsequently only $T=0$ and $T=$final samples were taken to limit sample extraction from the chambers. Oxygen uptake was then calculated as the difference between $T=$final and $T=0$ samples and the duration of the incubation.

C. Measuring Methane Efflux

Measurements of the efflux of CH_4 were made simultaneously to those GPP and ER, after Yvon-Durocher *et al.* (2010b). A single gas chamber was positioned at the water surface of each mesocosm on each sampling occasion. The chambers were made of polycarbonate and enclosed a headspace (300 mL) of ambient air at the air–water interface of the mesocosm [see Yvon-Durocher *et al.* (2010b) for details]. The lid of the gas chamber was equipped with a Teflon septum port, through which samples of gas (1 mL) were removed using a gas-tight syringe (2 mL VICI gas tight syringe) every 15 min for the first hour of the incubation, then hourly for up to 10 h thereafter. The samples were then transferred to water-filled gas-tight vials (3 mL, Exetainers; Labco, High Wycombe, UK) through a two-way valve venting through a narrow bore needle. The gas-tight vials were then stored upside down prior to analysis.

The concentration of CH_4 in the headspace of the sample was determined by gas chromatography as follows. Samples (50 μL) were withdrawn from the headspace of the sample vials and injected into a gas-chromatograph fitted with a flame ionising detector (GC/FID; Agilent Technologies, UK). Headspace concentrations of CH_4 were calculated from peak areas calibrated against known standards (Scientific and Technical gases, Staffs, UK) and the total amount of CH_4 in the gas tight vial (water plus headspace) was calculated using the appropriate solubility coefficients (Yamamoto *et al.*, 1976). The efflux of CH_4 across the water–air interface was calculated by ordinary least squares regression analysis of the change in concentration of CH_4 in the chamber headspace over time. Subsequently, 1 h was used as an appropriate duration for accurately estimating the flux of CH_4 (Lambert and Frechette, 2005). Regression slopes with a significance of $P > 0.05$ and/or an R-squared of below 0.9 were considered non-significant and were excluded from further analyses (9 from the 140 individual flux measurements).

D. Dissolved Methane

The concentration of dissolved CH_4 in the water column was measured by removing a water sample (30 mL in a gas-tight syringe) and gently transferring it to a gas tight vial (12.5 mL Exetainers, Labco, High Wycombe, UK), allowing it to overflow, fixing it with a bactericide (100 μL 50%, w/v, $ZnCl_2$) and sealing it. Samples were collected at hourly time intervals (in total 6–10 h depending on the time of year) over a day for each replicate on alternate months for 1 year (April 2007 to April 2008, $n = 1416$ individual measurements). Upon return to the laboratory, a headspace (2 mL analytical grade helium) was introduced to the gas-tight vial and the sample was shaken vigorously for 0.5 min and then allowed to stand for a further 30 min to allow for headspace equilibration, before analysis of the headspace concentration of CH_4 using a gas chromatograph as described above, see also (Sanders *et al.*, 2007; Yvon-Durocher *et al.*, 2010b). Finally, the 1416 measurements were pooled in each case to give an average daily pool of dissolved CH_4 for each pond (140 measures over the year for 70 heated and 70 ambient).

E. Statistical Analyses

All experimental data were checked for normality using the Shapiro Wilks test for normality and were natural log or $\log_{10}$ transformed prior to statistical analysis, where necessary. The activation energy of any metabolism is given by the slope of the relationship of an Arrhenius plot between ln(*x* flux) and $1/kT$, where k is Boltzmann's constant and T is absolute temperature (K). Here we replace the $1/kT$ with $1/k(1/T - 1/T_c)$: this transformation centres the inverse temperature data at zero to make the intercept of the linear model equal to the flux rate at T_c (here we have chosen 15 °C). This greatly reduces the correlation between the activation energy and the intercept and makes the intercept more biologically meaningful. The activation energy of ln(CH_4 efflux), ln(GPP) and ln(ER) was determined by linear mixed effects models using the *lme* function in R (R Development Core Team, 2006). In the models, treatment and sampling occasions were treated as fixed effects, while a random effects term in which pond identity was nested within sampling occasion, accounted for temporal pseudo-replication due to repeated sampling of the mesocosms over the year. The most parsimonious model to describe the temperature dependence of metabolism was obtained by first fitting the most complex model (i.e. different slopes for each treatment and sampling occasion) to test for statistical differences in the slopes of each of these relationships between treatments and sampling occasions; subsequently, non-significant terms were deleted to give the best model to describe the data. Model comparison was carried out using the Akaike Information Criterion (AIC). The activation energies of the literature compilation data for primary production and respiration were determined by ordinary least squares regression.

Between-treatment differences in the overall mean annual values of GPP, ER, CH_4 efflux, inorganic nutrients, gas transfer velocities, ER/GPP, CH_4 efflux /GPP and CH_4 efflux /ER were also analysed using the *lme* (linear mixed-effects model) function in R (R Development Core Team, 2006). In the models, treatment (heated or unheated) was treated as the fixed effect, and temporal pseudo-replication from repeated sampling of the mesocosms seasonally over the year was accounted for by including mesocosm identity nested with sampling occasion as random effects. The repeated measures model was used to test for overall statistical differences between treatments in mean annual values of the above parameters.

F. Literature Data Compilation and Meta-Analysis

Data for lake and stream primary production were extracted from Morin *et al.* (1999), and data for oceanic primary productivity were taken from the ocean primary productivity working group (OPPWG http://marine.rutgers.

edu/opp/Database/DB.html). Within this extensive data repository we used only data from the most comprehensive survey of oceanic primary production currently available, the MARMAP (1978–1982) survey, to minimise potential between-study methodological biases. Data for lake respiration were taken from Gudasz *et al.* (2010), and data for pelagic estuarine respiration data were from Hopkinson and Smith (2005). Marine pelagic bacterial respiration data were taken from a large database on bacterial metabolism (http://www.uea.ac.uk/env/people/facstaff/robinsonc) in Robinson (2008) and from Rivkin and Legendre (2001). To investigate the effects of resource acclimation on the temperature dependence of ER further we divided the data on marine bacterial respiration into those from coastal seas (i.e. within the continental shelf: e.g. the North Sea) which are likely to receive allochthonous carbon subsidies, and those from the open ocean (i.e. beyond the continental shelf; e.g. the North East Atlantic ocean) which are unlikely to receive allochthonous carbon subsidies from estuaries.

IV. RESULTS

A. Ecosystem-Level Carbon Fluxes: Experimental Tests

Data from our mesocosm experiment broadly supported our theoretical predictions. Rates of GPP were strongly related to temperature (Table 1) and a plot of ln(GPP) versus $1/k(1/T - 1/T_c)$ revealed that its activation energy was 0.45 eV (Figure 3A; 95% CI 0.38–0.53 eV). This value was steeper than predicted [0.32 eV; Allen *et al.* (2005)], based on assumptions made for C3 photosynthesis, and might reveal fundamental differences between aquatic and terrestrial photosynthesis. ER was also strongly related to temperature (Table 1) and its activation energy in a plot of ln(ER) versus $1/k(1/T - 1/T_c)$ was 0.62 eV (Figure 3B; 95% CI 0.55–0.69 eV). The activation energy for ER was statistically indistinguishable from the predicted value of 0.65 eV (Gillooly *et al.*, 2001) and revealed that our assumption of long-term non-steady-state dynamics in response to experimental warming was validated because the temperature dependence of ER was not constrained by GPP over the year (Table 2). Finally, ME was similarly strongly related to temperature (Table 1), and a plot of ln(ME) versus $1/k(1/T - 1/T_c)$ revealed that its activation energy (Figure 3C; 95% CI 0.64–1.02 eV) was also indistinguishable from the predicted value ($\sim$0.88 eV) based on the activation energy of methanogenesis (Yvon-Durocher *et al.*, 2010b). This finding also provides strong support for our model assumption that the temperature dependence of ME and CH_4 production (MP) are equivalent and that CH_4 oxidation has little effect on the temperature dependence of ME.

Table 1 Results from Linear Mixed Effects (*lme*) models

Relationship	d.f.	*F* ratio	*P* value
ln(CH_4 efflux) vs. $1/kT$	1,123	97.98	*<0.0001*
ln(GPP) vs. $1/kT$	1,123	146.6	*<0.0001*
ln(ER) vs. $1/kT$	1,123	294.85	*<0.0001*
Difference in slope of ln(CH_4 efflux) vs. $1/kT$ between treatments	1,123	1.23	0.23
Difference in intercept of ln(CH_4 efflux) vs. $1/kT$ between treatments	1,123	2.29	0.132
Difference in slope of ln(GPP) vs. $1/kT$ between treatments	1,123	0.23	0.82
Difference in intercept of ln(GPP) vs. $1/kT$ between treatments	1,123	2.55	0.11
Difference in slope of ln(ER) vs. $1/kT$ between treatments	1,123	0.56	0.46
Difference in intercept of ln(ER) vs. $1/kT$ between treatments	1,123	2.27	0.13
Difference in slope between ln(CH_4 efflux) vs. $1/kT$ and ln(GPP) vs. $1/kT$	1,254	21.61	*<0.0001*
Difference in slope between ln(CH_4 efflux) vs. $1/kT$ and ln(ER) vs. $1/kT$	1,254	8.36	*0.0042*
Difference in slope between ln(ER) $\times$ ln (GPP) vs. $1/kT$	1,254	3.2	*0.0015*

The first set of models test for relationships between ecosystem-level metabolic rates (GPP, ER or ME) and temperature $[1/k(1/T - 1/T_c)]$, parallelism between treatments, and differences between intercepts in the linear models. Metabolic rates are used as dependent variables, temperature $[1/k(1/T - 1/T_c)]$ as the independent variable, while treatment (warmed or ambient) was a fixed factor and mesocosm nested within sampling occasion were treated as random effects to account for temporal pseudoreplication. The second set of models tested for differences in the slope of the temperature dependence between metabolic rates (i.e. ER $\times$ GPP and ER $\times$ ME, ME $\times$ GPP). Here metabolic rate is used as the dependent variable, temperature ($1/kT$) as the independent variable and metabolic rate ID (e.g. NPP or ER) as the fixed factor. Significant P values are given in italics.

The three fundamental metabolic fluxes of the carbon cycle exhibited strong seasonal trends in the mesocosm experiment (Figure 4). For example, flux rates of GPP, ER and ME peaked in early summer and were lowest in winter, presumably reflecting the temperature dependence of metabolism. Rates of benthic respiration, however, exhibited slightly different seasonal dynamics with rates peaking in late summer and autumn (Figure 4D). For all metabolic fluxes, the mean annual values were significantly elevated in the warmed mesocosms (Table 3).

As predicted from our models, the key metabolic fluxes of the carbon cycle had sequentially greater activation energies (i.e. GPP = 0.45 eV < ER = 0.62 eV < ME = 0.85 eV) in the mesocosm experiment, that were predictable from the sub-cellular kinetics of their particular metabolic pathways.

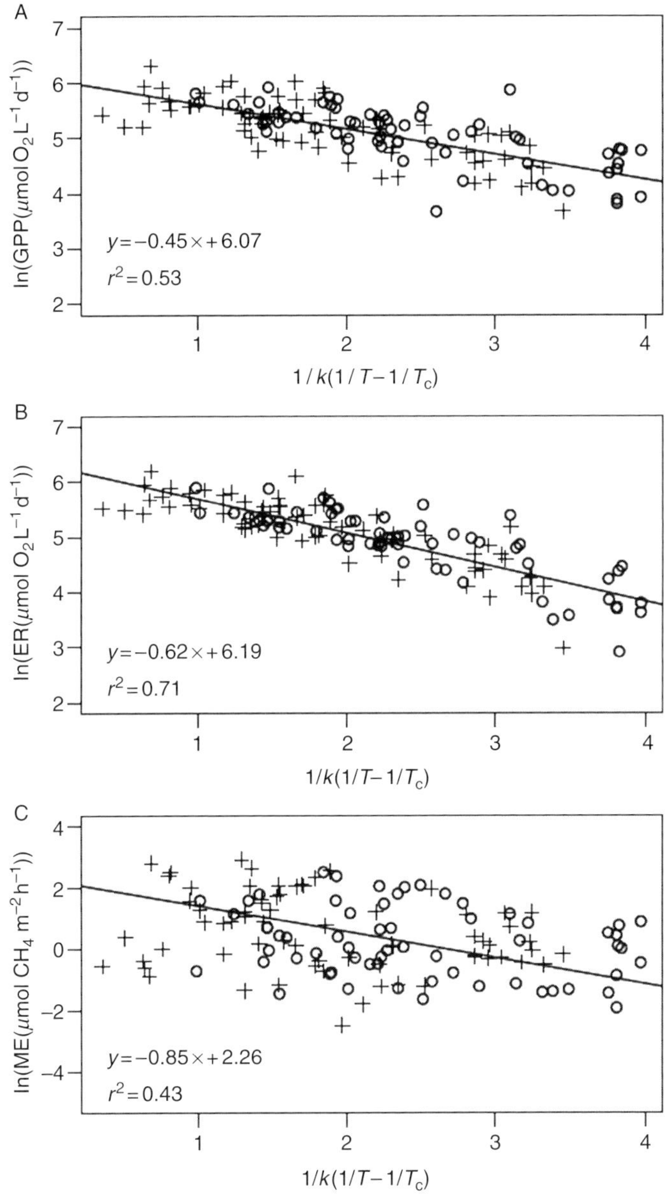

Figure 3 Temperature dependence of the three main ecosystem-level carbon fluxes in the mesocosm experiment. (A) Gross primary production (GPP), (B) Ecosystem respiration (ER), and (C) CH_4 efflux (ME). The slope of the temperature–metabolism relationship is equivalent to the activation energy of the metabolic process. Each data point corresponds to either the GPP, ER or CH_4 efflux from a single mesocosm on each of the seven sampling occasions ($n = 131$). Ambient treatments are denoted by circles and warmed treatments by plus signs. Note that the three fundamental metabolic fluxes in the carbon cycle have sequentially greater activation energies. Dashed lines are predicted from theory (a = 0.32 eV, b = 0.62 eV, c = 0.88 eV). Figure redrawn from Yvon-Durocher *et al.* (2010a,b).

Table 2 Annual activation energies for GPP and ER for each mesocosm derived from seasonal measures of GPP and ER

Pond	Annual activation energy GPP	Annual activation energy ER
1	−0.497	−0.642
2	−0.467	−0.608
3	−0.502	−0.673
4	−0.473	−0.639
5	−0.461	−0.623
6	−0.374	−0.535
7	−0.459	−0.633
8	−0.504	−0.673
9	−0.541	−0.645
10	−0.477	−0.621
11	−0.367	−0.553
12	−0.454	−0.634
13	−0.431	−0.580
14	−0.477	−0.632
15	−0.548	−0.634
16	−0.354	−0.473
17	−0.622	−0.762
18	−0.453	−0.676
19	−0.416	−0.522
20	−0.522	−0.645

Data reveal that the activation energy of ER is consistently higher than that of GPP suggesting that in all mesocosms ER was not at steady with GPP over the year, validating our theoretical assumptions.

This divergence among the temperature dependences of these metabolic pathways suggests that under future global warming these processes have the potential to go out of balance.

B. Ecosystem-Level Carbon Fluxes: Meta-Analysis of Field Survey Data

An extensive global compilation of data on rates of primary production in the ocean revealed that its temperature dependence in a plot of ln(PP) versus $1/k(1/T - 1/T_c)$ was 0.44 eV (95% CI 0.39–0.49 eV) (Figure 5A) which was again steeper than the expected value of 0.32 eV derived for terrestrial C3 plants (Allen *et al.*, 2005), but almost identical to the value of 0.45 eV measured in our mesocosm experiment (Figure 5A). Similarly, a global database of rates of primary production compiled for freshwater ecosystems (e.g. lakes and streams) demonstrated a temperature dependence identical to that observed in the ocean with a plot of ln(PP) versus $1/k(1/T - 1/T_c)$ revealing an activation energy of 0.48 eV (95% CI 0.35–0.62 eV) (Figure 5B). Again this value is

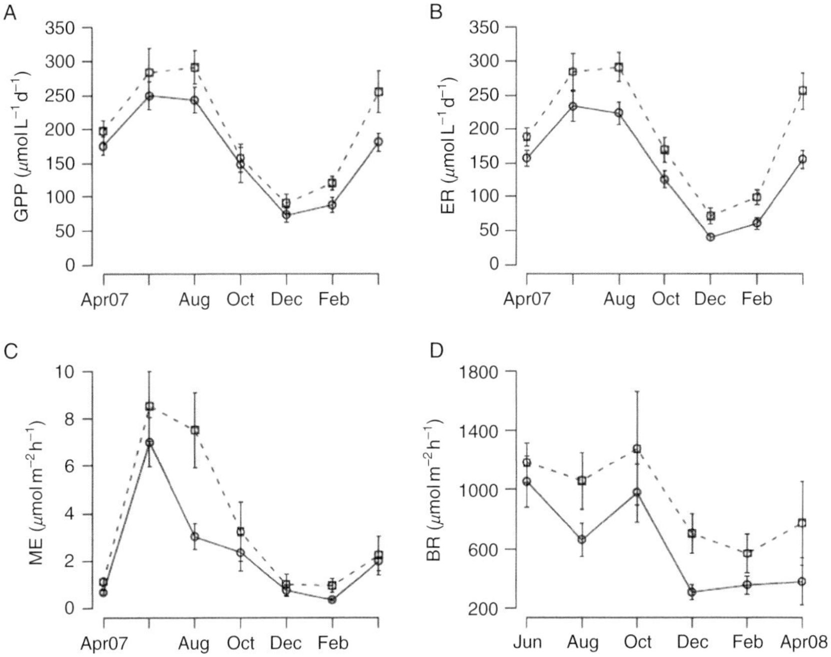

Figure 4 Seasonality of gross primary production (GPP; A), Ecosystem respiration (ER; B), CH_4 efflux (ME; C) and Benthic respiration (BR; D). Ambient treatments are denoted by solid lines and circles, while heated treatments are given by dashed lines and squares. Data points are the mean values (±SE) of each treatment on each month. Data reveal strong seasonal trends with rates peaking during early summer. Warmed treatments (dashed lines) are typically greater than the ambient treatments (solid line), as expected from the differences in their activation energies. Mean annual metabolic fluxes of each process were significantly elevated in the warmed mesocosms (Table 3). Note that data for ME, GPP and ER were standardised to μmol m^{-2} day^{-1} in the derivation of the carbon balance ratios in Figures 7 and 8. Figure redrawn from Yvon-Durocher *et al.* (2010a,b).

steeper than the 0.32 eV expected for terrestrial photosynthesis, but identical to that observed in our experiment and also in the ocean.

Data compiled on rates of community respiration, from a range of aquatic ecosystem types from across the globe revealed a similar coherence with our theoretical predictions (Figure 6). For example, in estuarine ecosystems, the activation energy of short-term pelagic community respiration was 0.58 eV (95% CI 0.46–0.7; Figure 6A), close to both the predicted value of 0.65 eV, and also the 0.62 eV observed in our experiments. Similarly, for the same estuaries, mean annual (i.e. long term) respiration had an activation energy of 0.54 eV (Figure 6B; 95% CI 0.41–0.8) which was statistically indistinguishable from the short-term flux data and the predicted theoretical value. Data compiled on rates of short-term and long-term sediment respiration in lakes across

Table 3 Linear mixed effects model analysis

Variable	d.f.	*F* ratio	*P* value
ER	1,123	31.0	*<0.0001*
GPP	1,123	42.9	*<0.0001*
ln(CH_4 efflux)	1,123	6.22	*0.014*
Benthic respiration	1,84	7.56	*0.0073*
ER/GPP	1,123	12.89	*0.0005*
ln(CH_4 efflux)/ln(GPP)	1,123	6.33	*0.013*
ln(CH_4 efflux)/ln(ER)	1,123	6.23	*0.0139*

Analysing differences between heated and unheated treatments in the annual means of gas transfer velocity [ln(*k* transfer)], ER, GPP, methane efflux [ln(CH_4 efflux)], the metabolic balance (ER/GPP), the ratio of methane efflux to GPP [ln(CH_4 efflux)/ln(GPP)], and the ratio of methane efflux to ER [ln(CH_4 efflux)/ln(ER)]. A linear mixed effects model was conducted with restricted maximum likelihood methods using the *lme* (linear mixed-effects model) function in R, treatment (heated or unheated) was the fixed effect, and temporal pseudo-replication from repeated sampling of the mesocosms over the year was accounted for by including mesocosm identity nested with sampling occasion as random effects. Significant *P* values are given in italic.

the arctic tundra was consistent with both the predicted values of 0.65 and 0.62 eV revealed by our experiment and exhibited an activation energies of 0.63 eV (Figure 6C; 95% CI 0.57–0.69) and 0.66 eV (Figure 6D; 95% CI 0.44–0.9), respectively. The activation energy of short-term bacterial respiration measured in coastal seas was 0.68 eV (Figure 6D; 95% CI 0.54–0.98 eV), again almost identical to the expected value of 0.65 eV. However, the activation energy of short-term bacterial respiration measured in the open ocean was 0.27 eV (Figure 6E; 95% CI 0.12–0.45) and was much shallower than that of heterotrophic metabolism, but closer to the activation energy of photosynthesis.

C. The Carbon Balance

The metabolic balance (i.e. ER/GPP) of the mesocosm experiment exhibited strong seasonal trends (Figure 7A). In 4 months of the annual study (June, August, October and April 2008) the metabolic balance was >1, indicating that ER $>$ GPP and that the warmed mesocosms were therefore net sources of CO_2 to the atmosphere. As expected from the differential activation energies of photosynthesis and respiration, the metabolic balance was significantly elevated in the warmed mesocosms (Table 3; Figure 7A). Moreover, the mean annual ratio of ER/GPP was elevated by 13% in the warmed mesocosms (Table 3). Similarly, the ratio of ME/GPP (Figure 7B) and ME/ER (Figure 7C) revealed strong seasonal trends with both carbon balances peaking in early summer and reaching seasonal lows in February. As predicted by the differential activation energies of these three metabolic

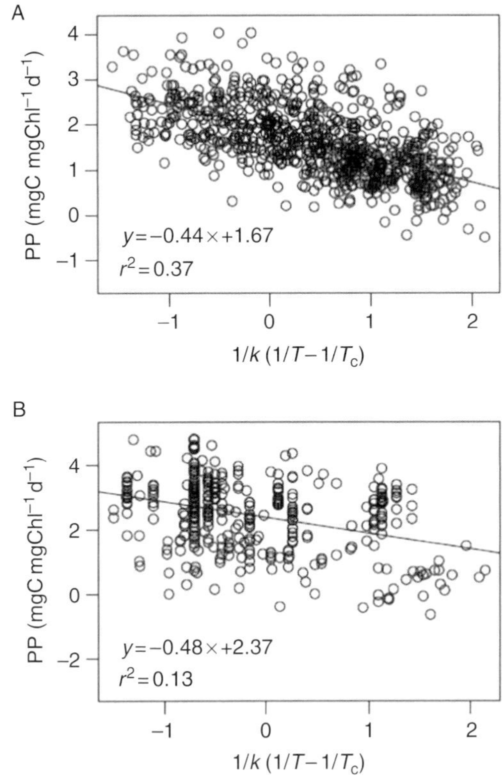

Figure 5 Literature data for the temperature dependence of ecosystem-level primary production in (A) the ocean (Data are from the OPPWG http://marine.rutgers.edu/opp/Database/DB.html), and (B) freshwater ecosystems [i.e. lakes and streams; Data from Morin *et al.* (1999)]. The slope of the temperature–metabolism relationships is equivalent to the activation energy of the metabolic process. These data reveal that the activation energies of ecosystem-level primary production in the ocean and in freshwater ecosystems are identical to the slope of GPP in the mesocosm experiment (Figure 3A), though steeper than that predicted by Allen *et al.* (2005) for terrestrial plants (i.e. 0.32 eV).

processes, both the ratio of ME/GPP and ME/ER were on average over the year, significantly elevated by 20% and 10%, respectively, in the warmed mesocosms (Table 3).

In line with their seasonal dynamics, the balance of the three key processes in the carbon cycle of the mesocosm experiment was related to temperature. As predicted by Eq. 13 a plot of ln(ER/GPP) and $1/k(1/T - 1/T_c)$ revealed that the temperature dependence of the metabolic balance in our experiment was 0.14 eV (95% CI 0.10 to 0.18 eV) and was statistically indistinguishable from the predicted value of 0.17 eV (Figure 8A) based on the empirically measured

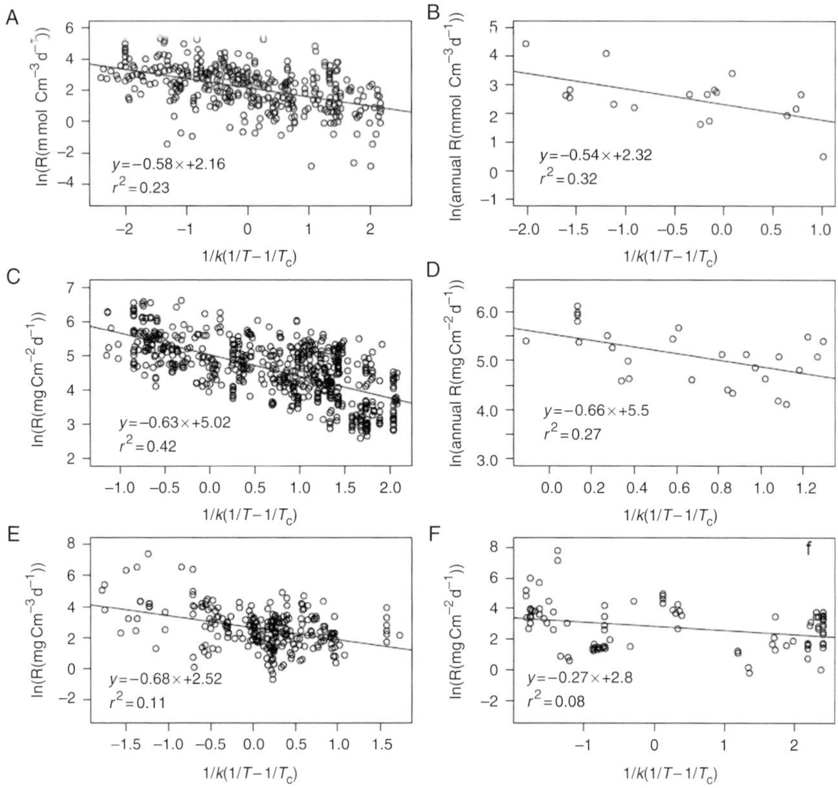

Figure 6 Literature data for the temperature dependence of community respiration. (A) Short-term pelagic estuarine ecosystems [data from Hopkinson and Smith (2005)]. (B) Long-term (i.e. mean annual) pelagic estuarine ecosystems [data from Hopkinson and Smith (2005)]. (C) Short-term sediment respiration in lakes [data from Gudasz *et al.* (2010)]. (D) Long-term (i.e. mean annual) sediment respiration in lakes [data from Gudasz *et al.* (2010)]. (E) Short-term bacterial respiration in the coastal ocean [data from Robinson (2008) and Rivkin and Legendre (2001)]. (F) Short-term bacterial respiration in the open ocean [data from Robinson (2008) and Rivkin and Legendre (2001)]. The slope of the temperature–metabolism relationships is equivalent to the activation energy of the metabolic process. These data reveal that the activation energies of ecosystem-level respiration in aquatic ecosystem are all close to the expected 0.62 eV predicted from our model and revealed in our mesocosm experiment, except for bacterial respiration in the open ocean which is likely constrained by GPP.

difference between the activation energies for GPP and ER – that is $E_r - E_p$. This finding suggests that the temperature dependence of the carbon sequestration capacity of aquatic ecosystems can be predicted from the temperature dependences of photosynthesis and respiration at the cellular level. Similarly, a plot of ln(ME/GPP) and $1/k(1/T - 1/T_c)$ revealed that its activation

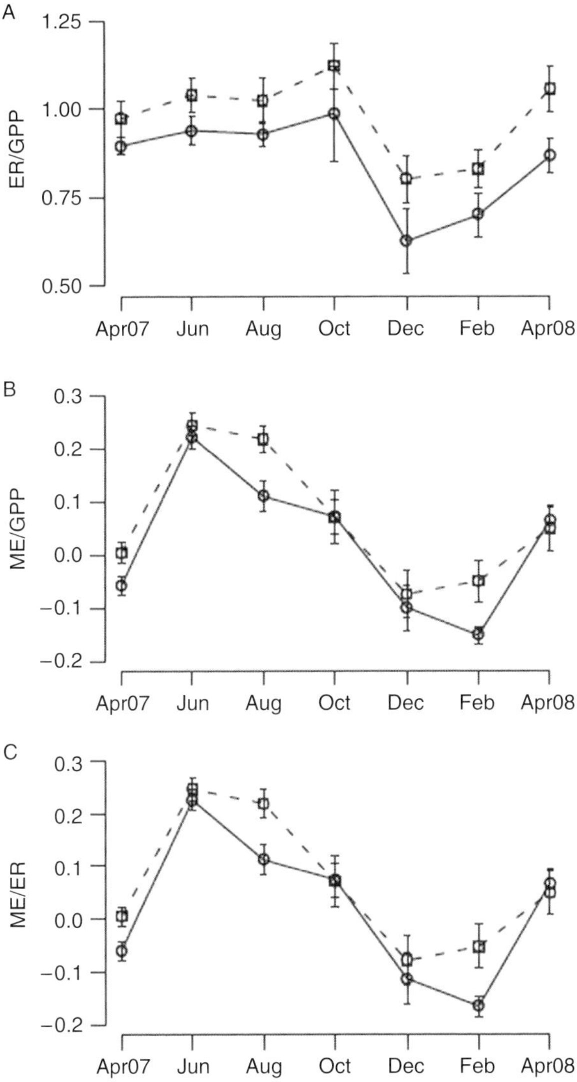

Figure 7 Seasonality of the carbon balance of the mesocosm experiment: (A) the metabolic balance (ER/GPP), (B) the balance between CH_4 efflux and gross primary production (ME/GPP), (C) the balance between CH_4 efflux and ecosystem respiration (ME/ER). Ambient treatments are denoted by solid lines and circles, while heated treatments are given by dashed lines and squares. Data points are the mean values ($\pm$SE) of each treatment on each month. Data reveal that the balance of the fundamental components of the carbon cycle are typically elevated in the warmed mesocosms as expected from the relative difference in their activation energies. The mean annual carbon balances were all significantly elevated in the warmed mesocosms (Table 3). Figure redrawn from Yvon-Durocher *et al.* (2010a,b).

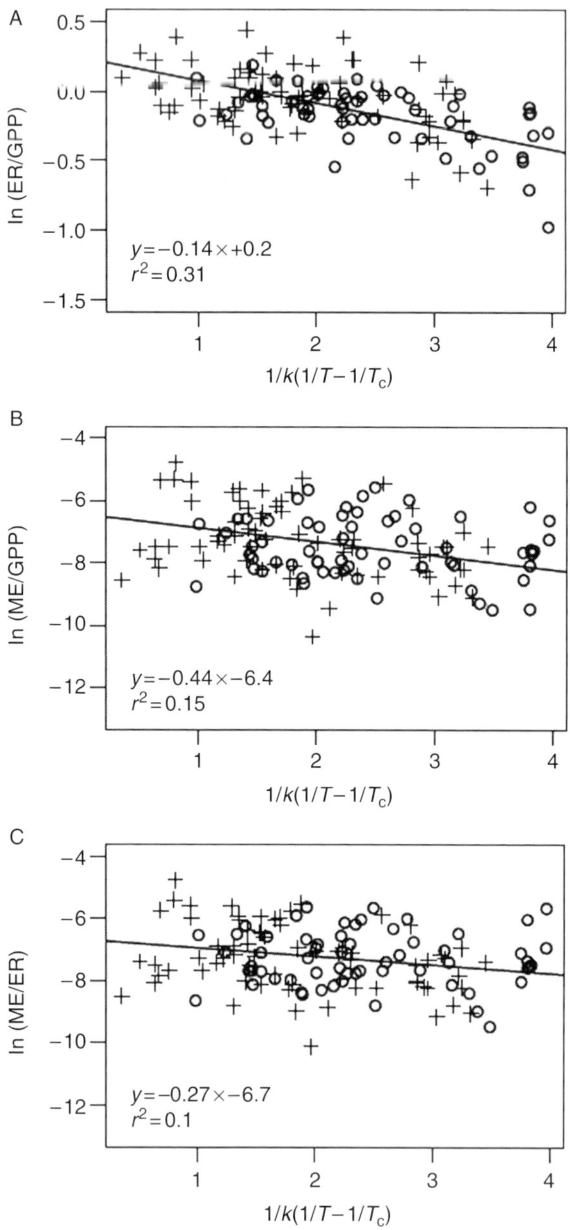

Figure 8 Temperature dependence of the carbon balance of the mesocosm experiment. (A) The metabolic balance (ER/GPP), (B) the balance between CH_4 efflux and gross primary production (ME/GPP), (C) the balance between CH_4 efflux and ecosystem respiration (ME/ER). Ambient treatments are denoted by circles and warmed treatments by plus signs. The slopes of each of these relationships are indistinguishable from the differences in their empirically derived activation energies and are predicted by Eqs. 13–15 (e.g. a, $E_r - E_p = 0.62 - 0.45 = 0.17$. b, $E_m - E_p = 0.85 - 0.45 = 0.4$. c, $E_m - E_r = 0.85 - 0.62 = 0.23$). Figure redrawn from Yvon-Durocher *et al.* (2010a,b).

energy was 0.44 eV (95% CI 0.26 to 0.63 eV) (Figure 8B) and supported our prediction (Eq. 14) that the temperature dependence of ME/GPP should be equivalent to the difference in the activation energies of ME and GPP – that is $E_m - E_r = 0.4$. Moreover, our prediction of the importance of the activation energies of metabolism for the temperature dependence of the carbon balance of aquatic ecosystems were further supported by a plot of ln(ME/ER) and $1/k(1/T - 1/T_c)$ which revealed that its temperature dependence was 0.23 eV (95% CI 0.09 to 0.45 eV) (Figure 8C) and was indistinguishable from the predicted value of 0.27 eV, based on the difference in the activation energies of ME and ER – that is $E_m - E_p$.

V. DISCUSSION

A. The Temperature Dependence of the Key Components of the Carbon Cycle

The three key components of the carbon cycle in aquatic ecosystems that we measured in our experiment and compiled from global empirical databases were all strongly related to temperature. Moreover, their temperature dependence was typically constrained by the average activation energy of their particular metabolic pathway and, as predicted by our models, GPP, ER and ME had sequentially greater temperature dependences (i.e. GPP < ER < ME). The temperature dependence of the carbon cycle also differed between aquatic and terrestrial ecosystems which could have important implications for the propagation of feedbacks between future warming and the carbon cycle in these systems. In the following sections, we start by discussing each of the key metabolic fluxes in the carbon cycle, and subsequently assess the level of agreement between our experiments, meta-analyses and theory. We then focus on the apparent similarities and differences of the temperature dependence of the metabolic flux of interest between aquatic and terrestrial ecosystems. Finally, we shift our focus to the balance between the three key metabolic fluxes in the carbon cycle and consider the coherence between our theoretical predictions and experimental findings.

1. Experiments, Data and Theory

In general, there was strong agreement between our models, experiment and our empirically determined temperature dependence for each of the three metabolic processes. For instance, the 'effective' activation energy for GPP in our experiment was 0.45 eV and was identical to that observed in our global data compilations of primary production in the ocean (0.44 eV) and in freshwater ecosystems (0.48 eV). Similarly, the temperature dependence of ER in our experiment had an activation energy of 0.62 eV, which was statistically indistinguishable from the average activation energy of heterotrophic metabolism [0.65 eV; Gillooly *et al.* (2001)] and supported our theoretical assumption of non-steady state between ER and GPP. The close coherence between the temperature dependence of ER observed in our experiment with the temperature dependence of lake (0.63 eV), estuarine (0.58 eV) and coastal bacterial respiration (0.68 eV) from the global data compilations again provide further support for the generality of our experimental findings in aquatic ecosystems. Rates of CH_4 efflux (ME) from the mesocosms were strongly dependent on temperature and, as predicted, had an activation energy (0.85 eV) which was almost identical to that of methanogenesis in pure cultures of methanogens (0.88 eV) (Yvon-Durocher *et al.*, 2010b). These data corroborate the assumption in our theoretical model, that although CH_4 oxidation can considerably influence the absolute amounts of CH_4 production emitted to the atmosphere it has a negligible effect on the overall temperature dependence of ME.

The agreement between our experimentally and empirically derived activation energies provides strong support to suggest that the findings of our mesocosm experiment may be applicable more widely to aquatic ecosystems in general. These findings also support the notion that the ecosystem-level fluxes in the carbon cycle might be relatively independent of species composition (Allen *et al.*, 2005; Enquist *et al.*, 2003; Manning *et al.*, 2006) because the temperature response of each of the metabolic fluxes in our experiment was identical to that observed in the ocean, estuaries and in freshwater ecosystems, each of which contrast markedly in their community structure and species composition. This point emphasises the potentially broad predictive power of our theoretical models based on biochemical kinetics (Allen *et al.*, 2005; Brown *et al.*, 2004) for understanding the temperature dependence of the aquatic carbon cycle and that these ecosystem-level processes might largely transcend seemingly contingent community-level differences between systems.

2. *Aquatic-Terrestrial Comparisons*

a. Primary production. The 'effective' activation energies of primary production in aquatic ecosystems documented in this study (experiment = 0.45 eV; Ocean = 0.44 eV; Freshwater = 0.48 eV) are steeper than the predicted value of 0.32 eV derived by Allen *et al.* (2005) based on assumptions for C3 photosynthesis in terrestrial plants [i.e. 1. internal CO_2 concentrations are about 70% of ambient; 2. co-limitation of photosynthesis by Rubisco; and 3. similar kinetic properties for Rubisco across species (Farquhar *et al.*, 1980)]. This 'effective' activation energy of ~0.32 eV has been verified for terrestrial net primary production in a global data compilation by Allen *et al.* (2005), and for individual level phytoplankton net photosynthesis in the ocean by Lopez-Urrutia *et al.*

(2006). However, the precision of the coefficient in the data reported in the latter study was low ($r^2 = 0.06$), so confidence in its absolute value is questionable. Our results suggest that the effective activation energy for aquatic photosynthesis might be greater than that of terrestrial photosynthesis, and this could have important implications for how these different ecosystem types may respond to future warming, - i.e. an elevated capacity for CO_2 fixation at higher temperature in aquatic ecosystems.

The potential divergence in the temperature dependence of photosynthesis between aquatic and terrestrial plants might reflect fundamental differences in the physiology and biochemistry of photosynthesis in their respective autotrophic groups. For example, in the derivation of $E_p = 0.32$ eV, Allen *et al.* (2005) assume, following Farquhar *et al.*'s (1980) well-established model of leaf photosynthesis, that the rate of C3 photosynthesis in terrestrial plants is co-limited by the initial step in carbon fixation – that is the Rubisco carboxylation reaction and the light-dependent RuBP (Ribulose-1,5-bisphosphate) regeneration step. They demonstrated that the maximum rate of Rubisco carboxylation has an activation energy of $E_c = 0.68$ eV. However, the overall weaker effect of temperature on terrestrial photosynthesis is due to a decrease in the ratio of this process to photorespiration (i.e. oxygen fixation) at high temperatures (Allen *et al.*, 2005). Oxygen and CO_2 compete at the binding site of Rubisco, which under conditions of low intracellular CO_2 concentrations, fixes O_2 rather than CO_2, thereby reducing the efficiency of photosynthesis (Falkowski and Raven, 1997). Moreover, intracellular CO_2 concentrations are strongly and negatively temperature dependent, increasing the prevalence of photorespiration and slowing photosynthesis at high temperatures in terrestrial plants (Berry and Bjorkman, 1980).

In aquatic systems, however, the rate-limiting step in photosynthesis tends to be governed by the diffusivity of CO_2 and/or HCO_3^- across cell membranes, and the delivery of CO_2 to Rubisco (Smith and Walker, 1980). This raises the possibility that rates of photosynthesis in aquatic autotrophs may, in part, be limited by diffusion and therefore governed by fundamental physical laws. Another, more biologically based mechanism, relates to the fact that many aquatic plants have carbon concentrating mechanisms, including the inducible activation of the protein carbonic anhydrase that enables many unicellular green algae and cyanobacteria to utilise HCO_3^- (Raven *et al.*, 2008). Carbonic anhydrase is an efficient scavenger of DIC and is not inhibited by O_2, unlike Rubisco. The conversion of DIC to CO_2 by carbonic anhydrase provides elevated CO_2 to Rubisco, thus enabling more efficient CO_2 fixation at low concentrations, whilst simultaneously decreasing the oxygenase activity of Rubisco (i.e. photorespiration) (Raven *et al.*, 2008). It seems possible then that carbon concentrating mechanisms in aquatic autotrophs might mitigate the constraints of elevated photorespiration to carboxylation ratios experienced by terrestrial plants at high temperatures. This putative mechanism might explain why our data for aquatic photosynthesis has a considerably greater effective activation energy than terrestrial C3 photosynthesis.

Another possibility is that there are fundamental differences in the kinetic properties of the Rubisco enzyme between plants adapted to aquatic and terrestrial realms. Recent evidence suggests that the subcellular ratio of CO_2:O_2 is an important determinant of Rubisco kinetics that can differ markedly between autotroph species (Tcherkez *et al.*, 2006). Furthermore, they find that the CO_2:O_2 specificity is inversely related to the maximal rate of Rubisco carboxylation. Thus, Rubiscos in the presence of high intracellular CO_2 concentrations (e.g. organisms with carbon concentrating mechanisms) tend to have low CO_2/O_2 specificity but high maximal carboxylation rates, which may reflect a biochemical evolutionary adaptation to the sub-cellular environment. Therefore, the greater prevalence of carbon concentrating mechanisms in aquatic autotrophs (Raven *et al.*, 2008) and the potential for elevated maximum Rubisco carboxylation rates might explain the stronger temperature dependence and higher effective activation energy of aquatic primary production reported in our experiment and global data compilations relative to those associated with terrestrial primary production.

b. Ecosystem respiration. For ER, our experimental findings, and those for aquatic ecosystems in general, appear to contrast with the model and global data compilation presented by Allen *et al.* (2005) for terrestrial ecosystems. In their model of the terrestrial carbon balance, Allen *et al.* (2005) make the assumption, for terrestrial ecosystems, of a steady state between GPP and ER over time periods of a year or greater. At steady state ER is constrained to equal GPP in the long term (i.e. 1 year or more), because GPP provides the substrate to fuel ER. Therefore, at steady state in a terrestrial ecosystem, Allen *et al.* (2005) predict that the temperature dependence of ER should be constrained to equal the effective activation energy of photosynthesis (i.e. $\sim$0.32 eV). They also predict that because $E_p < E_r$, if growing season temperatures increase to a new long-term average, respiration at a given temperature (i.e. the normalisation constant, $M_{TOT}^{r} \langle M^{1-\alpha} \rangle_r$ in Eq. 8) will decline. This is because M_{TOT}^{r} – that is the total biomass of heterotrophs – is predicted to decline with increases in temperature, because warming-induced increases in metabolism mean that fewer individuals can be supported per unit of GPP. These assumptions are intuitive for terrestrial ecosystems in which autochthonous production, and detritus derived from autochthonous production, is typically the dominant energy source that fuels heterotrophic food chains (Cebrian, 1999).

The results from our model, experiment and global data compilations suggest that the temperature dependence of ER in aquatic ecosystems might be constrained by a somewhat different set of mechanisms from those operating in terrestrial ecosystems. In our experiment, the temperature dependence of ER (0.62 eV) was significantly greater than that of GPP (0.45 eV), indicating that over the temporal scale of the experiment (i.e. 1 year) ER was not at steady state with respect to GPP (Yvon-Durocher *et al.*, 2010a). In both warmed and control mesocosms the carbon balance deviated from steady state because ER/GPP was <1 averaged over the year, suggesting that the mesocosms were accumulating carbon. Because the mesocosms were not at steady state (i.e. ER/GPP < 1), ER was not substrate limited by contemporary NPP, and heterotrophic metabolism was unconstrained by the weaker temperature dependence of GPP. Moreover, our experimental data revealed that the intercept of ln(ER) versus $1/k(1/T - 1/T_c)$ – that is $M^{r}_{TOT} \langle M^{1-\alpha} \rangle_r$ – was not significantly affected by long-term increases in temperature, contrary to the prediction of the Allen *et al.* (2005) steady-state model. Thus, because heterotrophic metabolism was not limited by organic carbon compounds from NPP, total heterotrophic biomass (M^{r}_{TOT}) did not decline in response to warming.

In our global data compilations, lake (0.63 eV), estuarine (0.58 eV) and coastal bacterial respiration (0.68 eV) were indistinguishable from one another and from the average activation energy of heterotrophic respiration [0.65 eV; Gillooly *et al.* (2001)], but greater than reported for terrestrial ecosystems [0.32 eV; Allen *et al.* (2005)]. This marked discrepancy between the temperature dependence of respiration in aquatic and terrestrial ecosystems could reflect fundamental differences in the equilibrium between autochthonous production, heterotrophic consumption and allochthonous carbon subsidies between these ecosystem types. As previously explained, the weaker temperature dependence of terrestrial ER arises from the fact that it is typically constrained by GPP over time periods greater than 1 year. However, in contrast, many aquatic ecosystems can receive considerable carbon subsidies from adjacent ecosystems that support heterotrophic metabolism beyond that which could be sustained solely by autochthonous primary production (Battin *et al.*, 2008; Cole and Caraco, 2001; Cole *et al.*, 2000, 2002; del Giorgio and Peters, 1994; del Giorgio *et al.*, 1997; Pace *et al.*, 2004; Ram *et al.*, 2007). For example, lakes and rivers often drain sizeable terrestrial catchments and in doing so can receive substantial inputs of dissolved organic carbon, originally synthesised in the terrestrial biosphere (Battin *et al.*, 2008; Cole and Caraco, 2001; Cole *et al.*, 2000, 2002; del Giorgio and Peters, 1994; Pace *et al.*, 2004). Similarly, estuaries are transitional ecosystems at the interface between riverine, terrestrial and marine ecosystems and can receive significant allochthonous subsidies of organic carbon (Ram *et al.*, 2007). Many marine ecosystems, particularly coastal seas, are often influenced by nutrient upwellings along continental shelves and can also receive substantial carbon inputs from the estuaries (del Giorgio *et al.*, 1997). The influence of carbon subsidies to aquatic ecosystems complicates the concept of steady state between ER and GPP because a significant proportion of heterotrophic metabolism in aquatic ecosystems can be supported by carbon synthesised externally (Cole *et al.*, 2000; del Giorgio and Peters, 1994; Pace *et al.*, 2004). Thus, the stronger temperature dependence of respiration in the aquatic ecosystems highlighted here might arise if heterotrophic metabolism is not limited by GPP for organic carbon compounds. There is, however, one important exception to this pattern that is borne out in our data compilation: the open ocean – the region beyond the continental shelf – is typically oligotrophic or hyper-oligotrophic and receives few, if any, external carbon subsidies. In line with the argument outlined above, the temperature dependence of bacterial respiration from these regions was considerably lower (0.27 eV) than in coastal seas (0.68 eV; e.g. in the North Sea or the Mediterranean) and far closer to that reported in terrestrial systems.

B. The Carbon Balance of Aquatic Ecosystems

The balance of the three carbon fluxes investigated here and their response to future warming may affect the strength of biotic-atmospheric feedbacks on a potentially global scale (Woodwell *et al.*, 1998). The strong correlation between temperature and each flux meant that overall rates of metabolism were significantly elevated in the warmed mesocosms over the course of the year (Yvon-Durocher *et al.*, 2010a,b). However, more importantly, the different temperature dependences of these three processes resulted in substantial shifts in the carbon balance of the warmed mesocosms (Yvon-Durocher *et al.*, 2010a,b). For example, the balance between ER and GPP was elevated by 13% on average over the year in the warmed mesocosms, resulting in a marked reduction in the capacity of the warmed mesocosms to sequester carbon. Similarly, the ratio between ME and GPP increased by 20%, while that of ME to ER increased by 9% on average over the year in the warmed mesocosms. These ratios determine the carbon balance (Whiting and Chanton, 2001; Yvon-Durocher *et al.*, 2010b) and also the relative greenhouse gas efflux potential of ecosystems. Because CH_4 has up to 21 times the radiative forcing potential of CO_2 over periods of up to 20 years (Rodhe, 1990), the increase in ME relative to GPP and ER in response to a 4 °C rise in temperature could represent a substantial, but until now unknown, positive feedback between global warming and the carbon cycle if true of freshwater ecosystems on a global scale. Clearly, considerable caution must be exercised when extrapolating results from mesocosm experiments to global scale phenomena (Benton *et al.*, 2007), but the close coherence between our theoretical models, experiments, and global data compilations suggest that our findings may be applicable to a wide range of natural ecosystems.

Our models predicted not only the direction but also the magnitude of the shift in the carbon balance of our mesocosms in response to warming. For example, our model, based on non-equilibrium dynamics, predicted that the ER/GPP ratio would be related to $1/k(1/T - 1/T_c)$ with a temperature dependence equal to the difference between the activation energy for respiration and that of Rubisco carboxylation ($E_r - E_p$). Data from our experiment strongly supported this prediction. Further, our model predictions for the temperature dependence of ME/GPP and ER/GPP were similarly corroborated by our experimental data and demonstrated that the temperature dependence of carbon balance could be predicted by the differences in the activation energies of these metabolic pathways (i.e. $E_m - E_p$ and $E_m - E_r$, respectively).

C. Conclusions, Caveats and Further Study

The divergence between the temperature dependence of the three key metabolic fluxes of the carbon cycle in our experiment resulted in significant imbalances between these biogeochemical processes at the ecosystem level and an overall reduction in the carbon sequestration capacity of the warmed mesocosms. Our models revealed that the constraints imposed by warming on individual metabolism were able to predict both the direction and magnitude of these experimentally observed shifts. Our data also suggest that the temperature dependence of primary production and respiration in aquatic ecosystems might be governed by a different set of mechanisms from those that constrain these processes in terrestrial ecosystems. For example, the temperature dependence of primary production in aquatic ecosystems was considerably stronger than that observed in terrestrial ecosystems, suggesting that aquatic photosynthesis might be more efficient at higher temperatures, relative to C3 photosynthesis in terrestrial plants. The prevalence of carbon-concentrating mechanisms in aquatic plants (Raven *et al.*, 2008), combined with differences in the kinetic properties of the enzyme Rubisco (Tcherkez *et al.*, 2006) might in part explain this phenomenon.

There are a number of important caveats that must also be considered when interpreting our results on the temperature dependence of primary production in both our experiment and in our global data compilations. For example, in Eqs. 2 and 3 and in our determination of the activation energy of primary production (Figures 3A and 5A) we assume that the body mass term (i.e. total biomass) is independent of environmental temperature. Co-variability of biomass and temperature may influence the observed activation energy of primary production in our experimental data and meta-analyses. In fact, in a recent study, we have demonstrated that on two of the seven sampling occasions in the mesocosm experiment, total phytoplankton biomass actually declined in response to warming (Yvon-Durocher *et al.*,

2010c). Similarly, a study in the open ocean has recently demonstrated that in the mesocosm experiment, total phytoplankton biomass tends to be lower in warmer, lower latitude regions (Moran *et al.*, 2010). In both cases declines in autotroph biomass as a function of temperature should elevate the observed activation energy of aquatic primary production and further emphasise the differences between aquatic and terrestrial photosynthesis. In a similar manner, light and temperature are also typically correlated in temperate latitudes that experience large seasonal fluctuations (Hessen *et al.*, 2005) and light is fundamental in controlling rates of photosynthesis in many aquatic ecosystems (Falkowski and Raven, 1997). Therefore, we cannot rule out that elevated light levels during the warmer months may at least in part explain the elevated temperature dependence of primary production in both our experiment and the global data compilations. If data were consistently available, rates of primary production should ideally have been corrected for latitudinal and seasonal differences in light levels before analysis. An additional caveat that relates to the MARMAP data set is that in the ocean, temperature and nutrient concentrations in the photic zone are often inversely correlated because the former determines the degree of stratification and hence the replenishment of nutrients from the deeper, more enriched waters (Finkel *et al.*, 2005, 2010). However, greater nutrient limitation at higher temperatures would (assuming all else is equal) be expected to increase the strength of the temperature dependence of primary production in the ocean and accentuate the difference between aquatic and terrestrial ecosystems if accounted for in the analysis. Further, even the theoretical value of 0.32 eV derived by Allen *et al.* (2005) is based on experimental parameterisation, and is therefore not strictly mechanistic, and open to methodological error. Clearly more research is needed in this area to rigorously test the potential divergences between aquatic and terrestrial photosynthesis we have highlighted in this study.

As with GPP, the temperature dependence of ER in aquatic ecosystems was also much stronger than in terrestrial ecosystems, and might be explained by key differences in autochthonous production, consumption and external carbon subsidies in aquatic and terrestrial ecosystems. Many aquatic ecosystems receive significant subsidies of organic carbon that fuel heterotrophic metabolism over and above that which could be sustained on autochthonous production (Cole and Caraco, 2001; Cole *et al.*, 2000, 2002; del Giorgio and Peters, 1994; del Giorgio *et al.*, 1997; Pace *et al.*, 2004; Ram *et al.*, 2007; Tcherkez *et al.*, 2006). Therefore, unlike terrestrial ecosystems, aquatic respiration is often not at steady state with autochthonous production (del Giorgio and Peters, 1994; del Giorgio *et al.*, 1997), even over time periods of a year or more, and is thus not constrained by the weaker temperature dependence of photosynthesis. These potentially crucial differences between aquatic terrestrial ecosystems could be important in predicting their respective responses to future global warming. For example, autochthonous production as a constraint on ER in terrestrial ecosystems is often cited as a negative feedback mechanism, causing acclimation over time periods (i.e. years) relevant to the feedbacks between warming and the carbon cycle (Dewar *et al.*, 1999; Gifford, 2003; Luo *et al.*, 2001). However, the potential lack of this ecosystem-level feedback in aquatic ecosystems might fuel the capacity of aquatic systems to remain net heterotrophic (i.e. net sources of CO_2 to the atmosphere) in a warmer climate long enough to seriously alter atmospheric chemistry.

ACKNOWLEDGEMENTS

We thank Brian Godfrey, Dan Perkins and the Freshwater Biological Association for their help with the experiment. Ricard Solé discussed ideas and provided comments on early drafts. We also thank Angel Lopez-Urrutia, Jon Cole and Jon Benstead for comments and suggestions that significantly improved the manuscript. G. Yvon-Durocher was supported by a Natural Environment Research Council studentship (NER/S/A2006/14029). J. Montoya was funded by the NERC Fellowship Scheme (NE/C002105/1), and a Ramon y Cajal Fellowship (RYC-2008-03664).

APPENDIX I. POTENTIAL CONFOUNDING VARIABLES AND SUPPLEMENTARY INFORMATION

A. Inorganic Nutrient Regime (Figure S1)

The overarching goal of this monograph was to build an understanding the effects of warming on the carbon cycle in aquatic ecosystems. Therefore, it was crucial to isolate the effects of warming *per se* from any other potentially confounding variables that might influence rates of carbon cycling. One such potentially confounding variable is the extent of inorganic nutrient limitation, because this can strongly influence rates of primary production (Woodward, 2007) and also the size structure of phytoplankton communities (Finkel *et al.*, 2005; Winder *et al.*, 2009). To determine whether the experimental treatment (i.e. warming) affected the extent of nutrient limitation in the mesocosms, we made detailed seasonal measurements of the major inorganic nutrients which might be expected to regulate primary production: water samples for measuring dissolved inorganic nutrient concentrations were collected from mid-depth in each mesocosm at 9 a.m. on each sampling occasion. Samples were filtered (Whatmann GF/F) and stored frozen (-20 °C) for subsequent determination of NO_3^-, NO_2^-, NH_4^+, PO_4^{3-} and Si ($Si(OH)_4$) using a segmented flow auto-analyser (Skalar, San++, Breda, Netherlands), according to Kirkwood (1996).

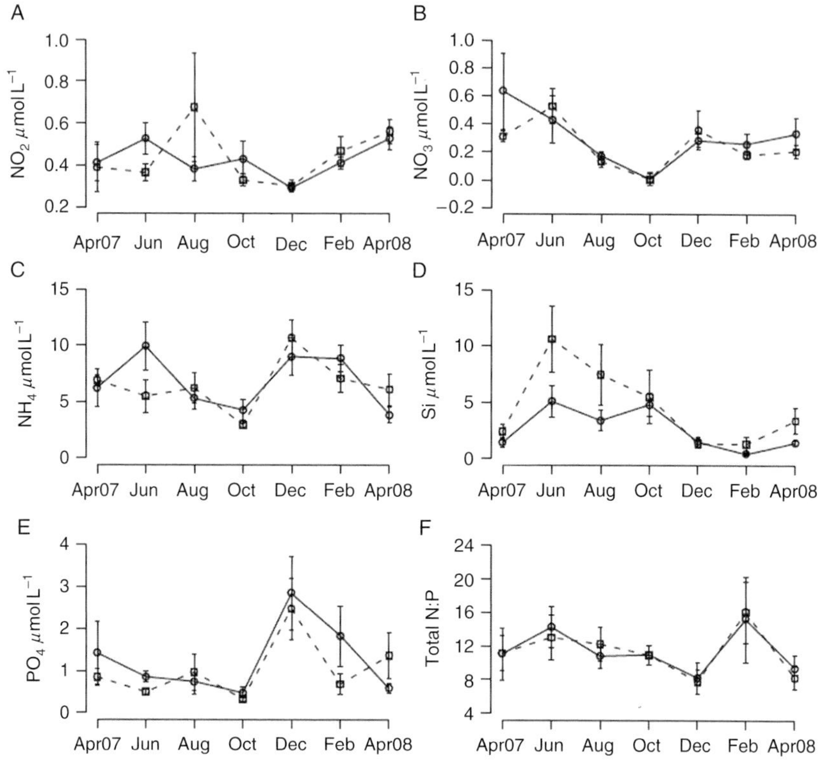

Figure S1 Seasonality of inorganic nutrients in the mesocosm experiment: (a) Nitrite, (b) Nitrate, (c) Ammonium, (d) Silicate, (e) Phosphate, (f) the stoichiometry of the inorganic nutrient pool. Warmed treatments are denoted by the dashed lines and squares, while ambient treatments are denoted by the solid lines and circles. Data highlight that inorganic nutrient regimes were identical in warmed and ambient treatments (see Table S1).

Inorganic nutrients (NO_3^-, NO_2^-, NH_4^+, PO_4^{3-} and Si) exhibited strong seasonal trends (Figure 3). For example, NO_3^- concentrations peaked in spring and declined progressively throughout the summer, and were depleted to $\sim$0.005 μmol l^{-1} by October, before remineralisation in the winter. Concentrations of NO_3^-, NO_2^-, NH_4^+ and PO_4^{3-} showed identical seasonal patterns in the warmed and ambient treatments, with no significant differences in the overall mean annual concentrations of these nutrients (Table S1). Furthermore, the stoichiometry of the inorganic nutrient pool exhibited remarkable similarity between treatments, with a mean annual ratio of total inorganic N to P of $\approx$11:1 in both heated and ambient mesocosms. The only inorganic nutrient which differed markedly between treatments was Si. Consistency of the dissolved inorganic nutrient concentrations between

Table S1 Results of the linear mixed effects model testing for differences in the concentration of inorganic nutrients between heated and ambient mesocosms

Inorganic nutrient	d.f.	*F*	*P*
NO_2^-	1,120	0.06	0.812 (NS)
NO_3^-	1,120	0.65	0.420 (NS)
NH_4^+	1,120	0.23	0.632 (NS)
Si	1,120	6.08	*0.015*
PO_4^{3-}	1,120	0.68	0.412 (NS)
Total inorganic N to P	1,120	0.009	0.922 (NS)

A linear mixed effects model was conducted with restricted maximum likelihood methods using the *lme* (linear mixed-effects model) function in R, treatment (heated or unheated) was the fixed effect, and temporal pseudo-replication from repeated sampling of the mesocosms over the year was accounted for by including mesocosm identity nested with sampling occasion as random effects. *P* values are given in italic.

experimental treatments meant that this potentially confounding variable, because it represents the potential resource supply rate for primary production, could be discounted from further analyses: that is, any changes in carbon cycling in the mesocosms could be ascribed to the effects of warming *per se*.

B. Air–Water Gas Exchange Due to Advection and Diffusion (Figure S2)

Apart from biological metabolic activity, gas exchange with the atmosphere due to diffusion and advection is an additional factor that might affect the concentration of dissolved gases (Cole and Caraco, 1998) and may also affect the interpretation of my results. Gas flux across the air–water interface is dependent on the concentration gradient between the water and the overlying air, and the gas transfer velocity, *k* (otherwise known as the piston velocity). Gas flux across the air–water interface can be described by the following equation (Cole and Caraco, 1998):

$$f = k(C_{\text{water}} - C_{\text{eq}}) \tag{1}$$

where k is the gas transfer velocity (cm h^{-1}), and Cwater – Ceq is the concentration gradient of the gas between the water and the concentration that would be at equilibrium with the atmosphere. In running waters characterised by turbulent flow, reaeration due to physical processes is typically a crucial determinant of the concentration of dissolved gases, and must therefore be accounted for in calculation of metabolism from changes in the concentration of dissolved gases (Marzolf *et al.*, 1994; Mulholland *et al.*, 2001). In still waters, however, k is typically determined by wind velocity (Cole and

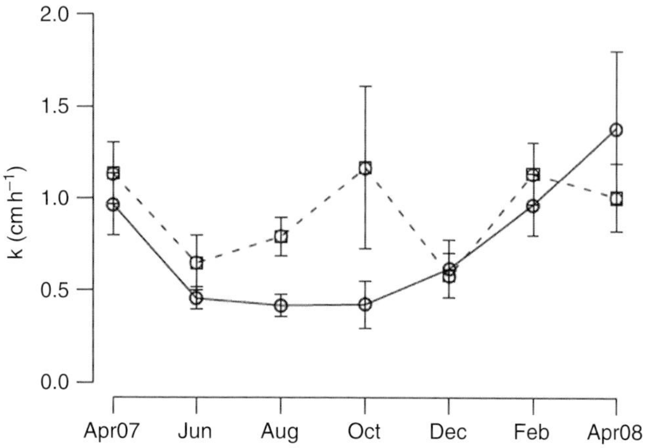

Figure S2 Seasonal trends in the gas transfer velocity (k), between heated (dashed lines, squares) and unheated treatments (solid lines, circles). The gas transfer exhibited little seasonal variability and was not systematically affected by the heating of the mesocosms. Data were natural log transformed for statistical analysis but presented here untransformed for ease of interpretation.

Caraco, 1998) which determines surface water turbulence. In the present study measured wind velocities were typically very low (average 0.53 m s^{-1}), with only 2.22% of measurements above 3 m s^{-1}. Importantly, k is largely independent of wind velocity at wind speeds less than ~3 m s^{-1} (Cole and Caraco, 1998) therefore, enhanced gas exchange due to the turbulence created by wind was not considered in the calculations of ecosystem metabolism or CH4 efflux. However, advective processes which also determine k at low winds might still have been influenced by the heating of the mesocosms (i.e. through convection). To determine whether experimental warming systematically altered the gas transfer velocity we estimated k from simultaneous measurements of the efflux of CH4 and dissolved CH4 (detailed above) from:

$$k = f/(C_{\text{water}} - C_{\text{eq}}) \tag{2}$$

where f is the measured efflux of CH_4 across the air–water interface, $C_{\text{water}} - C_{\text{eq}}$ is the concentration gradient of the gas in the water and the concentration in the water at equilibrium with the atmosphere (C_{eq}). C_{eq} was calculated using the equations of Yamamoto *et al.* (1976) and the measured mixing ratio for CH_4 in the air and temperature of the water on each occasion.

The gas transfer velocity, k, exhibited no clear seasonal variability (i.e. it was independent of seasonal changes in temperature; Figure S2) and was not significantly different between treatments on average over the course of the experiment ($F_{1,\ 123} = 3.46$, $P = 0.068$). This evidence suggests that the physical influence of heating the mesocosms by ~4 °C had little discernable effect on advective processes. Consequently, this potentially confounding variable was also discounted from further analyses of the biogeochemical processes in the experiment, and changes in the concentration of oxygen and methane in the water column and the efflux of methane across the air–water interface can be ascribed to biological metabolism.

REFERENCES

Allen, A.P., Gillooly, J.F., and Brown, J.H. (2005). Linking the global carbon cycle to individual metabolism. *Funct. Ecol.* **19**, 202–213.

Atkin, O.K., and Tjoelker, M.G. (2003). Thermal acclimation and the dynamic response of plant respiration to temperature. *Trends Plant Sci.* **8**, 343–351.

Badger, M.R., and Collatz, G.J. (1977). Studies on the kinetic mechanism of Ribulose-1,5-bisphosphate carboxylase and oxygenase reactions, with particular reference to the effect of temperature on kinetic parameters. *Annu. Rep. Director Dep. Plant Biol.* **76**, 355–361.

Baird, D., and Ulanowicz, R.E. (1989). The seasonal dynamics of the Chesapeake bay ecosystem. *Ecol. Monogr.* **59**, 329–364.

Bales, J.D., and Nardi, M.R. (2007). Automated routines for calculating whole stream metabolism. In: *Theoretical Background and Users Guide: US Geological Survey Techniques and Methods 4-C2* (Ed. by J.D. Bales and M.R. Nardi). US Geological Survey.

Battin, T.J., Kaplan, L.A., Findlay, S., *et al.* (2008). Biophysical controls on organic carbon fluxes in fluvial networks. *Nat. Geosci.* **1**, 95–100.

Bellamy, P.H., Loveland, P.J., Bradley, R.I., Lark, R.M., and Kirk, G.J.D. (2005). Carbon losses from all soils across England and Wales 1978-2003. *Nature* **437**, 245–248.

Benton, T.G., Solan, M., Travis, J.M.J., and Sait, S.M. (2007). Microcosm experiments can inform global ecological problems. *Trends Ecol. Evol.* **22**, 516–521.

Berry, J., and Bjorkman, O. (1980). Photosynthetic response and adaptation to temperature in higher plants. *Annu. Rev. Plant Physiol. Plant Mol. Biol.* **31**, 491–543.

Brown, J.H., Gillooly, J.F., Allen, A.P., Savage, V.M., and West, G.B. (2004). Toward a metabolic theory of ecology. *Ecology* **85**, 1771–1789.

Cebrian, J. (1999). Patterns in the fate of production in plant communities. *Am. Nat.* **154**, 449–468.

Clarke, A. (2004). Is there a Universal Temperature Dependence of metabolism? *Funct. Ecol.* **18**, 252–256.

Cole, J.J., and Caraco, N.F. (1998). Atmospheric exchange of carbon dioxide in a low-wind oligotrophic lake measured by the addition of SF6. *Limnol. Oceanogr.* **43**, 647–656.

Cole, J.J., and Caraco, N.F. (2001). Carbon in catchments: Connecting terrestrial carbon losses with aquatic metabolism. *Mar. Freshw. Res.* **52**, 101–110.

Cole, J.J., Pace, M.L., Carpenter, S.R., and Kitchell, J.F. (2000). Persistence of net heterotrophy in lakes during nutrient addition and food web manipulations. *Limnol. Oceanogr.* **45**, 1718–1730.

Cole, J.J., Carpenter, S.R., Kitchell, J.F., and Pace, M.L. (2002). Pathways of organic carbon utilization in small lakes: Results from a whole-lake C-13 addition and coupled model. *Limnol. Oceanogr.* **47**, 1664–1675.

Cox, P.M., Betts, R.A., Jones, C.D., Spall, S.A., and Totterdell, I.J. (2000). Acceleration of global warming due to carbon-cycle feedbacks in a coupled climate model. *Nature* **408**, 184–187.

Davidson, E.A., and Janssens, I.A. (2006). Temperature sensitivity of soil carbon decomposition and feedbacks to climate change. *Nature* **440**, 165–173.

del Giorgio, P.A., and Peters, R.H. (1994). Patterns in planktonic P/R ratios in lakes: Influence of lake trophy and dissolved organic carbon. *Limnol. Oceanogr.* **39**, 772–787.

del Giorgio, P.A., Cole, J.J., and Cimbleris, A. (1997). Respiration rates in bacteria exceed phytoplankton production in unproductive aquatic systems. *Nature* **385**, 148–151.

R Development Core Team (2006). *R Foundation for Statistical Computing*. R Development Core Team, Austria, Vienna.

Dewar, R.C., Medlyn, B.E., and McMurtrie, R.E. (1999). Acclimation of the respiration photosynthesis ratio to temperature: Insights from a model. *Global Change Biol.* **5**, 615–622.

Enquist, B.J., Economo, E.P., Huxman, T.E., Allen, A.P., Ignace, D.D., and Gillooly, J.F. (2003). Scaling metabolism from organisms to ecosystems. *Nature* **423**, 639–642.

Enquist, B.J., Kerkhoff, A.J., Huxman, T.E., and Economo, E.P. (2007). Adaptive differences in plant physiology and ecosystem paradoxes: Insights from metabolic scaling theory. *Global Change Biol.* **13**, 591–609.

Falkowski, P.G., and Raven, J.A. (1997). Aquatic Photosynthesis. Blackwell Science, Oxford.

Farquhar, G.D., von Caemmerer, S., and Berry, J.A. (1980). A biochemical model of photosynthetic CO2 assimilation in leaves of C3 species. *Planta* **149**, 78–90.

Finkel, Z.V., Katz, M.E., Wright, J.D., Schofield, O.M.E., and Falkowski, P.G. (2005). Climatically driven macroevolutionary patterns in the size of marine diatoms over the cenozoic. *Proc. Natl. Acad. Sci. USA* **102**, 8927–8932.

Finkel, Z.V., Beardall, J., Flynn, K.J., Quigg, A., Rees, T.A.V., and Raven, J.A. (2010). Phytoplankton in a changing world: Cell size and elemental stoichiometry. *J. Plankton Res.* **32**, 119–137.

Friedlingstein, P., Cox, P., Betts, R., *et al.* (2006). Climate-carbon cycle feedback analysis: Results from the (CMIP)-M-4 model intercomparison. *J. Clim.* **19**, 3337–3353.

Gifford, R.M. (2003). Plant respiration in productivity models: Conceptualisation, representation and issues for global terrestrial carbon-cycle research. *Funct. Plant Biol.* **30**, 171–186.

Gillooly, J.F., Brown, J.H., West, G.B., Savage, V.M., and Charnov, E.L. (2001). Effects of size and temperature on metabolic rate. *Science* **293**, 2248–2251.

Gudasz, C., Bastviken, D., Steger, K., Premke, K., Sobek, S., and Tranvik, L.J. (2010). Temperature-controlled organic carbon mineralization in lake sediments. *Nature* **466**, 478–481.

Hessen, D.O., van Donk, E., and Gulati, R. (2005). Seasonal seston stoichiometry: Effects on zooplankton in cyanobacteria-dominated lakes. *J. Plankton Res.* **27**, 449–460.

Hopkinson, C.S., and Smith, E.S. (2005). Estuarine respiration: An overview of benthic, pleagic and whole system respiration. In: *Respiration in Aquatic Ecosystems* (Ed. by P.A. del Giorgio and P.J.le.B. Williams), pp. 122–146. Oxford University Press, Oxford.

Hutchinson, G.E. (1959). A homage to Santa Rosalia or why are there so many kinds of animals? *Am. Nat.* **93**, 145.

IPCC (2007). Climate change 2007: The physical sciences basis. In: *Contribution of Working Group I to the Fourth Assessment Report of the Intergovernmental Panel on Climate Change* Parry, M.L., Canziani, O.F., Palutikof, J.P., van der Linden, P.J. and Hanson, C.E. (Eds.), pp. 7–22. Cambridge University Press, Cambridge.

Jones, J.I., Young, J.O., Eaton, J.W., and Moss, B. (2002). The influence of nutrient loading, dissolved inorganic carbon and higher trophic levels on the interaction between submerged plants and periphyton. *J. Ecol.* **90**, 12–24.

Kirkwood, D.S. (1996). *Nutrients: Practical notes on their determination in seawater*. ICES, Copenhagen.

Kirschbaum, M.U.F. (2000). Will changes in soil organic carbon act as a positive or negative feedback on global warming? *Biogeochemistry* **48,** 21–51.

Knorr, W., Prentice, I.C., House, J.I., and Holland, E.A. (2005). Long-term sensitivity of soil carbon turnover to warming. *Nature* **433**, 298–301.

Kozlowski, J., and Konarzewski, M. (2005). West, Brown and Enquist's model of allometric scaling again: The same questions remain. *Funct. Ecol.* **19**, 739–743.

Lambert, M., and Frechette, J.-L. (2005). Analytical techniques for measuring fluxes of CO2 and CH4 from hydroelectric reservoirs and natural water bodies. In: *Greenhouse Gas Emission – Fluxes and Processes* (Ed. by A. Tremblay, L. Varfalvy, C. Rohem and M. Garneau), pp. 37–60. Springer, Berlin, Germany.

Laws, E.A., Falkowski, P.G., Smith, W.O., Ducklow, H., and McCarthy, J.J. (2000). Temperature effects on export production in the open ocean. *Global Biogeochem. Cycles* **14**, 1231–1246.

Lindeman, R.L. (1942). The trophic-dynamic aspect of ecology. *Ecology* **23**, 399–418.

Lopez-Urrutia, A., San Martin, E., Harris, R.P., and Irigoien, X. (2006). Scaling the metabolic balance of the oceans. *Proc. Natl. Acad. Sci. USA* **103**, 8739–8744.

Lovelock, J.E. (1972). Gaia as seen through the atmosphere. *Atmos. Environ.* **6**, 579.

Luo, Y.Q., Wan, S.Q., Hui, D.F., and Wallace, L.L. (2001). Acclimatization of soil respiration to warming in a tall grass prairie. *Nature* **413**, 622–625.

Makarieva, A.M., Gorshkova, V.G., and Li, B.L. (2005). Revising the distributive networks models of West, Brown and Enquist (1997) and Banavar, Maritan and Rinaldo (1999): Metabolic inequity of living tissues provides clues for the observed allometric scaling rules. *J. Theor. Biol.* **237**, 291–301.

Manning, P., Newington, J.E., Robson, H.R., *et al.* (2006). Decoupling the direct and indirect effects of nitrogen deposition on ecosystem function. *Ecol. Lett.* **9**, 1015–1024.

Marzolf, E.R., Mulholland, P.J., and Steinman, A.D. (1994). Improvements to the diurnal upstream-downstream dissolved-oxygen change techniques for determining whole stream metabolism in small streams. *Can. J. Fish. Aquat. Sci.* **51**, 1591–1599.

Mastepanov, M., Sigsgaard, C., Dlugokencky, E.J., Houweling, S., Strom, L., Tamstorf, M.P., and Christensen, T.R. (2008). Large tundra methane burst during onset of freezing. *Nature* **456**, 628–630.

McKee, D., Atkinson, D., Collings, S.E., *et al.* (2003). Response of freshwater microcosm communities to nutrients, fish, and elevated temperature during winter and summer. *Limnol. Oceanogr.* **48**, 707–722.

Medlyn, B.E., Dreyer, E., Ellsworth, D., *et al.* (2002). Temperature response of parameters of a biochemically based model of photosynthesis. II. A review of experimental data. *Plant Cell Environ.* **25**, 1167–1179.

Melillo, J.M., Steudler, P.A., Aber, J.D., *et al.* (2002). Soil warming and carbon-cycle feedbacks to the climate system. *Science* **298**, 2173–2176.

Montoya, J.M., and Raffaelli, D. (2010). Climate change, biotic interactions and ecosystem services. *Philos. Trans. R. Soc. Lond. Ser. B-Biol. Sci.* **365**, 2013–2018.

Montoya, J.M., Pimm, S.L., and Sole, R.V. (2006). Ecological networks and their fragility. *Nature* **442**, 259–264.
Moran, X.A.G., Lopez-Urrutia, A., Calvo-Diaz, A., and Li, W.K.W. (2010). Increasing importance of small phytoplankton in a warmer ocean. *Global Change Biol.* **16**, 1137–1144.
Morin, A., Lamoureux, W., and Busnarda, J. (1999). Empirical models predicting primary productivity from chlorophyll a and water temperature for stream periphyton and lake and ocean phytoplankton. *J. North Am. Benthol. Soc.* **18**, 299–307.
Mulholland, P.J., Fellows, C.S., Tank, J.L., *et al.* (2001). Inter-biome comparison of factors controlling stream metabolism. *Freshw. Biol.* **46**, 1503–1517.
Pace, M.L., Cole, J.J., Carpenter, S.R., *et al.* (2004). Whole-lake carbon-13 additions reveal terrestrial support of aquatic food webs. *Nature* **427**, 240–243.
Parmesan, C., and Yohe, G. (2003). A globally coherent fingerprint of climate change impacts across natural systems. *Nature* **421**, 37–42.
Perkins, D.M., McKie, B.G., Malmqvist, B., Gilmour, S.G., Reiss, J., and Woodward, G. (2010) Environmental warming and biodiversity-ecosystem functioning in freshwater microcosms: Partitioning the effects of species identity, richness and metabolism. *Adv. Ecol. Res.* **43**, 177–209.
Ptacnik, R., Moorthi, S.D., and Hillebrand, H. (2010) Hutchinson reversed, or why there need to be so many species. *Adv. Ecol. Res.* **43**, 1–43.
Purdy, K.J., Hurd, P.J., Moya-Laraño, J., Trimmer, M., and Woodward, G. (2010) Systems biology for ecology. *Adv. Ecol. Res.* **43**, 87–149.
Ram, A.S.P., Nair, S., and Chandramohan, D. (2007). Bacterial growth efficiency in a tropical estuary: Seasonal variability subsidized by allochthonous carbon. *Microb. Ecol.* **53**, 591–599.
Raven, J.A., Cockell, C.S., and La Rocha, C.L. (2008). The evolution of inorganic carbon concentrating mechanisms in photosynthesis. *Philos. Trans. R. Soc. B-Biol. Sci.* **363**, 2641–2650.
Reiss, J., Bailey, R.A., Cássio, F., Woodward, G., and Pascoal, C. (2010a) Assessing the contribution of micro-organisms and macrofauna to biodiversity-ecosystem functioning relationships in freshwater microcosms. *Adv. Ecol. Res.* **43**, 151–176.
Reiss, J., Forster, J., Hirst, A.G, Cássio, F., Pascoal, C., Stewart, R., and Hirst, A.G. (2010b) When microscopic organisms inform general ecological theory. *Adv. Ecol. Res.* **43**, 45–85.
Rivkin, R.B., and Legendre, L. (2001). Biogenic carbon cycling in the upper ocean: Effects of microbial respiration. *Science* **291**, 2398–2400.
Robinson, C. (2008). Heterotrophic bacterial respiration. In: *Microbial Ecology of the Oceans* (Ed. by D.L. Kirchman). 2nd ed. Wiley, New York.
Rodhe, H. (1990). A comparison of the contribution of various gases to the greenhouse-effect. *Science* **248**, 1217–1219.
Sanders, I.A., Heppell, C.M., Cotton, J.A., Wharton, G., Hildrew, A.G., Flowers, E.J., and Trimmer, M. (2007). Emission of methane from chalk streams has potential implications for agricultural practices. *Freshw. Biol.* **52**, 1176–1186.
Savage, V.M., Gillooly, J.F., Woodruff, W.H., West, G.B., Allen, A.P., Enquist, B.J., and Brown, J.H. (2004). The predominance of quarter-power scaling in biology. *Funct. Ecol.* **18**, 257–282.
Schroter, D., Cramer, W., Leemans, R., *et al.* (2005). Ecosystem service supply and vulnerability to global change in Europe. *Science* **310**, 1333–1337.
Schutz, H., Holzapfelpschorn, A., Conrad, R., Rennenberg, H., and Seiler, W. (1989). A 3 year continuous record on the influence of daytime, season, and fertilizer

treatment on methane emission rates from an Italian rice paddy. *J. Geophys. Res.-Atmos.* **94**, 16405–16416.
Segers, R. (1998). Methane production and methane consumption: A review of processes underlying wetland methane fluxes. *Biogeochemistry* **41**, 23–51.
Smith, F.A., and Walker, N.A. (1980). Photosynthesis by aquatic plants: Effects of unstirred layers in relation to assimilation of CO2 and HCO3 – and to carbon isotopic discrimination. *New Phytol.* **86**, 245–259.
Tcherkez, G.G.B., Farquhar, G.D., and Andrews, T.J. (2006). Despite slow catalysis and confused substrate specificity, all ribulose bisphosphate carboxylases may be nearly perfectly optimized. *Proc. Natl. Acad. Sci. USA* **103**, 7246–7251.
Trumbore, S. (2000). Age of soil organic matter and soil respiration: Radiocarbon constraints on belowground C dynamics. *Ecol. Appl.* **10**, 399–411.
Ventura, M., Liboriussen, L., Lauridsen, T., Sondergaard, M., and Jeppesen, E. (2008). Effects of increased temperature and nutrient enrichment on the stoichiometry of primary producers and consumers in temperate shallow lakes. *Freshw. Biol.* **53**, 1434–1452.
Walther, G.R., Post, E., Convey, P., *et al.* (2002). Ecological responses to recent climate change. *Nature* **416**, 389–395.
Westbrook, G.K., Thatcher, K.E., Rohling, E.J., *et al.* (2009). Escape of methane gas from the seabed along the West Spitsbergen continental margin. *Geophys. Res. Lett.* **36**, 5.
Whiting, G.J., and Chanton, J.P. (2001). Greenhouse carbon balance of wetlands: Methane emission versus carbon sequestration. *Tellus B-Chem. Phys. Meteorol.* **53**, 521–528.
Winder, M., Reuter, J.E., and Schladow, S.G. (2009). Lake warming favours small-sized planktonic diatom species. *Proc. R. Soc. B-Biol. Sci.* **276**, 427–435.
Woodward, F.I. (2007). Global primary production. *Curr. Biol.* **17**, R269–R273.
Woodward, G., Perkins, D.M., and Brown, L.C. (2010a). Climate change and freshwater ecosystems: Impacts across multiple levels of organization. *Philos. Trans. R. Soc. B* **365**, 2093–2106.
Woodward, G., Benstead, J.P., Beveridge, O.S., Blanchard, J., Brey, T., Brown, L., Cross, W.F., Friberg, N., Ings, T.C., Jacob, U., Jennings, S., Ledger, M.E., *et al.* (2010b). Ecological networks in a changing climate. *Adv. Ecol. Res.* **42**, 71–138.
Woodwell, G.M., Mackenzie, F.T., Houghton, R.A., Apps, M., Gorham, E., and Davidson, E. (1998). Biotic feedbacks in the warming of the earth. *Clim. Change* **40**, 495–518.
Yamamoto, S., Alcauskas, J.B., and Crozier, T.E. (1976). Solubility of methane in distilled water and seawater. *J. Chem. Eng. Data* **21**, 78–80.
Yvon-Durocher, G., Jones, J.I., Woodward, G., Trimmer, M., and Montoya, J.M. (2010a). Warming alters the metabolic balance of ecosystems. *Philos. Trans. R. Soc. Lond. Ser. B-Biol. Sci.* **365**, 2117–2126.
Yvon-Durocher, G., Montoya, J.M., Jones, J.I., Woodward, G., and Trimmer, M. (2010b). Warming increases the proportion of primary production emitted as methane from freshwater ecosystems. *Global Change Biol.* doi: 10.1111/j.1365-2486.2010.02289.x.
Yvon-Durocher, G., Trimmer, M., Woodward, G., and Montoya, J.M. (2010c). Warming alters the size spectrum and the distribution of body size in aquatic ecosystems. *Global Change Biol.* doi: 10.1111/j.1365-2486.2010.02321.x.

Index

Advances in Ecological Research Volume 1–43

Cumulative List of Titles

ERRATUM

Chapter 1 — Hutchinson Reversed, or Why There Need to Be So Many Species

Page 8: Figure 1 - A portion of the figure legend is missing in the chapter. Please refer the correct figure legend below.

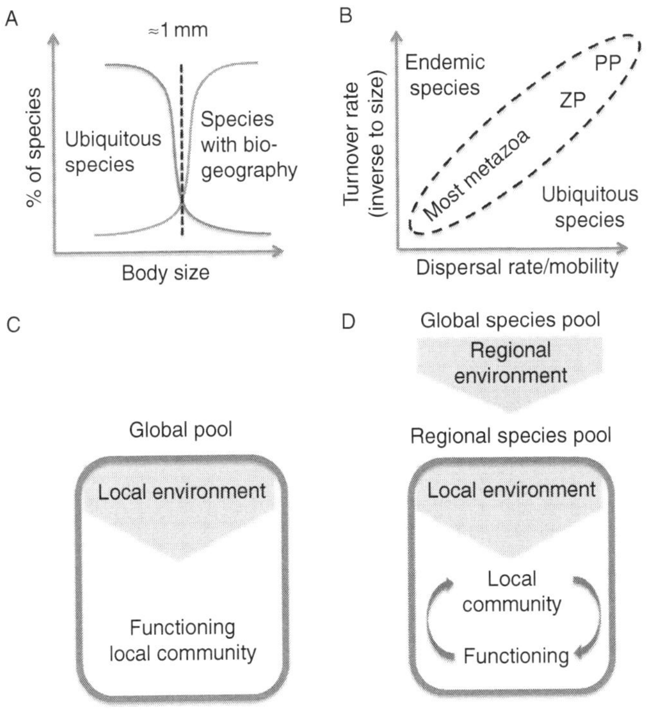

Figure 1 Illustration of the ongoing paradigm shift in plankton ecology. (A) Organisms <1mm were assumed to be globally dispersed (modified from Finlay, 2002). (B) Recent evidence suggests that both dispersal rate and community turnover rate affect the expression of regional meta-populations (Ptacnik et al., 2010). Since both processes scale with body size, most organisms are assumed to be found along the diagonal, that is, to have meta-populations. Combinations of low dispersal with fast turnover (endemic species) or vice versa (globally dispersed taxa) should be the exception. The traditional view implied that local communities were selected from a global species pool (C), and regarded the local community a result of local dynamics. Taking regional species pools into account (D), the local community is constrained by regional dynamics and becomes itself a driver of local system dynamics.